W9-BJV-622

The
Transformation
of American Law,
1780–1860

The
Transformation
of American Law,
1780–1860

MORTON HORWITZ

New York   Oxford
OXFORD UNIVERSITY PRESS
1992

Oxford University Press

Oxford   New York   Toronto
Delhi   Bombay   Calcutta   Madras   Karachi
Kuala Lumpur   Singapore   Hong Kong   Tokyo
Nairobi   Dar es Salaam   Cape Town
Melbourne   Auckland

and associated companies in
Berlin   Ibadan

Library of Congress Cataloging-in-Publication Data
Horwitz, Morton J., 1938–
The transformation of American law, 1780–1860 / Morton J. Horwitz.
p. cm.   Originally published: Cambridge, Mass.
Harvard University Press, 1977.
Includes bibliographical references and index.
ISBN 0-19-507829-2
1. Law—United States—History. I. Title.
KF366.H6   1992   349.73'09'034—dc20
[347.300934]   92-3464

9 8 7 6 5 4 3 2 1

Printed in the United States of America
on acid-free paper

For Sandra

Acknowledgments

C HAPTERS I, II, and VIII and portions of Chapter VI have previously appeared in *Perspectives in American History*, the *University of Chicago Law Review*, the *American Journal of Legal History*, and the *Harvard Law Review*, respectively.

I wish to gratefully acknowledge the following institutions for allowing me to use and quote from their manuscript collections: the Harvard Law School Library, the New York Historical Society, the Swem Library of the College of William and Mary, the Massachusetts Historical Society. Erika Chadbourn, Edith Henderson, and Margaret Moody of the Harvard Law School Library have assisted me in many ways. Charles Cullen, of *The Papers of John Marshall*, has also been very helpful.

My secretaries, Mary Malone and Susan Levin, and my research assistants, John Fisher, Robert Rosen and Stephen Yeazell, were all extraordinarily giving of their time and energies.

Several of my colleagues generously read portions of the manuscript: John P. Dawson, Andrew Kaufman, and Robert Keeton. Duncan Kennedy, Alfred Konefsky, and William Nelson made searching criticisms at various stages of this work. Professor Nelson also graciously shared with me the fruits of his own research into the Massachusetts court records. Stanley N. Katz, editor of this series, has given me constant support and help for many years.

Thanks are due for generous research support from the Russell Sage Foundation, the Charles Warren Fund of the Harvard Law School, the American Philosophical Society, and the National Endowment for the Humanities. I also wish to express my debt and gratitude to Dean Albert M. Sacks of the Harvard Law School for unwavering support of my scholarly efforts.

Finally, I wish to acknowledge the help given by the Harvard University Press; by Aida Donald, Joan Ryan, my editor, and Nancy Donovan, who compiled the index.

Contents

Introduction

MY first aspiration in this book is to make the history of technical and obscure areas of American law accessible to professional historians and to other nonlegally trained scholars. Yet, every scholar who writes the history of a specialized discipline becomes sensitive to the problems of addressing a multiple audience. Whether the field be the history of science or economic or legal history, the historian is brought face to face with the problems of being faithful to the internal technical structure of a discipline while at the same time providing a more general perspective from which to measure its significance. Thus, one is constantly faced with choices about how technical to get. A completely nontechnical exposition not only deprives other specialists of the necessary data with which they can either challenge or build upon one's work; it also deprives the nonspecialist of both the essential texture and the structure of historical change within the discipline. To exclude such material may mislead the reader into believing he can ignore the ways in which the internal technical life of a field generates autonomous forces that determine its history.

Every specialist ought to aspire to render his own "mysterious science" less mysterious. This goal usually requires that he attempt to translate the technical vocabulary and concepts of the specialist into more general and accessible categories.

It has been my ardent desire to reach the general historian. I am aware, nevertheless, that there are many points in this book which general readers will find too technical for their own purposes. I can only say that I have tried to confine to the notes many technical questions that I thought might present unnecessary difficulties for nonspecialists. Yet, there remains a residue which I regard as indispensable for anyone who wishes seriously to understand legal history.

When I began writing this book, my purpose was to study the relationship between private law (tort, contract, property, commercial law) and economic change in the nineteenth century. Constitutional law, I felt, had been overstudied both in terms of its impact on the development of the American economy and in terms of its representative character. Not only do traditional constitutional histories include a large number of atypical "great cases" but constitutional cases are also unrepresentative either as intellectual history or as examples of social control. Indeed, constitutional law in America represents episodic legal intervention buttressed by a rhetorical tradition that is often an unreliable guide to the slower (and often more unconscious) processes of legal change in America.

But another, more crucial, distortion has been introduced by the excessive equation of constitutional law with "law." Because of the peculiar intellectual and institutional background of judicial review, the study of constitutional law focuses historians on the naysaying function of law and, more specifically, on the rather special circumstances of judicial intervention into statutory control. Yet judicial promulgation and enforcement of common law rules constituted an infinitely more typical pattern of the use of law throughout most of the nineteenth century. By thus focusing on private law we can study the more regular instances in which law, economy, and society interacted.

Nevertheless, I have not written about all areas of private law nor even about all clearly relevant areas in the period studied. One criterion of exclusion turned on whether a particular field had already been well covered in the literature. To the extent that the existing studies revealed not only great sensitivity to the relationship between law and society but were also able to penetrate the difficult technical problems that regularly impede real understanding of legal questions, I tended to rely upon them. So, for example, corporation law and labor law, of which there are a number of excellent studies, are not dealt with in any systematic way.

Another criterion influenced the selection process. I was drawn to study those areas of the law that historians (and indeed most lawyers) have tended to regard as neutral from a standpoint of policies concerning economic growth or the distribution of wealth or political power. In most general historical studies, fields like torts and contracts, when they are examined at all, are treated as if they stand apart from content or policy. Even sophisticated lawyers, who regularly address themselves to the policies imbedded in contemporary legal rules, tend to treat the historical study of law with an arid formalism that is striking and surprising.

Still, no historian of law can fail to recognize that legal consciousness in any particular period is not simply the sum of those contemporary social forces that impinge upon law. Law is autonomous to the extent that ideas are autonomous, at least in the short run. In addition, the internal needs of professional interest and ideology have enabled the legal professional to serve as a buffer or filter between social forces and the law. That process ought not to be confused with the self-justificatory claims of lawyers to mediate between opposing social forces in the interest of a politically neutral law. Rather, it simply acknowledges that lawyers too have usually had interests in creating the appearance of a neutral, apolitical legal system.

This study attempts to challenge certain features of the "consensus" history that has continued to dominate American historiography since the Second World War. Within that tradition, several major assumptions or conclusions have marked recent historical writing about the particular period (1780-1860) that this study focuses upon. First, most of these studies were written during or immediately after the New Deal in the midst of charges that New Deal regulation had radically departed from a well-established noninterventionist American political tradition. These studies implicitly sought to demonstrate that far from representing a novel departure from a consistent laissez-faire past, the regulatory state in fact only represented a return to an earlier pattern of governmental interventionism. The studies of the Handlins and Hartz, for example, successfully demonstrated a postrevolutionary pattern of systematic governmental regulation of the economy as well as a much later (post-1850) development of a laissez-faire ideology which for the first time erected theoretical objections to governmental activity. Other writers also showed that a pattern of mixed ownership of, for example, the newly developed canal system belied any early principled objection to governmental ownership of enterprise. The lesson presumably to be drawn from this demonstration was that there was no historical norm of laissez-faire and that, if anything, laissez-faire ideology itself represented an aberrational strand in defining the legitimate relationship between government and economy. Through this process, the New Deal was provided its own historical pedigree.

My own study of the antebellum period does nothing to overthrow these basic conclusions, and, indeed, they are confirmed at virtually every point. My point of disagreement instead focuses on the questions these historians asked and the resulting structure of analysis that they forged. By and large, the New Deal historians were much more concerned with finding evidence of governmental intervention than they were in asking in whose interest these regulations were forged. To a surprisingly great extent, they treated all instances of state intervention as equally proving their point, as indeed such evidence would tend to do, given the qualitatively undifferentiated questions they tended to ask.

One of the most important consequences of this approach is that the historical writing of the last generation tended to ignore all questions about the effects of governmental activity on the distribution of wealth and power in American society. They tended to assume that virtually all regulation was in the public interest without ever providing any real criteria for such a conclusion. And, by and large, they accepted at face value the rhetorical public interest justifications offered by the proponents of governmental activity. Out of this approach, there often emerged striking contradictions, as when Leonard Levy invoked the Handlins' conception of a Commonwealth (essentially a state operating to promote the public interest) to explain the Chief Justiceship of Lemuel Shaw (1830-1860) during a period when the Handlins themselves sought to show that the Commonwealth ideal was disintegrating. But Levy was just as successful as the Handlins in collecting utterances by Shaw and his contemporaries which demonstrated (not unexpectedly) that Shaw also believed he was operating in the public interest.

Another example of how the political categories of the earlier generation determined the character of their history relates to promotion of economic development. To the extent that we are interested equally in all forms of governmental intervention, it hardly matters whether canals, for example, were financed through debt or through taxation, since both forms of financing equally demonstrate governmental promotion of economic development. But to the extent that financing through government bonds was regarded as a means of avoiding taxation (with subsequent liquidation of the debt through tolls) it may be of great significance for the resulting distribution of wealth which form actually was adopted. The fact that the Massachusetts budget was held constant at $133,000 for over a generation led to governmental promotion of enterprise through forms not involving cash outlays, such as franchises and monopolies. But, again, while either mode of promotion may indicate governmental activism, it is clear that different modes affected the distribution of wealth and power quite differently. So, for example, we know almost nothing about what the distributive effects of promotion through taxation would have been because this set of questions was seemingly irrelevant.

In this book, I seek to show that one of the crucial choices made during the antebellum period was to promote economic growth primarily through the legal, not the tax, system, a choice which had major consequences for the distribution of wealth and power in American society.

Another main departure in this study concerns use of the concept of laissez-faire. Beginning with the Progressive historians, this notion has generally stood for objections to governmental regulation of the economy, and governmental regulation has usually referred to the absence of legislative (statutory) or administrative intervention in the economy. Since few historians have been primarily concerned with the common law power of judges, they have not thought through the problems of using this concept in a common law context. Strictly speaking, there could never be a laissez-faire regime unless judges refused to enforce all contracts and refused to compensate for all injuries to persons and property. Yet, Willard Hurst's famous phrase "release of energy" correctly suggests that under certain circumstances even judicial refusal to act could be motivated by developmental goals. In this sense, the contractarian ideology of nineteenth century judges was both instrumental (in the sense of promoting economic development) and laissez-faire (in the traditional sense of being hostile to legislative or administrative regulation). In short, when it comes to analyzing the activities of private law judges deciding disputes concerning tort, contract, and property, the category laissez-faire is often not useful primarily because it does not distinguish between developmental and distributional goals. It also ignores the political significance of leaving the task of governmental regulation primarily to judges, which was often the real goal of advocates of laissez-faire.

The ambiguities in the term laissez-faire involve, however, a still deeper problem in the historiography of this period. Were the New Deal historians successful in casting the question of governmental activity in a consensus framework? What evidence did they have for indiscriminately characterizing most forms of governmental activity as a promotion of the public interest?

From the perspective of our own time, it seems inconceivable that anyone could have ever so casually equated governmental activity with promotion of some abstract public interest. But the consensus historians were not so naive as all that. In principle, they recognized that powerful interest groups could convert their own particular interests into governmental policies. But their view of American society made any such eventuality unlikely. Whether they worked from a view of a relatively homogeneous, conflict-free society or instead from a belief in a benevolent equilibrium between conflicting groups of relatively equal power, they never seriously doubted that law and policy could be characterized as an accurate reflection of the undifferentiated needs of society at large.

Study of antebellum society with particular focus upon legal conflict tempts one to characterize these earlier histories as ideological. During the eighty years after the American Revolution, a major transformation of the legal system took place, which reflected a variety of aspects of social struggle. That the conflict was turned into legal channels (and thus rendered somewhat mysterious) should not obscure the fact that it took place and that it enabled emergent entrepreneurial and commercial groups to win a disproportionate share of wealth and power in American society.

The transformed character of legal regulation thus became a major instrument in the hands of these newly powerful groups. While they often used the rhetoric of promotion of the public interest — and what self-interested group does not? — one ought to remain skeptical about their claims.

In one sense, their claims are plausible. If the sole criterion of the public interest is the maximization of economic growth, a case can be made for the fact that the American legal system after the Revolution was transformed successfully to promote developmental goals. But if we look at the resulting distribution of economic wealth and power — at the legal expropriation of wealth or at the forced subsidies to growth coerced from the victims of the process — it is difficult to characterize it as codifying some consensus on the objective needs of the society.*

---

*During the past ten years, under the influence of R. H. Coase's seminal article, "The Problem of Social Cost," 3 *J. of Law & Econ.* 1 (1960), a large literature has grown up dealing with an economic analysis of legal rules. Coase has shown that where there are no "transaction costs" between parties — that is, in situations where the parties are able to engage in costless bargaining — it does not matter from the standpoint of efficiency which of the parties is saddled with legal liability. Coase's theorem thus cautions the legal historian to take care when accounting for legal change in terms of the goal of economic efficiency.

On the other hand, we have only begun to realize that Coase's theorem is based on a static model. Since legal rules do determine the distribution of wealth, there are various allocationally efficient outcomes that relate to different wealth distributions. See, *e.g.*, Baker, "The Ideology of the Economic Analysis of Law," 5 *Philos. & Public Aff.* 3 (1975); E. J. Mishan, "Pareto Optimality and the Law," 19 (N.S.) *Oxford Economic Papers* 225 (1967) and *Cost-Benefit Analysis* 121-137 (1971).

More concretely, the dynamic goal of economic growth, so important to nineteenth century judges, was clearly understood in terms of the assumption that different sets of legal rules would have differential effects on economic growth depending both on the distribution of wealth they produced and the level of investment they encouraged. See A. Hirschman, *The Strategy of Economic Development* 55-61 (1961). In short, despite the Coase theorem, it seems clear that legal historians are correct in ascribing to nineteenth century jurists their own frequently stated strategy of adjusting legal rules to promote economic growth.

The
Transformation
of American Law,
1780–1860

I   The Emergence of an Instrumental Conception of Law

E VEN the most summary survey of developments in American
law during the nineteenth century confirms Daniel Boorstin's
view that it represented one of those truly "creative outbursts of
modern legal history."[1] And while other historians have characterized the period as a "formative era" or "golden age" in American
law,[2] it has never been entirely clear why at this particular time the
legal system should have taken on such an innovative and transforming role. The most fundamental change undoubtedly involves the
function of the common law. In eighteenth century America, common law rules were not regarded as instruments of social change;
whatever legal change took place generally was brought about
through legislation. During this period, the common law was conceived of as a body of essentially fixed doctrine to be applied in
order to achieve a fair result between private litigants in individual
cases. Consequently, American judges before the nineteenth century rarely analyzed common law rules functionally or purposively,
and they almost never self-consciously employed the common law as
a creative instrument for directing men's energies toward social
change.

What dramatically distinguished nineteenth century law from its
eighteenth century counterpart was the extent to which common law
judges came to play a central role in directing the course of social
change. Especially during the period before the Civil War, the common law performed at least as great a role as legislation in underwriting and channeling economic development. In fact, common
law judges regularly brought about the sort of far-reaching changes
that would have been regarded earlier as entirely within the powers

of the legislature. During the antebellum period, Mark DeWolfe Howe observed, "The legislative responsibility of lawyers and judges for establishing a rule of law was far more apparent than it was in later years. It was as clear to laymen as it was to lawyers that the nature of American institutions, whether economic, social, or political, was largely to be determined by the judges. In such a period questions of private law were seen and considered as questions of social policy."[3] Indeed, judges gradually began to shape common law doctrine with an increasing awareness that the impact of a decision extended far beyond the case before them and that their function had expanded beyond the necessity merely of doing justice in the individual case. Courts "appear[ed] to take the opportunity which each case afforded, not only of deciding that case, but of establishing rules of very general application."[4] In short, by 1820 the process of common law decision making had taken on many of the qualities of legislation. As judges began to conceive of common law adjudication as a process of making and not merely discovering legal rules, they were led to frame general doctrines based on a self-conscious consideration of social and economic policies.

During the last fifteen years of the eighteenth century, one can identify a gradual shift in the underlying assumptions about common law rules. For the first time, lawyers and judges can be found with some regularity to reason about the social consequences of particular legal rules. For example, it would have been very unusual a decade earlier to hear a lawyer argue, as the attorney general of South Carolina did in 1796, that the state Supreme Court should not require compensation for land taken for road building because it would "thwart and counteract the public in the exercise of this all-important authority for the interest of the community."[5] Similarly, jurists began to frame legal arguments in terms of "the importance of the present decision to the commercial character of our country,"[6] or of the necessity of deciding whether adherence to a particular common law rule will result in "improvement in our commercial code."[7] By the first decade of the nineteenth century, it was common for judges to argue that "to admit a party to a negotiable note to come forward as a witness to impeach that note would greatly embarrass trade and commerce, and almost entirely prevent the circulation of this species of paper."[8]

One of the best instances of the emerging legal mentality occurred in the New York case of *Palmer v. Mulligan,* decided in 1805. There the court was called upon to decide whether the plaintiff, a downstream mill owner, could recover damages for obstruction of the flow of water by the defendant, who had erected his own upstream dam. Not only did the court refuse to apply the common law rule, which had always allowed a downstreamer to recover for any obstruction of the natural flow, but the judicial opinions were filled with functional analyses of the common law test. Under the common law, one judge observed, "He who could first build a dam or mill on any public or navigable river, would acquire an exclusive right, at least for some distance." The result of allowing the plaintiff to prevail would be that "the public, whose advantage is always to be regarded, would be deprived of the benefit which always attends competition and rivalry."[9]

This increasing preoccupation with using law as an instrument of policy is everywhere apparent in the early years of the nineteenth century. Two decades earlier it would have been impossible to find an American judge ready to analyze a private law question by agreeing "with Professor *Smith,* in his 'Wealth of Nations,' . . . that distributing the burthen of losses, among the greater number, to prevent the ruin of a few, or of an individual, is most conformable to the principles of insurance, and most conducive to the general prosperity of commerce."[10] It would have been equally improbable that a judge, as in the *Philadelphia Cordwainers Case* of 1806, would "consider the effect" that a labor conspiracy "would have upon the whole community" so that, as a result of conspiracy, no "man can calculate . . . at what price he may safely contract to deliver articles."[11]

In a whole variety of areas of law, ancient rules are reconsidered from a functional or purposive perspective, often before new or special economic or technological pressure for change in the law has emerged. Only in the nineteenth century, for example, did American judges begin to argue that the English law of waste "is inapplicable to a new, unsettled country" because of its restraint on improvement of land, even though the problem appears to have been central to eighteenth century concerns as well.[12] Similarly, the English rule of allowing dower on unimproved lands was routinely followed by the Massachusetts court at the end of the eighteenth century but rejected early in the nineteenth century because "it would operate as a clog upon estates, designed to be the subject of transfer." Yet, the goal of free alienation of land in order to encourage economic improvement was regarded as equally important by eighteenth century Americans as it was by their successors.[13] It is only in the nineteenth century that American courts abandon the marine insurance rule of strict construction of warranties in deciding cases under fire insurance policies on the ground that it "would render them so inconvenient as wholly to prevent them."[14]

In short, an instrumental perspective on law did not simply emerge as a response to new economic forces in the nineteenth century. Rather, judges began to use law in order to encourage social change even in areas where they had previously refrained from doing so. My task, then, is to explain why it was not until the nineteenth century that the common law took on its innovating and transforming role in American society. This in turn forces one to ask whether an explanation for this fundamental shift in the conception of law between 1780 and 1820 can be found in more general changes in the political theory of the period.

THE EIGHTEENTH CENTURY CONCEPTION OF THE COMMON LAW

The generation of Americans who made the American Revolution had little difficulty in conceiving of the common law as a known and determinate body of legal doctrine. After more than a decade of insistence by political writers that the "grand basis of the common law" was "the law of nature and its author,"[15] it is not surprising that the first Continental Congress in 1774 should maintain that Americans were "entitled to the common law" as well as to English statutes existing at the time of colonization.[16] In a similar move, the Virginia Convention of 1776 adopted as "the rule of decision" "the common law of England" as well as all English statutes "of a general nature" passed before 1607, while the New Jersey Constitution of the same year declared that "the common law of England as well as so much of the statute law, as have been heretofore practiced in this Colony, shall still remain in force."[17] Between 1776 and 1784, eleven of the thirteen original states adopted, directly or indirectly, some provisions for the reception of the common law as well as of limited classes of British statutes.[18]

In light of the attacks on the common law that began to appear within the next generation, it is remarkable that the revolutionary generation saw no difficulty in establishing the common law as the rule of decision in legal controversies. None of the persistent hostility to English legislation that prevailed throughout the colonial period seems to have influenced their commitment to common law doctrines. Nor did anti-British attitudes, which later were to take the form of opposition to the English common law, result in such opposition at the moment of the break from Great Britain. Although there was considerable fear of judicial discretion voiced both during and after the colonial period, there is nevertheless little evidence in the earlier period—unlike the later one—that this fear was identified with the uncertain nature of common law rules. For example, Massachusetts Chief Justice Hutchinson could declare in 1767 that "laws should be established, else Judges and Juries must go according to their Reason, that is, their Will." It was also imperative "that *the Judge* should never be the *Legislator:* Because, then the Will of the Judge would be the Law: and this tends to a State of Slavery." Unless the laws were "known" and "certain" so that "People may know how to act," he concluded, "they depend upon the arbitrary Opinion of Another." Nevertheless, he saw no contradiction in maintaining that where there was no statutory rule, "the Common Law of England is the Rule."[19] The great danger of judicial discretion for the colonists arose not from common law adjudication but in connection with judicial construction of statutes, "for whenever we leave Principles and clear and positive Laws," John Adams observed, "and wander after Constructions, one Construction or Consequence is piled up upon another until we get an immense distance from Fact, Truth and Nature."[20]

Hutchinson's statement illustrates the "inevitable and rapid reception of the body of English common law" in the colonies during the eighteenth century. The persistent appeals to the common law in the constitutional struggles leading up to the American Revolution "created a regard for its virtues that seems almost mystical." As a result, by the end of the eighteenth century, lawyers regarded the "concept of the common law as a body of principles," which "encouraged uninhibited use of English precedents by the legal profession in the federal courts."[21] And while Americans always insisted on the right to receive only those common law principles which accorded with colonial conditions, most of the basic departures were accomplished not by judicial decision but by local statute, so that by the time of the American Revolution one hears less and less about the unsuitability of common law principles to the American environment.

By contrast, the colonial attitude toward British statutes was summed up in the typical revolutionary constitutional or legislative provision confining statutory reception to the period before colonization or else limiting the force of British statutes to those previously adopted. Indeed, this sharp distinction between statute and common law serves to underline the fundamentally different conceptions that eighteenth century jurists applied to the two bodies of law. Until around 1720 both colonial and English discussions of common law reception continued to be dominated by Sir Edward Coke's distinction between lands conquered from Christians and those taken from infidels.[22] Resisting the claim that the American plantations were "conquered" territory and as such not entitled to the benefits of English law, the colonists largely succeeded after 1720 in establishing the proposition that they carried the English common law with them.

Between 1712 and 1718, the legislatures of South and North Carolina, and Pennsylvania, as well as the Maryland proprietor, claimed the benefits of the common law but of only limited categories of British statutes.[23] And while Coke's dichotomy continued to influence the debate over the authority of British statutes right up to the Revolution, it largely dropped from view after 1720 on the question of common law reception. In that year English Attorney General Richard West acknowledged that "the common law of England is the common law of the [American] plantations and all statutes in affirmance of the common law, passed in England antecedent to the settlement of a colony, are in force in that colony, unless there is some private act to the contrary; though no statutes, made since those settlements, are thus in force unless the colonists are particularly mentioned. Let an Englishman go where he will, he carries as much of law and liberty with him as the nature of things will bear."[24] Although there were occasional attempts before the American Revolution to deny the applicability of the common law to the colonies,[25] West's statement came to represent the overwhelming view as to the authority of common law rules,[26] and his formula, distinguishing between statutory and common law reception, was almost uniformly adopted by American legislatures at the time of the Revolution.

The fundamentally different conceptions of the sources of common and statute law in the eighteenth century can also be illustrated by the frequent contention of colonial lawyers that the retroactive application of statutes was illegitimate. "It is a general observation," Virginia Attorney General John Randolph argued in 1768, "that laws are unjust and unreasonable, when by retrospection they attempt to direct those that were prior." To retroactively apply a statute limiting entails of slaves under the guise of merely clarifying a prior law, he urged, "would fall heavy on purchasers who understood the former law in the common sense, and who saw not the hidden meaning which the legislature have thought it necessary to unfold by a subsequent law. In criminal cases they have been branded with an infamy well merited; and why should the safety of our property stand on a basis less firm than that of our persons?"

While Randolph was voicing a familiar complaint about the retroactive application of a statute, there is no indication that any colonial lawyer expressed a similar concern about an equivalent application of common law rules. Randolph himself acknowledged that a different theory would apply to nonstatutory crimes and, presumably, to common law civil rules as well. He conceded that "a law founded on the law of nature may be retrospective, because it always existed, and a breach of it was criminal, though not forbidden by any human law: but that an institution, merely arbitrary and political, if retrospective, is injurious, since, before it existed, it could not be broken nor could any person foresee that such an action would, at a future day, be made criminal."[27] In short, common law doctrines were derived from natural principles of justice, statutes were acts of will; common law rules were discovered, statutes were made.

An excellent manifestation of this conception appears in *Anderson v. Winston,* a Virginia case decided in 1736. In an action of debt on a Virginia statute barring usurious interest, the defendant maintained that, since he had contracted with the plaintiff before the act was passed, it was "against natural justice to punish any man for an action, innocent in itself with respect to human laws, by a law made *ex post facto.*" But his argument did not conclude there, for he understood that if the statute was merely declaratory of natural law, it hardly mattered whether or not the statute applied to his case. It was therefore essential also to demonstrate at great length that "most of the writers upon the law of nature agree" that natural law principles did not establish any preexisting bar against usury.[28]

Believing, as the elder Daniel Dulany wrote in 1728, "That the Common Law, takes in the Law of Nature, the Law of Reason and the revealed Law of God; which are equally binding, at All Times, in All Places, and to All Persons,"[29] Americans had little cause to fear judicial discretion or the retroactive application of legal doctrine in common law adjudication. Where common law rules were conceived of as *"founded in principles, that are permanent,* uniform *and* universal," and where common law and natural law were interchangeably defined as "the Law which every Man has implanted in him," jurists would be unlikely to think of the common law as deriving its legitimacy from the will of a lawmaker.[30] It was precisely this reasoning that encouraged the colonists to receive those English statutes merely declaratory of the common law while refusing to be bound by those that changed common law rules. If, for example, it could be shown that "the [English] Act of Settlement created no innovation of the ancient constitution" with respect to tenure of judges and that judicial independence "is not of a late date but part of the ancient constitution," it seemed to follow that such a doctrine was an "inseparably inherent" right of individuals. There would be no difficulty in retrospectively applying that rule, it was argued, for it would "introduce no new law" but was merely an "affirmance of the old law, that which was really law before."[31]

In much of the prerevolutionary legal writing about common law, we see this identification of natural law principles with customary usage. The determinations of English courts "for so many ages past," one judge maintained, "shew, not only what the common law . . . is, but that these rules of the common law, are the result of the wisdom and experience of many ages." Indeed, John Adams was prepared to argue—contrary to his own political desires—that the common law had not provided for life tenure of judges because "*custom and nothing else prevails, and governs in all those cases.*" "General customs," he declared, ". . . form the common law . . . ; they have been used time out of mind, or for a time whereof the memory of man runneth not to the contrary." Thus, the only question was whether a particular office provided for life tenure "by custom, that is, immemorial usage, or common law."[32]

The equation of common law with a fixed, customary standard meant that judges conceived of their role as merely that of discovering and applying preexisting legal rules. Because a "Usage had been uninterrupted," proclaimed the judges of the Massachusetts high court in 1765, "the Construction of the Law [is] thereby established" and they "therefore would make no Innovation." Are common law courts free to depart from the known rules of the common law, one judge asked. "No, surely, unless they take it upon them to be wiser than the Law." Should there be a discretion in Equity to modify the legal rule concerning jointly held property? No, a Virginia lawyer argued in 1737, for it would "overturn the most ancient and established Rules of Law" and "render Right and Property very precarious." "Instead of being determined by fixed and settled Rules and Principles," he concluded, "Law and Right [would] depend upon arbitrary Decisions which are ever fluctuating & contradicting one another."[33]

The result of all this was a strict conception of precedent. While lawyers would occasionally argue, as did James Otis, that it is "Better to observe the known Principles of Law than any one Precedent," the overwhelming fact about American law through most of the eighteenth century is the extent to which lawyers believed that English authority settled virtually all questions for which there was no legislative rule. In fact, this state of mind continued for a time after the American Revolution. Writing in 1786, the young James Kent continued to emphasize the colonial dichotomy between statute and common law. While he regarded any "Admission" that statutes enacted after settlement were binding as "subversive in Effect of our independent Rights," he continued to believe that the common law "can only be discovered & known by searching into the Decisions of the English Courts," which "are regarded with us as *authentic Evidence* of the Common Law & therefore are cited as *Precedents binding* with us even down to the year 1776." Thus, the rule of strict precedent was derived from the very conception of the nature of common law principles as preexisting standards discoverable by judges. Since "Judges cannot make law," the younger Daniel Dulany maintained in 1767, ". . . if they take upon themselves to frame a regulation, *in prospect*, which is to govern in [the] future, and which they have so framed, not to determine a case existing before them, but for the determination of cases that may happen, they essentially assume a power legislative." Just as the framing of general rules to guide future conduct was conceived of as a legislative function, judicial innovation itself was regarded as an impermissible exercise of will. This fundamental distinction between the very bases of legitimacy of common law and legislation determined the modest role the former played in the process of social change. In short, for the jurist of the eighteenth century, it was important, as Supreme Court Justice James Iredell declared in 1793, "that the distinct boundaries of law and Legislation [not] be confounded," since "that would make Courts arbitrary, and in effect *makers of a new law*, instead of being (as certainly they alone ought to be) *expositors of an existing one*."[34]

COMMON LAW CRIMES: THE BREAKDOWN OF THE EIGHTEENTH CENTURY CONCEPTION OF LAW

One of the most interesting manifestations of the breakdown of a unitary conception of the common law is the remarkable attack on the doctrine of federal common law crimes that emerged in the last years of the eighteenth century. Unheard of before 1793 and, with one exception, uniformly rejected by American judges during the first two decades of the Republic, the opposition to common law crimes nevertheless prevailed when in 1812 the United States Supreme Court declared that it had been "long since settled in public opinion" that a person could not be convicted of a federal crime without a statute.[35]

What was the nature of the attack on federal common law crimes? Basically, there were two arguments, the first deriving from states rights constitutional theories, the second reflecting a changing conception of the nature of law. Growing out of Jeffersonian hostility to criminal indictments of American citizens for pro-French activities,[36] the constitutional objection to common law crimes boiled down to the assertion that if the federal judiciary possessed jurisdiction to impose criminal sanctions without a statute it would be able to obliterate all constitutional limitations on the federal government. "[I]f the principle were to prevail," Thomas Jefferson wrote in 1800, "of a common law being in force in the United States" it would "possess . . . the general government at once of all the powers of the State Governments and reduce . . . us to a single consolidated government."[37]

Although it was reiterated many times over, it is difficult to understand precisely what the Jeffersonian argument was all about. In spite of his Jeffersonian loyalties, James Sullivan of Massachusetts understood that the question of common law jurisdiction involved no special constitutional difficulties, for all that it required was that federal common law jurisdiction be limited to those substantive crimes over which Congress had legislative power.[38] In short, if Congress could not constitutionally make an activity criminal, the courts could not impose common law criminal sanctions. There would be no greater danger of a federal court punishing activity beyond the scope of federal power than of Congress passing a statute exceeding those same limits.

There are still greater mysteries associated with the Jeffersonian constitutional position on common law crimes. Most thoughtful representatives of that position agreed that Congress could legitimately confer common law power on the courts to punish activities within the reach of federal constitutional power.[39] All that was lacking, they maintained, was that Congress in fact had not done so in the Judiciary Act of 1789.[40] Yet, they were thereby conceding that if Congress had passed an act similar to those enacted after the Revolution by state legislatures adopting the common law as the rule of decision, there would be no special constitutional problem. Indeed, this approach laid bare the proposition that there was no real issue concerning the proper allocation of powers between national and state governments, and, in fact, that the only respectable constitutional question involved the separation of powers between legislature and judiciary. And it was here that the campaign against common law crimes revealed the changing conception of the nature of law.

Beginning sometime in the 1780s and reaching a high point in the first decade of the nineteenth century, American jurists succeeded in dethroning the common law from the unchallenged place it had occupied in the jurisprudence of the revolutionary generation. It would be "in vain," St. George Tucker wrote in his famous 1803 edition of Blackstone, that we should "attempt, by a *general theory*, to establish an uniform authority and obligation in the common law of England, over the American colonies, at any period between the first migrations to this country, and that epoch, which annihilated the sovereignty of the crown of England over them." Emphasizing the differences between rules of the English common law and those of the American colonies as well as the diversity among the various colonies themselves, he argued that "it would require the talents of an Alfred to harmonize and digest into one system such opposite, discordant, and conflicting municipal institutions, as composed the codes of the several colonies of the period of the revolution."[41]

Tucker's attack on the conception of a unitary common law constituted the foundation of a more general assault on the uncertainty and unpredictability of common law adjudication. The diversity of the past, he emphasized, threatened uncertainty in the future. If the common law did not represent a known and discoverable body of legal rules, how could it provide either for the liberty of the citizen or for the protection of property? Indeed, these jurisprudential concerns as much as any question of constitutional allocation of power led the Virginia General Assembly in 1800 to instruct its United States senators that the assertion of federal common law criminal jurisdiction "opens a new code of sanguinary criminal law, both obsolete and unknown, and either wholly rejected or essentially modified in almost all its parts by State institutions. . . . It subjects the citizen to punishment, according to the judiciary will, when he is left in ignorance of what this law enjoins as a duty, or prohibits as a crime."[42]

Due to an excessive preoccupation with the political and constitutional dimensions of the struggle over common law crimes, historians have not fully appreciated the underlying change in the conception of law that laid the foundation for the constitutional struggle. For it is clear that the decline in the authority of the common law was neither coextensive with party loyalty nor with constitutional philosophy. For example, the leading target of Jeffersonian attacks on the judiciary, Federalist Justice Samuel Chase, was also the first federal judge to hold that there were no federal common law crimes. In *United States v. Worrall* (1798) — delivered one year before the question burst forth as a political issue—Chase declared that, since "the whole of the common law of England has been no where introduced" in America and since "the common law . . . of one State, is not the common law of another," there could be no general common law of the United States. Dismissing a common law indictment for bribery of a federal official, he argued that it was "essential, that Congress should define the offenses to be tried, and apportion the punishments to be inflicted." Even if Congress had declared and defined the offense, he added, it would be improper for a judge to exercise discretion in prescribing punishments. Thus, even where there was no need to discover and apply common law rules, Chase maintained that only legislative standards could legitimize judicial discretion.[43]

Although fear of judicial discretion had long been part of colonial political rhetoric, it is remarkable that before the last decade of the eighteenth century it was not associated with attacks on the common law jurisdiction of the judiciary. As late as 1786, Pennsylvania Admiralty Judge Francis Hopkinson attacked an amendment of the Commonwealth's penal laws on the ground that it conferred discretion on judges to impose punishments within certain limits. "This is," he protested, "in fact vesting them with legislative authority within those limits. It is a distinguishing mark of a free government, that the people shall know before hand, the penalty which the laws annex to every offense; and, therefore, such a system is called a government of laws, and not of men. But within the scope of the present bill, no offender can tell what his punishment is to be, till after conviction. The *quantum*, at least, is to be determined by the particular state of mind the judge happens to be in at the time of passing sentence." Nevertheless, even though he wrote two lengthy articles largely anticipating the challenge to common law crimes during the next decade, Hopkinson never so much as mentioned the problem of convicting an individual without a statute.[44] Common law standards, it appears, were just not conceived of as allowing a judge discretion.

Similarly, the eighteenth century debates over the obligation of English statutes in the colonies invariably brought forth concern over judicial discretion. As early as 1701, an anonymous Virginian wrote that the doctrine that only those Acts of Parliament were binding "where the Reason is the same here, that it is in England" was opposed by some as "leaving too great a latitude to the Judge." Without some guide as to the authority of British statutes, he concluded, "we [are] left in the dark in one of the most considerable

Points of our Rights, and the Case being so doubtful, we are too often obliged to depend upon the Crooked Cord of a Judge's Discretion in matters of the greatest moment and value."[45] Yet, there was no indication that the writer saw any problem of judicial discretion in applying common law as opposed to legislative rules. In spite of a continuing colonial preoccupation with the "uncertainty" of statutory rules, there is no evidence that before the Revolution Americans ever thought that the reception of common law principles endowed judges with the power to be arbitrary.

One of the few suggestions that colonials had any qualms about imposing criminal punishments without a statute occurs in a 1712 Connecticut case. After a trial of one Daniel Gard, the jury returned a special verdict. Finding certain facts true, the jury declared that "whether . . . Daniel Gard is guilty of murther or manslaughter we leave to the discretion of the court." The judges, in turn, adjourned the case and inquired of the Assembly whether they might determine the nature of the crime "by the rules of the common law, there being not so particular direction for the resolution of that point contained in our printed laws." In addition, if they should decide that the defendant had committed only manslaughter, they wished to know "what directions the Judges ought to have reference to in determining the punishment and giving sentence." The Assembly replied that the Judges might, "in the case above proposed, determine by the rules of the common law."[46]

It is, of course, quite difficult to make much out of this sparse record. Why did the judges see any difficulty in applying the common law? Was it that they anticipated some modern conception of the primary legitimacy of statute law? If indeed they did believe that the legislature was the only legitimate agency for defining punishments, it still does not seem that it was because they believed that the rules of the common law were uncertain or unascertainable. More likely, they were merely reflecting the burden under which all colonial jurists labored in the early years of the eighteenth century: uncertainty as to whether they were entitled at all to use and apply common law rules in this newly settled land. It seems therefore that, at most, the judges were looking to the legislature for authority to impose common law standards in criminal cases. Whatever the judges thought, however, it is clear that the Connecticut Assembly had no difficulty in believing that a direction to "determine" the case "by the rules of the common law" conveyed a clear and intelligible guide. Indeed, after this legislative direction, the judges themselves were able to determine the appropriate level of guilt.

Whatever one makes of the Gard case, it is plain that the specific attack upon common law crimes emerged from a distinctively post-revolutionary conviction that the common law was both uncertain and unpredictable. For example, even before the issue was turned into a constitutional question at the national level it began to surface in the states as well shortly after the Revolution. In his "Dissertation on the Act Adopting the Common and Statute Laws of England" (1793), Vermont Chief Justice Nathaniel Chipman "lay[s] it down as an unalterable rule that no Court, in this State, ought ever to pronounce sentence of death upon the authority of a common law precedent, without the authority of a statute." And two years later, in his treatise on Connecticut law, Zepheniah Swift, soon to be that state's chief justice, indicated that he too was troubled by the doctrine "that every crime committed against the law of nature may be punished at the discretion of the judge, where the legislature has not appointed a particular punishment." Distinguishing between "crimes which are expressly defined by statute or common law" and those actions over which "courts of law have assumed a discretionary power of punishing," he warned that judges "ought to exercise [the latter] power with great circumspection and caution," since "the supreme excellency of a code of criminal laws consists in defining every act that is punishable with such certainty and accuracy, that no man shall be exposed to the danger of incurring a penalty without knowing it." It would be unjust, he continued, for "a man [to] do an act, which he knows has never been punished, and against which there is no law, yet upon a prosecution for it, the court may by a determination subsequent to the act, judge it to be a crime, and inflict on him a severe punishment." "This mode of proceeding," he concluded, "manifestly partakes of the odious nature of an *ex post facto* law."[47] While Swift clearly did not abandon the natural law framework within which common law crimes traditionally had been understood, he was no longer prepared to assume that even the first judicial pronouncement of a legal rule was merely a declaration of some known and preexisting standard of natural law. Indeed, his entire discussion assumed the inability of individuals to know their legal duties without some express legislative or judicial pronouncement. Though it was left to others to extend Swift's analysis to the whole of common law crimes, his preoccupation with the unfairness of administering a system of judge-made criminal law was a distinctly postrevolutionary phenomenon, reflecting a profound change in sensibility. For the inarticulate premise that lay behind Swift's warnings against the danger of judicial discretion was a growing perception that judges no longer merely discovered law; they also made it.

Even at the national level, the problem soon transcended the constitutional rhetoric with which it had first been associated. In 1813 the question arose whether without a statute a person could be punished in admiralty for murder on the high seas. What distinguished this case from its predecessors was that there was no doubt that the Constitution had conferred exclusive jurisdiction in admiralty on the federal courts; thus, unlike earlier cases, no question could be raised of the federal courts' exceeding their constitutional powers. Even so, Supreme Court Justice William Johnson, on circuit, held that there could be no punishment. "[I]t was a favorite object" of the framers of the Constitution, he wrote, "to leave no one to search for the road of safety or the *Dii Limini* between crime and innocence, anywhere but in the Statute Book of the Legislative body they were about creating." In the colonies as well, he argued, "the adoption of the Common Law, depended upon the voluntary act of the legislative power of the several States" and not, as had previously been assumed, on some inherent obligation deriving from natural law or custom. Denying that the whole of the common law had been received in any state before the Revolution, or that the use of common law terms in statutes gave any "more validity to the Common Law than use of terms peculiar to the Civil or Cannon [sic] Law would to the latter," he concluded that the recognition of common law crimes even in Admiralty would invite "judicial discretion." Such a system "would . . . increase the oddity of the state of things," since "the judiciary would have then to decide what system they would adopt as their guide on the Law of Nations, or be left at large to be governed by their own views on the Fitness of things."[48]

As the attack on common law crimes became more and more dissociated from its originally narrow constitutional basis, there appeared in 1819 a remarkable and influential book articulating the problem entirely in terms of a new conception of law. In *Historical Sketches of the Principles and Maxims of American Jurisprudence,* Ohio lawyer John Milton Goodenow inquired for the first time whether common law crimes could be in force even in an individual state. Beginning his inquiry with a general analysis of the nature of legal obligation, his initial axioms reveal how far many Americans had moved from eighteenth century natural law assumptions:

[L]aws, which man creates from himself, in his social state, are not the emanations of Divine Will, nor yet the pure institutions of nature and reason;

but changeable and arbitrary in their formation; they are necessarily of a *positive, local existence*; made, declared and *published* in a shape and character clear and unequivocal to all to whom they are directed; otherwise, they could never become obligatory: because they are not of intuition, discoverable by the eye of reason.

Confining the obligations of natural law entirely to "what may be called the right of conscience," and asserting that for violations of natural law man "is accountable for error in judgment to none but his GOD," Goodenow concluded that the principles of natural law "may be disregarded and lie for ages buried beneath the rubbish of human invention."

Our ears are often saluted in our courts of justice, with a sort of text to all common law arguments, that "law is the perfection of reason—that a law against reason is void—that no man is bound to obey a municipal law which infringes the law of nature." If such a doctrine were in fact to prevail, the blood of half mankind would flow in its execution.

From these premises Goodenow moved on to argue that "all human laws for the punishment of crimes, are mere matters of social policy; diverse in different states as their political purposes, forms and ends of government, are diverse." Criminal laws depended "no more upon man's natural reason, no more consonant therewith, than the political shade and texture of the time in which they are made." It followed that the only legitimate authority that could impose criminal sanctions was the legislature by means of statutory enactment. "The judge . . . of a criminal tribunal," Goodenow concluded, "is *governed* himself by *positive* law, and executes and inforces the will of the supreme power, which is the will of THE PEOPLE."[49]

With Goodenow's *Historical Sketches* we see most clearly the always implicit relationship between the attack on common law crimes and more general changes in political theory and conceptions of the nature of law. Our task now is to see more precisely how these changes came about and, finally, to determine how these transformations in legal theory led ultimately to an instrumental conception of common law.

The Emergence of an Instrumental Conception of Law

Before the American Revolution common law and statute law were conceived of as two separate bodies of law, and the authority of judges and legislators was justified in terms of the special category of law that they administered. This dichotomy between the nature of the two forms of law was itself a fairly recent creation, reflecting the beginnings of the modern conception of sovereignty.[50] For example, when Lord Coke, in deciding *Calvin's Case* in 1608, held that English law was not automatically received in conquered territories, there was no suggestion of a distinction between statute and common law, for statutes were still largely conceived of as an expression of custom. We begin to see a different treatment accorded to the two forms of law early in the eighteenth century as represented by Attorney General West's 1720 statement confining the problem of reception to parliamentary acts. As statutes began to be understood as a command of the sovereign, the formula governing reception thus took on a function entirely different from Coke's formulation, raising the question of the basis of obedience to rules not derived from natural law principles. Finally, as the principle of sovereignty emerged full blown in English law by the second half of the century, Blackstone was again able to maintain, with an eye toward the emerging constitutional struggles in the colonies, that both statute and common law were not received in America.[51]

It is not generally appreciated to what extent the American experience after the Revolution recapitulated that of the English. As much as they attacked the Blackstonian conception of a single and indivisible sovereignty, Americans after the Revolution began widely to accept the modern theory of law underlying that conception. While they disputed the supremacy of parliament, they simultaneously argued that written constitutions were legitimate because they embodied the "will" of the people. And as they sought to redefine the basis of legal obligation in terms of popular sovereignty, they tended to assert the ultimate primacy of the legislature and of statute law. The result was that the original natural law foundation of common law rules began to disintegrate.

The demand for codification of all laws, based on the principle that "in republics, the very nature of the constitution requires the judges to follow the letter of the law," emerged with particular vigor after the Revolution. If judges "put such a construction on matters as they think most agreeable to the spirit and reason of the law," the author of *People the Best Governors* argued, "they assume what is in fact the prerogative of the legislature, for those that made the laws ought to give them a meaning when they are doubtful." In fact, courts "must take the law as it is," another commentator argued, "and by all due and proper means execute it, without any pretense to judge of its right or wrong."[52]

What underlay these demands for codification was a new conviction that much of the English common law itself was a product of the whim of judges. Of great importance was the reforming work of Lord Mansfield in England, which convinced many Americans that judges could not be depended on merely to apply existing law. More than any other factor, it appears, Mansfield's decisions convinced Thomas Jefferson that a check need be established on the common law powers of judges. While "the object of former judges ha[d] been to render the law more & more certain," Jefferson wrote in 1785, Mansfield had sought "to render it more uncertain under pretence of rendering it more reasonable. No period of English law of what ever length it be taken, can be produced wherein so many of it's [sic] settled rules have been reversed as during the time of this judge." As a result, he concluded, Mansfield's "accession to the bench should form the epoch, after which all recurrence to English decisions should be proscribed." In some of his work, in fact, Jefferson went even further. Reporting on the efforts of the Virginia codifiers, Jefferson declared that "the common law of England, by which is meant that part of the English law which was anterior to the date of the oldest statutes extant," was "made the basis" of the code. He was thus prepared to limit common law reception to rules existing in the thirteenth century, if not before![53]

In the course of several decades, the destruction of a conception of a fixed and determinate common law would lead to the position ultimately articulated by Robert Rantoul in his famous plea for codification in 1836. "Why," he asked, "is an *ex post facto* law, passed by the legislature, unjust, unconstitutional, and void, while judge-made law, which, from its nature, must always be *ex post facto,* is not only to be obeyed, but applauded."[54] By the time Rantoul asked the question, judges and jurists had begun to convert the old natural law justification of the common law into a theory of legal "science," a development I will examine in Chapter VIII. In the immediate postrevolutionary period, however, the defense of common law was far different, representing an effort to find a unitary foundation for both statute and common law.

This special crisis of common law adjudication after the American Revolution can be seen most dramatically in Supreme Court Justice James Wilson's "Lectures on Law" delivered in 1791. The central purpose of these lectures, he pointed out, was to demonstrate that Blackstone's assertion that law is the will of a superior was "an improper principle," since "the sole legitimate principle of obedience to human laws is human consent." In the process, however,

Wilson revealed the extent to which he had come under the spell of modern conception of law as a sovereign command. "It is agreed, on all hands," he acknowledged, "that, in every state, there must be somewhere a power supreme, arbitrary, absolute, uncontrollable." The only dispute concerned "where does this power reside."

While Wilson continued to acknowledge the obligations derived from natural law, unlike his predecessors he reduced them to private questions of conscience. For the fundamental question was "whether a man can be bound by any human authority, except his own consent?"

Let us suppose, that one demands obedience from me to a certain injunction, which he calls a law. . . . I ask him, why am I obliged to obey it? He says it is just that I should do it. Justice, I tell him, is a part of the law of nature; give me a reason drawn from human authority . . . he assails me on a very different quarter; and softening his accents, represents how generous, nay how humane, it would be, to do as he desires. Humanity is a duty; generosity a virtue; but neither is to be referred to human authority.

Thus, the basis of obedience to law was set entirely within the modern framework of a will theory of law, however much Wilson argued over whose will ultimately legitimized legal commands. This definition of the basis of obligation in terms of popular will was a far cry from the eighteenth century conception of obligation derived from the inherent rightness or justice of law.

The result was a distinctly postrevolutionary phenomenon: an attempt to reconstruct the legitimacy not simply of statutes, but of common law rules, on a consensual foundation. Wilson, for example, insisted that custom was "of itself, intrinsick evidence of consent."

How was a custom introduced? By voluntary adoption. How did it become general? By the instances of voluntary adoption being increased. How did it become lasting? By voluntary and satisfactory experience, which ratified and confirmed what voluntary adoption had introduced. In the introduction, in the extension, in the continuance of customary law, we find the operations of consent universally predominant.

Thus, he concluded, the "origin of [the] obligatory force" of the common law rested "on nothing else, but free and voluntary *consent*."[55]

That Wilson had gone a long way toward the Blackstonian position can be seen by contrasting his argument with that of St. George Tucker, who was equally desirous of resisting Blackstone's propositions concerning sovereignty. While Blackstone "presupposes an act of the legislature in every case whatsoever," Tucker maintained, "there is more ingenuity than truth in the idea . . . that all the unwritten rules of law are founded upon some positive Statute, the memory of which has been lost." But as he moved further from Blackstone's definition of sovereignty, Tucker tended to emphasize "immemorial Custom, and Usage" as sources of legitimacy opposed to sovereignty. Though he succeeded to some extent in resisting the pressure of Blackstone, he exposed even more the vulnerability of a system of common law adjudication under a regime of popular sovereignty.[56]

If the principle of popular sovereignty seemed to some to lead logically to complete legislative codification, orthodox legal writers like James Wilson sought to show instead that the common law power of judges was entirely compatible with the sovereignty principle. The emphasis in postrevolutionary legal thought on the consensual foundation of the common law was thus designed to demonstrate that common law judges actually constituted the "trustees" or "agents" of the sovereign people.

The problem of fitting the common law into an emerging system of popular sovereignty became the central task of judges and jurists at the turn of the century. One response and one of the most subtle shifts in the theory underlying the legitimacy of common law rules can be seen in Jesse Root's short essays, "The Origin of Government and Laws in Connecticut" and "On the Common Law of Connecticut," published in 1798 as an introduction to the first volume of Root's Connecticut Reports. Root sought to expose the "ignorance of those who are clamorous for a new constitution" that would end the authority of English law, emphasizing the "mistake of those who suppose that the rules of the common law of England are the common law of Connecticut, until altered by a statute." Pointing to the feudal origins of English law, he insisted that the citizens of Connecticut had long followed their own indigenous common law. "Their common law," he declared, "was derived from the law of nature and of revelation; those rules and maxims of immutable truth and justice, which arise from the eternal fitness of things, which need only to be understood, to be submitted to; as they are themselves the highest authority."

Not only does Root's analysis at this point seem highly traditional, but if that were all there were it would be quite remarkable. One just does not find Americans at the turn of the century who ultimately rest upon immutable natural law principles while insisting at the same time on a separate and distinct body of indigenous American legal rules. But true to form, Root also found it necessary to explain the American deviation by the unifying principle of consent. There was, he wrote, "another branch of common law . . . derived from certain usages and customs, universally assented to and adopted in practice by the citizens at large . . ." which "courts of justice take notice of . . . as rules of right, and as having the force of laws."

But how does custom become law? No longer able to equate custom with obligatory natural law, Root's answer demonstrated the impact of the Blackstonian will theory of law. "That these customs and usages must have existed immemorially, and have been compulsory, in order to their being recognized to be law; seems to involve some degree of absurdity — that is, they must have the compulsory force of laws, before they can be recognized to be laws, when they can have no compulsory force till the powers of government have communicated it to them by declaring them to be laws." It was not enough that customs became law as soon as their origins were forgotten; "this may be necessary in arbitrary governments, but in a free government like ours," Root continued, there was a "better reason." Just as statutes are binding because they have been enacted with the consent of the legislature, he concluded, "so these unwritten customs and regulations . . . have the sanction of universal consent and adoption in practice" and for that reason courts of justice may declare them to be obligatory. "The reasonableness and utility of their operation, and the universality of their adoption" were the best evidence "of their having the general consent and approbation." Before he had finished, therefore, Root was forced to retreat from the traditional natural law framework with which he had begun. By insisting that America had its own common law, he managed ultimately to justify the compulsory force both of statute and common law by the single legitimating principle of consent.[57]

While Root sought to save the common law by appealing to an indigenous body of common law principles to which the people of Connecticut had consented, there were others, like James Sullivan of Massachusetts, who attempted to rest the old conception of a universal law on a new foundation of popular sovereignty. While rules concerning real estate, Sullivan pointed out, had always rested on local statutes or customs, legal doctrines governing "personal estate [are] not fixed to any place or country, and contracts depend on the *jus gentium* (the general law of nations) for their origin and their expositions, rather than on any municipal regulations of particular countries. . . . As personal contracts are founded in commerce, they cannot rest on the particular laws of one country only; but ought to be the subject of those principles of the general law of nations, which are acknowledged by the world." But Sullivan was no longer content to rest his appeal to a universal law on the eighteenth century trust in natural law principles. While these legal rules were "transfer[red] . . . with certainty from age to age" through the "written institutes" and "reports" of the common lawyers, "it may be asked, where do these books derive an authority to become the law of the land?" "The answer," Sullivan concluded, equating statute and common law, "is that the most solemn act of legislation is no more than an expression of the public will." These principles had been "received" as "principles, to which the people of the country have, for a long time together, submitted, considering them as proper and useful."

If this explanation still sounded too much like the mere submission to custom which Root had found prevalent "in arbitrary governments" Sullivan restated his conclusion in a somewhat different way. "It may be inquired," he wrote, "how are the English institutes and reporters evidence of the will of the American nation, or of this Commonwealth? The answer is, that the will of the people is expressed in the constitution, which they have of their own authority established for their Commonwealth." Since "the voice of the people has established this [common law] system, so far only as it had been before adopted and practiced upon," it was "the judges . . . who [were] to decide . . . whether a principle, urged as law, had heretofore been in practice."[58]

The result of this transformation in the underlying basis for the legitimacy of the common law was that jurists began to conceive of the common law as an instrument of will. Delivering his charge to the jury in the 1806 prosecution of the Philadelphia Cordwainers for common law conspiracy, Mayor's Court Judge Moses Levy declared that the jury was "bound to conform" to the common law rule "even though we do not comprehend the principle upon which it was founded." "We are not to reject it because we do not see the reason of it. It is enough, that it is the will of the majority. It is law because it is their will—if it is law, there may be good reasons for it though we cannot find them out."[59]

Nevertheless, this view of law was decidedly a two-edged sword. While nineteenth century judges were satisfied to limit the discretion of juries by reminding them that "unwritten custom" and "written law . . . are both equally the act of the legislature," their own behavior reflected James Sullivan's view that they had been given a

popular charter to mold legal doctrine according to broad conceptions of public policy. In their influential lectures at the Litchfield
Law School, Judges Tapping Reeve and James Gould reflected this
changed conception of law. By 1813 they were instructing students
in the Blackstonian definition of municipal law as "a rule of civil
conduct prescribed by the supreme power of the state commanding
what is right and prohibiting what is wrong." Not only had all the
earlier qualifications of the Blackstonian argument that had preoccupied James Wilson and St. George Tucker completely disap-
peared, but with their disappearance came a candid recognition of
the new basis of the common law. "Theoretical[ly] courts make no
law," they declared, "but in point of fact they are legislators." And
after citing cases where courts had made law, they inquired: "How
then could these laws have been prescribed by a supreme power in a
state? By the acquiescence of the legislature, they impliedly consented to these laws, and it is immaterial whether this consent be
subsequent or antecedent to there [sic] birth." Finally, with a dash
of irony, they laid to rest the old conception of law.

How can [common law rules] be said to have existed from time immemorial, when there [sic] origin is notorious. The Judges of the Courts of Judicature are considered as the depositories of the common law. Therefore
when they lay down a new, and before unheard of rule of law, they are supposed to take from this repository, where it had laid dormant and unwanted
from time immemorial.[60]

As judges began to conceive of themselves as legislators, the criteria by which they shaped legal doctrine began to change as well. The
principle upon which judges ought to decide whether to adhere to a
series of decisions, one jurist noted, "resolves *itself into a question of
expediency*." "But what shall be the test of . . . error" in prior decisions? he asked. "*I know nothing but the mind that is to judge*." As a
result, during the first two decades of the nineteenth century judges
began to conceive of themselves as the leading agents of legal
change. Since "a Legislature must establish a general unbending
rule," Zephaniah Swift argued in 1810, while "courts possess a discretion of shaping their rules to every possible variety of circumstance," there "is a vital principle, inherent in the constitution of the
judiciary . . . furnishing remedies according to the growing wants,
and varying circumstances of men, . . . without waiting for the slow
progress of Legislative interference."[61]

One of the most dramatic manifestations of the new role of courts
is a Pennsylvania statute of 1807 empowering the judges of the Su-

preme Court to decide which English statutes were in force in the Commonwealth. While the delegation of so explicit and self-conscious a legislative function to judges would have been inconceivable even two decades earlier, it was completely in tune with newly emerging perceptions of law as will. Reflecting this change, Pennsylvania Supreme Court Justice Hugh Henry Brackenridge strongly argued in 1808 for the abolition of the common law and the introduction of statutory codification. By 1814, however, he was prepared to deplore "the sullenness: or affected timidity of English judges, in the narrowness of their construction of powers given." Judges bind themselves to precedent, he observed, only because they have "a dread of innovation," while in fact "*departure* from rule can be justified only by *success*." The two greatest obstacles to change were "attachment to decisions" and "timidity of mind in effecting a reform." Judges, he insisted, are the architects of the legal system. The bulk of judges "trudge on through the slough as Hodge did *even after the bridge was built*; so far are they from attempting *to build a bridge*. Such may have the praise of being what are *sound lawyers*; but they must be contented with this, and cannot be called *great judges*." A great judge, he argued, "who has traversed all space of legal science," can be a reformer if "such a mind happened to be at the head of the highest court." While "no one but a *skilful architect* . . . who can have the whole edifice in his mind," ought to undertake the task of legal change, he concluded, reform is preeminently the judge's task, for "the legislature can act only in detail, and in particulars, whereas the able judge can remove at once, or alter, what was originally faulty or has become disproportioned in the building."[62]

As Brackenridge's observations indicate, one of the most universal features of postrevolutionary American jurisprudence was an attack on the colonial subservience to precedent. Even conservative jurists like Vermont Chief Justice Nathaniel Chipman complained that the legal profession had followed precedents "with too great veneration." These precedents, he noted in his "Dissertation of the Act Adopting the Common and Statute Laws of England" in 1793, "were made at a time when the state of society, and of property were very different from what they are at present." Moreover, many common law doctrines were formulated "in an age when the minds of men were fettered in forms [and] when forms were held to be substances, and abstractions real entities." As a result, at common law "technical reasoning and unmeaning maxims . . . frequently supplied the place of principles." In America, he emphasized, substance is more important than form. Consequently, Blackstone's rule of strict adherence to precedent might suffice in England, where "by too easy departure, Judges might unwarily disturb property acquired on the faith of such precedents." But to follow "arbitrary rules" or "arbitrary decisions" without understanding that they "arose out of [a different] state of society" would be "certainly contrary to the principles of our government and the spirit of our laws."[63]

Chipman's "Dissertation" reflects the beginnings of a postrevolutionary functionalism in American law. His emphasis on "principle" and "reason" in law and his rejection of the "arbitrary" authority of customary rules were derived from a new conception of common law as a self-conscious instrument of social policy. "Instead of entertaining a blind veneration for ancient rules, maxims, and precedents," he insisted, "we [sh]ould learn to distinguish between those which are founded on the principles of human nature in society, which are permanent and universal, and those which are dictated by the circumstances, policy, manners, morals and religion of the age." Similarly, the future chief justice of Connecticut, Zephaniah Swift, argued in 1795 that since the settlers of that state were "unfettered by the shackles of forms, customs, and precedent," they "have attended to enlarged, and liberal views of policy." Though he acknowledged that most colonial changes had been brought about by statute, these laws established "a variety of legal principles" by which judges would guide their own conduct. Thus, courts were "not absolutely bound by the authority of precedents."

If a determination has been founded on mistaken principles, or the rule adopted by it be inconvenient, or repugnant to the general tenor of the law, a subsequent court assumes the power to vary from it or contradict it. In such cases they do not determine the prior decisions to be bad law; but that they are not law. Thus in the very nature of the institution, is a principle established which corrects all errors and rectifies mistakes.[64]

Swift thus came as close as any jurist of the age to maintaining that law is what courts say it is. While no judge of the period acknowledged himself to be free from the restraints of "reason" and "principle" in formulating legal doctrine, one cannot survey the judicial output of the period without seeing the new policy orientation that judges brought to their work. Swift himself recognized in his 1810 treatise on the law of evidence what no jurist would have perceived a quarter of a century earlier: that "the rules of evidence are of an artificial texture, not capable in all cases of being founded on abstract principles of justice. They are positive regulations founded on policy." In short, "reason" and "principle" came to be understood not as rules or doctrines to be discovered, not as customary norms to be applied through precedent, but as a body of prudential regulations framed, as Swift himself saw, from the perspective of "enlarged and liberal views of policy." The result was that by the turn of the century, judges often explained in functional terms why they were free to disregard the authority of prior cases. In a typical case involving marine insurance decided in 1799, for example, a New York Supreme Court judge declared that if courts rigidly adhere to precedent "we must hope of little improvement in our commercial code. A single decision, though founded on mistake, would become of binding force, and by repetition, error might be continued, or heaped on error, until the common sense of mankind, and the necessity of the case oblige us to return to first principles, and abandon precedents."[65]

As courts were declaring their freedom from strict eighteenth century conceptions of precedent, they also began to see the problem of order and uniformity in the legal system in a new light. For the eighteenth century American jurist, the problem of legal certainty was essentially defined in political terms. Without "known" and "certain" laws, Massachusetts Chief Justice Hutchinson argued in 1767, "the Will of the Judge would be Law: and this tends to a State of Slavery," for citizens would come to "depend upon the arbitrary Opinion of another." In the nineteenth century, by contrast, legal certainty was conceived of as important primarily to the extent that it enabled individuals to plan their affairs more rationally. Swift, for example, remarked on the relationship between "uniformity of decision" in England and "the immense wealth, and unparalleled commercial prosperity of that nation." On the other hand, as adherence to precedent began to be understood as just one of a number of techniques for allowing men to order their affairs with regularity, judges were thus prepared to abandon precedent in order to create substantive doctrines that would themselves assure greater predictability of legal consequences. For example, in *Lee* v. *Boardman* (1807) the Massachusetts Supreme Court departed from the English rule that an insured party could not recover for total loss upon abandoning a captured ship. For the ship owner to await the outcome, Chief Justice Parker declared, would create "uncertainty," since it would be "out of the power of the assured to calculate with any certainty upon the extent of his funds: his commercial enterprise will be checked, and his plans embarrassed and defeated." In short, adherence to precedent was thought of as necessary only to the extent it allowed private parties to calculate in advance on the consequences of particular courses of conduct. Even where a judge acknowledged that it would not be "materially disadvantageous to commerce" to settle a marine insurance question "in either way, contended for in this cause,"—an agnostic attitude itself inconceivable two decades earlier—it was "of utmost importance, that the point should be clearly decided and settled in one or the other way; that merchants may know, and accommodate their affairs to the decision." As common law rules were conceived of as made and not discovered, precedent was no longer regarded as a technique for assuring that judges would apply a preexisting law. Instead, courts began to classify categories of legal rules in terms of the extent to which innovation would have "retrospective influence" on existing private arrangements. Thus in a "question of commercial concern, the determination of which can have no retrospective influence, nor affect pre-existing rights," a court regarded itself as "less restrained by the authority of existing cases."[66]

The retreat from precedent went hand in hand with a new definition of the role of the judge in formulating legal rules. One of the most radical departures occurred in a series of cases at the beginning of the nineteenth century involving the deference owed to decisions of English admiralty courts. The facts were all essentially the same: suits by ship owners to recover under marine insurance policies for ships that had been seized by the British for supposed violations of neutrality. Since the policies invariably covered only neutral property, the question was whether American courts were bound by the decision of the British admiralty holding that the property was nonneutral. The deeper question which confronted American judges was whether the traditional conception of courts discovering and impartially applying a fixed law of nations could coexist with clashing national policies at a time when critics were maintaining that the law of nations suffered from "the same uncertainty, which is felt so oppressively in the uncertainty of the common law." In a path breaking decision, the New York High Court of Errors in 1802 held that New York courts were not bound by prior English judgments. "There is not any uniform law by which these [English] courts govern themselves," one judge declared. "They listen more to instructions from the sovereign, than to the injunctions of the law of nations." It was important that American courts not establish rules interpreting insurance contracts "that would tend to embarrass commerce, or injure the assured; but adopt such a construction as will most promote the important objects in view." Since there was no clear requirement in international law, "we are at liberty to adopt such a construction as shall most subserve the solid interests of this growing country." The perception by American courts that the English admiralty courts were "governed . . . by ideas of political expediency" soon led American judges to see that it was necessary to adopt legal doctrines which in turn best promoted their own "solid interests."[67]

One of the most important consequences of the increased instrumentalism of American law was the dramatic shift in the relationship between judge and jury that began to emerge at the end of the eighteenth century. Although colonial judges had developed various techniques for preventing juries from returning verdicts contrary to law, there remained a strong conviction that juries were the ultimate judge of both law and facts.[68] And since the problem of maintaining legal certainty before the Revolution was largely identified with preventing political arbitrariness, juries were rarely charged with contributing to the unpredictability or uncertainty of the legal system. But as the question of certainty began to be conceived of in more instrumental terms, the issue of control of juries took on a new significance. To allow juries to interpret questions of law, one judge declared in 1792, "would vest the interpretation and declaring of laws, in bodies so constituted, without permanences, or previous means of information, and thus render laws, which ought to be an uniform rule of conduct, uncertain, fluctuating with every change of passion and opinion of jurors, and impossible to be known till pronounced." Where eighteenth century judges often submitted a case to the jury without any directions or with contrary instructions from several judges trying the case, nineteenth century courts became preoccupied with submitting clear directions to juries. Indeed, only when the judge-jury problem began to emerge at the end of the eighteenth century did Americans insist on published reports in order to relieve "the uncertainty and contradiction attending judicial decisions."[69]

Not until the nineteenth century did judges regularly set aside jury verdicts as contrary to law. At the same time, courts began to treat certain questions as "matters of law" for the first time. For example, at the turn of the century it was typical for a court in even a complicated marine insurance case to suggest a rule for measuring damages while conceding that "unless the defendant could shew a better rule, the Jury might adopt that, or such other as would do justice." By 1812, however, in a decision that expressed the attitude of nineteenth century judges on the question of damages, Justice Story refused to allow a damage judgment on the ground that the jury took account of speculative factors that "would be in the highest degree unfavorable to the interests of the community" because commercial plans "would be involved in utter uncertainty." As part of this tendency, legislatures began to take the question of damages entirely away from juries in eminent domain proceedings on the ground that, as one canal company maintained, this mode of assessing damages was "injurious and expensive, and . . . justice requires some amelioration" of the law's provisions. Finally, as part of the expanding notion of what constituted a "question of law" courts for the first time ordered new trials on the ground that a jury verdict was contrary to the weight of the evidence, despite the protest that "not one instance . . . is to be met with" where courts had previously reevaluated a jury's assessment of conflicting testimony.[70]

If an increasingly instrumentalist view of law led courts to narrow the province of the jury, this trend in turn encouraged judges to conceive of law in still more substantive terms, for a sharp distinction between law and facts provides a necessary inducement for courts to think of law in functional terms. Consequently, one of the most important symptoms of the emerging instrumentalism of nineteenth century law was the decline in the significance of technicalities in determining the outcome of cases. Following the lead of Nathaniel Chipman and Zephaniah Swift, who in the last decade of the eighteenth century sought to rescue American law from rules developed in "an age when the minds of men were fettered in forms," courts began to show contempt for lawyers who relied on technicalities to win cases. Formal pleading, the New York Supreme Court declared in 1805, "ought to be discountenanced, as being calculated to mislead magistrates, and involve proceedings . . . in all the technical niceties of special pleadings." In 1809 Massachusetts Chief Justice Parsons accused a defendant's lawyer of attempting to use his superior knowledge of pleading to evade "the apparent merits of the case." "All needless refinements," he concluded, "ought to be rejected, and all finesse intended to *ensnare* should be avoided." An important result was the increasing frequency with which courts allowed plaintiffs to amend imperfect pleadings instead of throwing them out of court.[71]

This reaction against form was part of a deeper change in thought. Not only were lawyers and jurists far less likely to analyze problems in relation to a static system of common law writs, but they were also led to think less and less in terms of self-contained analogies to inherent categories of the legal system. When a Massachusetts lawyer in 1810 successfully argued in favor of the unprecedented position that a corporation should be liable in tort, he no longer relied primarily on analogy to existing common law actions. Rather, he contended that "this form of action should be maintainable" because "corporations were anciently not common, and the necessity did not exist." However, since "they are now multiplying among us beyond all former examples, . . . individuals may frequently sustain injuries from their neglect." Similarly, when in 1793 Nathaniel Chipman reflected on whether promissory notes were negotiable, he freely conceded points that earlier lawyers had taken as decisive barriers to negotiability. Even though, he acknowledged, negotiability did not exist at common law and Vermont had not received the English statute permitting negotiability, it could still "be maintained . . . on principles of right, arising from the nature of the transaction itself." Once the actual understanding and purpose of the parties to the transaction were identified, he insisted, the legal system should provide the mechanism for enabling individuals to accomplish their own desires. The ascendancy of substance over form had begun to emerge.[72]

By 1820 the legal landscape in America bore only the faintest resemblance to what existed forty years earlier. While the words were often the same, the structure of thought had dramatically changed and with it the theory of law. Law was no longer conceived of as an eternal set of principles expressed in custom and derived from natural law. Nor was it regarded primarily as a body of rules designed to achieve justice only in the individual case. Instead, judges came to think of the common law as equally responsible with legislation for governing society and promoting socially desirable conduct. The emphasis on law as an instrument of policy encouraged innovation and allowed judges to formulate legal doctrine with the self-conscious goal of bringing about social change. And from this changed perspective, American law stood on the verge of what Daniel Boorstin has correctly called one of the great "creative outbursts of modern legal history."

II The Transformation in the Conception of Property

T HE productive development of land and natural resources at the beginning of the nineteenth century drew into question many legal doctrines formulated in an agrarian economy. In the eighteenth century, the right to property had been the right to absolute dominion over land, and absolute dominion, it was assumed, conferred on an owner the power to prevent any use of his neighbor's land that conflicted with his own quiet enjoyment.[1] Blackstone, in fact, asserted that even an otherwise lawful use of one's property could be enjoined if it caused injury to the land of another, "for it is incumbent on a neighboring owner to find some other place to do that act, where it will be less offensive."[2] Not until the nineteenth century did it become clear that, because this conception of ownership necessarily circumscribed the rights of others to develop their land, it was, in fact, incompatible with a commitment to absolute dominion. Logical difficulties had been easily concealed by experience, since the prevailing ideal of absolute property rights arose in a society in which a low level of economic activity made conflicts over land use extremely rare. As the spirit of economic development began to take hold of American society in the early years of the nineteenth century, however, the idea of property underwent a fundamental transformation—from a static agrarian conception entitling an owner to undisturbed enjoyment, to a dynamic, instrumental, and more abstract view of property that emphasized the newly paramount virtues of productive use and development. By the time of the Civil War, the basic change in legal conceptions about property was completed. This chapter examines the process by which the change took place.

PROPERTY RIGHTS IN THE NINETEENTH CENTURY: THE GENERAL VIEW

Two potentially contradictory theories of property rights underlay eighteenth century legal doctrines for resolving conflicts over uses of property. The first, an explicitly antidevelopmental theory, limited property owners to what courts regarded as the natural uses of their land, and often "natural" was equated with "agrarian." For example, in cases involving the conflicting claims of two riparian owners, courts usually gave precedence to appropriation of water not only for domestic purposes but often for agriculture and husbandry as well.[3]

Natural uses of land were probably favored also by strict liability in tort: any interference with the property of another gave rise to liability; only the lowest common denominator of noninjurious activity could avoid a suit for damages. The frequency with which eighteenth century courts solemnly invoked the maxim *sic utere tuo, ut alienum non laedas*[4] is a significant measure of their willingness to impose liability for injury caused by any but the most traditional activities.

The second theory of property rights on which courts drew in the eighteenth century, though it appeared in a variety of legal forms, amounted to a rule that priority of development conferred a right to arrest a future conflicting use. Sometimes this rule was simply stated by the long-standing maxim "first in time is first in right." More refined formulations required that the first user be engaged in his activity for a period of time sufficient to ripen into a prescriptive property right against interfering activities.

At first glance, the rule of priority seems more compatible with economic development, since it gives at least the first user freedom to develop his land as he wishes. By contrast, doctrines based on natural use confer on all landowners equal power to maintain the traditional order of things and thereby to impose a continuing pattern of nondevelopment. Before the nineteenth century, however, the theory of priority was harnessed to the common antidevelopmental end. Where two neighboring parcels of land were underdeveloped, each owner could claim a right, based on priority, to prevent further development. Thus, depending on the level of economic development from which one begins to measure priority, the consequences of the theories of natural and prior use may be the same; since the lowest level of development is also the earliest, each party acquires a prior right to the land in its natural state.

Furthermore, just as the theory of priority could be reduced to one of natural use, so could the natural use doctrine claim to enforce a rule of priority. If the starting point for judgment is not the first introduction of a new use on the land but rather the prior state of inactivity before the new use appears, then once again the doctrines of priority and natural use yield the same result. Indeed, when in the name of economic development the inevitable attack on eighteenth century property doctrine begins, these two are regularly lumped together by their opponents as part of one theory.

Though the two theories can be merged, they can also be made to have profoundly different consequences. If priority is measured not from a common denominator of natural use but from the time that a new technology appears, the theory of natural use continues to enforce its antidevelopmental premises, but a rule of priority now confers an exclusive property right on the first developer.

The potential for conflict between the two theories first began to surface in the nineteenth century. There are, for example, no cases before then dealing with conflicts over use of water in which an English or American court acknowledges that different consequences follow from a rule of "natural flow" as opposed to one of "prior appropriation." Courts were induced to distinguish between the two rules in such cases only when the judges began trying to break away from the antidevelopmental consequences of common law doctrine.

Once priority and natural use had taken on different operational meanings, the common law had moved into the utilitarian world of economic efficiency. Claims founded on natural use began to recede into a dim preindustrial past and the newer "balancing test" of efficiency came into sharp focus. As priority came to take on a life of its own distinct from doctrines of natural use, it was put forth not as a defense against encroachments of modernity but as an offensive doctrine justified by its power to promote economic development. In a capital-scarce economy, its proponents urged, the first entrant takes the greatest risks; without the recognition of a property right in the first developer—and a concomitant power to exclude subsequent entrants—there cannot exist the legal and economic certainty necessary to induce investors into a high-risk enterprise.

Though the strength of its hold varied among particular areas of the law, in general, priority became the dominant doctrine of property law in the early stages of American economic growth. Its development paralleled that of pervasive state-promoted mercantilism in the early nineteenth century American economies; while it was displaced almost immediately in some areas of the law, in others it continued to stand firm well into the century.

The attack on the rule of priority reveals the basic instability of utilitarian theories of property. As property rights came to be justified by their efficacy in promoting economic growth, they also became increasingly vulnerable to the efficiency claims of newer competing forms of property. Thus, the rule of priority, wearing the mantle of economic development, at first triumphed over natural use. In turn, those property rights acquired on the basis of priority were soon challenged under a balancing test or "reasonable use" doctrine that sought to define the extent to which newer forms of property might injure the old with impunity. Priority then claimed the status of natural right, but only rarely did it check the march of efficiency. Nor could a doctrine of reasonable use long protect those who advanced under its banner, since its function was to clear the path for the new and the efficient. Some of its beneficiaries eventually reclaimed the doctrine of priority, this time asserting the efficiency of "natural monopoly" and the inevitability of a standard of priority.

Viewed retrospectively, one is tempted to see a Machiavellian hand in this process. How better to develop an economy than initially to provide the first developers with guarantees against future competitive injury? And once development has reached a certain level, can the claims of still greater efficiency through competition be denied? By changing the rules and disguising the changes in the complexities of technical legal doctrine, the facade of economic security can be maintained even as new property is allowed to sweep away the old.

The plan that the historian sees in retrospect, however, was not what the participants in this process saw. They were simply guided by the conception of efficiency prevailing at the moment. Practical men, they may never have stopped to reflect on the changes they were bringing about, nor on the vast differences between their own assumptions and those of their predecessors.

WATER RIGHTS AND ECONOMIC DEVELOPMENT

The extensive construction of mills and dams at the end of the eighteenth and beginning of the nineteenth centuries gave rise to the first important legal questions bearing on the relationship of property law to private economic development, and it was here that the antidevelopmental doctrines of the common law first clashed with the spirit of economic improvement.[5] As a result, the evolving law of water rights had a greater impact than any other branch of law on the effort to adapt private law doctrines to the movement for economic growth.

Most of the legal controversies over water rights were of three types. The first, and by far the most important, involved an action by a downstream riparian landowner against his upstream neighbor, either for diverting the stream for agricultural purposes or for obstructing the natural flow of water in order to raise an upstream mill dam. As dams grew larger, a second set of cases dealt with the suit of an upstream mill owner against the downstream mill owner for throwing the water back so far as to impede the wheels or impair the fall of the upper mill. In a third group of cases, arising under the mill acts passed by most states, a neighboring landowner sued the proprietor of a mill who had flooded his land by raising a dam. Since the far-reaching impact of the mill acts is discussed later in this chapter,[6] the discussion here will focus on the widespread changes in the rules regulating the exploitation of water resources during the first half of the nineteenth century. This branch of law is important not only because of its direct influence on the course of early economic growth but also because the problems it was forced to confront and the legal categories developed for dealing with them reveal the basic structure of thought about all forms of property in the nineteenth century.

Without diminution or obstruction

Two basic assumptions determined the approach of the common law to conflicts over water rights. First, since the flow of water in its natural channel was part of nature's plan, any interference with this flow was an "artificial," and therefore impermissible, attempt to change the natural order of things. Second, since the right to the flow of a stream was derived from the ownership of adjacent land, any use of water that conflicted with the interests of any other proprietor on the stream was an unlawful invasion of his property. A late eighteenth century New Jersey case clearly expressed the prevailing conception:

In general it may be observed, when a man purchases a piece of land through which a natural water-course flows, he has a right to make use of it in its natural state, but not to stop or divert it to the prejudice of another. *Aqua currit, et debet currere* is the language of the law. The water flows in its natural channel, and ought always to be permitted to run there, so that all through whose land it pursues its natural course, may continue to enjoy the privilege of using it for their own purposes. It cannot legally be diverted from its course without the consent of all who have an interest in it. . . . I should think a jury right in giving almost any valuation which the party thus injured should think proper to afix to it.[7]

The premise underlying the law as stated was that land was not essentially an instrumental good or a productive asset but rather a private estate to be enjoyed for its own sake. The great English gentry, who had played a central role in shaping the common law conception of land, regarded the right to quiet enjoyment as the basic attribute of dominion over property. Thus, the New Jersey court regarded the legitimate uses of water as those that served domestic purposes and husbandry, requiring insignificant appropriations of the water's flow. All other interferences with the natural flow of water, including both diversion and obstruction, were illegal "without the consent of all who have an interest in it." Exploitation of water resources for irrigation or mill dams, which necessarily required significant interference with the natural flow of water, was thus limited to the lowest common denominator of noninjurious development, just as conflicts over the use of land were invariably resolved in favor of economic inactivity.

When American judges first attempted to resolve the tension between the need for economic development and the fundamentally antidevelopmental premises of the common law, the whole system of traditional rules was threatened with disintegration. Some courts went so far as virtually to refuse to recognize any right to prevent interference with the flow of water to a mill. Connecticut courts, for example, for a while limited the lower proprietor to the rights to prevent waste and to receive enough water to satisfy "necessary purposes."[8] And, even after discarding this doctrine in 1818, Connecticut jurists continued to disagree over whether a right to water was based on the common law rule of natural flow or was gained only by a long-standing pattern of appropriation.[9] The Supreme Judicial Court of Massachusetts in *Shorey v. Gorrell* (1783)[10] held that without long usage sufficient to confer a prescriptive right, there was no legal basis for preventing a newcomer from obstructing a stream. As a result, there were Massachusetts cases that denied any relief against even the substantial diversion of a stream for the purpose of irrigation.[11] Although it appears that the Massachusetts courts soon succeeded in eroding the force of *Shorey v. Gorrell* by treating the plea in prescription as a mere matter of form,[12] efforts to escape from the restrictive consequences of common law doctrines continued into the early nineteenth century as claims for relief against obstruction by upstream dams became more common. Some judges maintained that the common law action for diversion was simply not applicable to the temporary obstruction of water by upstream dams.[13] Others sought to modify the common law definition of legal injury in order to permit extensive, uncompensated use of water for business purposes.

The most important challenge to the common law doctrine was the so-called reasonable use or balancing test. Though it did not ultimately prevail until the second quarter of the nineteenth century, a handful of decisions early in the century had already laid the ground for its eventual triumph. In the earliest case, *Palmer v. Mulligan*[14] (1805), a divided New York Supreme Court for the first time held that an upper riparian landowner could obstruct the flow of water for mill purposes. The common law action for interference with the flow of water, Judge Brockholst Livingston said, "must be restrained within reasonable bounds so as not to deprive a man of the enjoyment of his property." Courts, Livingston argued, must be prepared to ignore "little inconveniences" to other riparian proprietors resulting from obstruction of the natural flow. Otherwise, he reasoned: "He who could first build a dam or mill on any public or navigable river, would acquire an exclusive right, at least for some distance, whether he owned the contiguous banks or not; for it would not be easy to build a second dam or mound in the same river on the same side, unless at a considerable distance, without producing some mischief or detriment to the owner of the first."[15]

Palmer v. Mulligan represents the beginning of a gradual acceptance of the idea that the ownership of property implies above all the right to develop that property for business purposes. Livingston understood that a rule making all injuries from obstruction of water compensable would, in effect, confer an exclusive right of development on the downstream property. The result, he concluded, would be that "the public, whose advantage is always to be regarded, would be deprived of the benefit which always attends competition and rivalry." Again, in *Platt v. Johnson*[16] (1818), in which the court held in favor of an upstream mill owner whose dam occasionally detained the flow of water for a number of days, the court observed that the sacred common law maxim *sic utere* "must be taken and construed with an eye to the natural rights of all." The court revealed a fundamentally new outlook on the question of conflicting rights to property: "Although some conflict may be produced in the use and enjoyment of such rights, it cannot be considered, in judgment of law, an infringement of right. If it becomes less useful to one, in consequence of the enjoyment by another, it is by accident, and because it is dependent on the exercise of the equal rights of others."[17]

These two cases marked a turning point in American legal development.[18] Anticipating a widespread movement away from property theories of natural use and priority, they introduced into American common law the entirely novel view that an explicit consideration of the relative efficiencies of conflicting property uses should be the paramount test of what constitutes legally justifiable injury. As a consequence, private economic loss and judicially determined legal injury, which for centuries had been more or less congruent, began to diverge.

Change in common law doctrine, however, is rarely abrupt, especially when a major transformation in the meaning of property is involved. Common lawyers are more comfortable with a process of gradually giving new meanings to old formulas than with explicitly casting the old doctrines aside. Thus, it is not surprising that, in periods of great conceptual tension, there emerges a treatise writer who tries to smooth over existing stresses in the law. Some such writers try to nudge legal doctrine forward by extracting from the existing conflict principles that are implicit but have not yet been expressed. Story's work in equity or commercial law comes to mind as an example. Others seek to banish novelty and to return the law to an earlier and simpler past. Joseph Angell was of the latter variety.

In his treatise, *Watercourses,* published in 1824, Angell reaffirmed the common law view that all diversion and obstruction of the natural flow of water was actionable. His only concession to a rule of equal or proportionate use was to observe that, since the common law rules "are undoubtedly liable to a rational and liberal construction," they would not allow "a right of action for every trivial and insignificant deprivation."[19] Although in *Palmer v. Mulligan* the New York court had attempted to justify its break with the past by showing that the injury to the lower mill owner was slight, Angell attacked the decision as "certainly contrary to the authorities, and obviously unjust."[20] In short, Angell's treatise asserted the traditional common law view that the only test for a reasonable use of water was the absence of all but the most trivial injury.

Though Angell sought to consolidate the past not a moment too early, his efforts hardly counted. The common law is especially cruel to those whom it casts aside. It either ignores them, soon forgetting that they ever existed, or, more usually, uses them as authority for propositions they did not accept. Angell's fate lay somewhere in between. In 1827 Justice Story wrote his influential opinion in *Tyler v. Wilkinson,*[21] citing all manner of contradictory authority, including, of course, Angell's treatise. Thereafter, Story was taken to have merely restated Angell's position and the clear inflexibility of the latter's analysis was forever replaced by the soothing, oracular quality of Story's formulation. Whenever Angell is cited afterwards, though the words be his, the meaning is supplied by Story.

Story's opinion is the classically transitional judicial opinion, filled with ambiguities sufficient to make any future legal developments possible.[22] It opens with a reaffirmation of traditional doctrine: the riparian owner "has a right to use of the water . . . in its natural current, without diminution or obstruction."[23] But even as he stated this principle, Story seemed to perceive its harsh antidevelopmental tendencies and he attempted to qualify its rigor. "I do not mean to be understood," he wrote, "as holding the doctrine that there can be no diminution whatsoever, and no obstruction or impediment whatsoever . . . for that would be to deny any valuable use. . . . The true test of the principle and extent of the use is, whether it is to the injury of the other proprietors or not." Some "diminution in quantity, or a retardation or acceleration of the natural current" is permissible if it is "not positively and sensibly injurious."[24]

By insisting, after all, that the true test of the riparian doctrine was reasonable use, but that reasonableness meant simply the absence of injury, Story managed to finesse the pressing problem of determining the extent to which conflicting and injurious uses of property could be regulated in the interest of economic development. Despite his invocation of the reasonable use formula in *Tyler,* however, it is clear from later opinions[25] that he, like Angell, wished to perpetuate the principle of natural flow.[26]

Tyler v. Wilkinson, which spawned a line of decisions opposed to all diversion or obstruction of water regardless of any beneficial consequences,[27] marks the nineteenth century high point in articulating the traditional conception of property that had already come under attack. Not only did it express a preproductive view of property as entailing the right to undisturbed ownership free from all outside interference, but proceeding from eighteenth century conceptions of absolute property rights, it condemned all conflicting injurious uses of land without consideration of whether such exploitation would maximize total economic welfare.

Yet, by acknowledging that the utilitarian criterion of valuable use was the ultimate source of legal rules, Story's reasonable use standard became almost immediately an open-ended formula through which common law judges could implement their own conceptions of desirable social policy. As a result, *Tyler v. Wilkinson* is cited more often during the second quarter of the nineteenth century to support than to condemn the reasonableness of a mill's interference with the flow of water.

The effort to free property law from its antidevelopmental premises was still very much a struggle at mid-century. As late as 1852 Massachusetts Chief Justice Shaw still found it necessary to argue that the law did not bar all obstructions of a watercourse "without diminution, acceleration, or retardation of the natural current," else "no proprietor could have any beneficial use of the stream, without an encroachment on another's right."[28] By the time of the Civil War, however, most courts had come to recognize a balancing test, making "reasonable use" of a stream "depend on the extent of detriment to the riparian proprietors below."[29]

The usages and wants of the community

It is important to appreciate the central role that the refashioning of American water law to the needs of industrial development played in the more general transformation of the law of property in the nineteenth century. Between 1820 and 1831 the productive capacity of American cotton mills increased sixfold.[30] Joseph Angell noted that more cases on the subject of water rights had been decided in the years between 1824 and 1833, when the first and second editions of his *Watercourses* appeared, than during the entire previous history of the common law.[31] Under the powerful influence of this rapid development, judges began to understand that the traditional rights of property entailed the power to exclude would-be competitors and that some injurious use of property was an inevitable consequence of any scheme of competitive economic development. They sought to free the idea of property from its exclusionary bias by enlarging the range of noncompensable injuries. The increasing frequency with which courts appealed to the idea of *damnum absque injuria*[32] seems to have occured in direct proportion to their recognition that conflicting and injurious uses of property were essential to economic improvement.

The most dramatic departure from common law riparian principles took place in Massachusetts, where, ever since the colonial mill acts, it had been the practice to confer privileges on mill owners in order to promote the growth of industry. As mills proliferated, a new set of technological considerations began to upset conventional legal doctrine. Since the amount of water power that a mill could generate depended largely on the fall of water, an increase in the height of a lower dam often reduced the fall of water from an upper dam. At the same time, the construction of large integrated cotton mills after 1815 unleashed such an enormous demand for water power that, as one observer noted in 1829, "in very many cases, only one of many proprietors can, in fact, improve a [stream] because the occupation of one mill site may render the others useless. Which proprietor shall, in such case, be preferred?"[33]

Chief Justice Shaw pondered this question in *Cary v. Daniels*[34] (1844). "One of the beneficial uses of a watercourse," he began, "and in this country one of the most important, is its application to the working of mills and machinery; a use profitable to the owner, and beneficial to the public." Proceeding from this new utilitarian orthodoxy, Shaw stated a legal doctrine strikingly different from Story's earlier formulation. Not only did the law require "a like reasonable use by the other proprietors of land, on the same stream, above and below," but it also took account of the "usages and wants of the community" and "the progress of improvement in hydraulic works." It required that "no one can wholly destroy or divert" a stream so as to prevent the water from flowing to the proprietor below, nor "wholly obstruct it" to the disadvantage of the proprietor above.[35] Thus, despite its invocation of "reasonable use," Shaw's formulation tended to erode a standard of proportionality: a mill owner who did not "wholly" obstruct a stream might claim that "the needs and wants of the community" justified his using more than a proportionate share of the water.

That Shaw intended this result is clear from *Cary v. Daniels* itself, in which the Chief Justice expressly rejected proportionality under the circumstances "growing out of the nature of the case."[36] Under manufacturing conditions then existing, he observed, beneficial uses of water were often, of necessity, mutually exclusive. Where the power needs of particular manufacturing establishments were such that maximum exploitation of limited water resources required a monopoly, "it seems to follow, as a necessary consequence from these principles, that . . . [the] proprietor who first erects his dam for such a purpose has a right to maintain it, as against the proprietors above and below; and to this extent, prior occupancy gives a prior title to such use."[37]

Shaw's opinion is premised on the desirability of maximizing economic development even at the cost of equal distribution. There was, of course, no reason why he could not have held that the permissible limits of economic growth were reached at the point at which exploitation of water resources was equal, that large cotton mills had to pay their own way to the extent that their operation exceeded the limits of proportional appropriation. Indeed, judges could also have demanded, a half century earlier, that equality of use be achieved through compensation of existing riparian owners. In the nineteenth century, however, there were few limits to the dominant mentality of maximization. When proportionate use was regarded as more efficient than priority, proportionality became the standard of reasonable use. When, in turn, proportionality stood in the way of the efficient use of water resources, the law returned to priority as the standard of reasonable use.

As *Cary v. Daniels* demonstrated, the doctrine of reasonable use could assimilate its historic antagonist—the rule of priority—and thereby make monopoly reasonable once again. Once the question of reasonableness of use became a question of efficiency, legal doctrine enabled common law judges to choose the direction of American economic development. By the time of *Cary v. Daniels,* they were so captivated by the spirit of improvement that they were willing to manipulate the concept of property to conform to their own notions of the needs of industrialization. In the seventeen years after *Tyler v. Wilkinson,* the direction of the law had turned around entirely.

PROPERTY RIGHTS AND MONOPOLY POWER

In this century there has been considerable discussion whether, before *Tyler v. Wilkinson,* the common law of water rights was a regime of riparian ownership or one based on principles of prior appropriation.[38] As we have seen, however, it was only after judges in the early nineteenth century began to respond to the felt need to modify the common law doctrine of diversion as it applied to mills and dams that the theory of prior appropriation emerged as an independent factor in American water law. By mid-century it had outlived its usefulness and, along with the related law of prescription, was discarded.

The benefit which always attends competition

In their effects on economic development, the common law principle of natural flow and a rule of prior appropriation were at first indistinguishable. Early cases relied on the two theories as alternative grounds for preventing the exploitation of water.[39] Only when a mill owner justified obstruction or diversion on the basis of prior appropriation, while the downstream plaintiff claimed a natural right to prevent all interference with the flow regardless of priority, could a choice between the two theories be of consequence. Indeed, it was only in the nineteenth century that a theory of priority was used offensively in order to maintain a right to obstruct the flow.[40]

The new circumstance that encouraged this change was the advent of large dams. While the original action for diversion placed lower mill owners at a relative advantage, since there was no one below them to injure, the erection of large dams introduced the possibility of equivalent injury by forcing water back upstream and causing damage to upper mills. Thus, the theoretical potential of the natural flow rule to check all development was made actual. This change forced courts to confront one of the great, and essentially unsettled, issues in nineteenth century America: how much certainty and predictability the law would guarantee to those who invested in economic development. Should the first person to build a mill be assured that his water supply would not be diminished by a subsequent upstream occupant? Could a guarantee be given without conferring a monopoly on the first developer?

There were two basic responses to this problem. The first was to recognize a right to interfere with the flow of water based on priority of appropriation. This view, though often coupled with the natural flow doctrine, began to take on a life of its own in the first quarter of the century.[41] The second response was a reasonable use or balancing test. Chancellor Kent, who managed not only to defend the natural flow doctrine but to approve a rule of priority as well, was also sympathetic to some form of reasonable use test. The natural flow doctrine, he wrote, following Story, "must not be construed literally, for that would be to deny all valuable use of the water to the riparian proprietors," since "rivers and streams of water would become utterly useless, either for manufacturing or agricultural purposes."[42]

While either a priority or a reasonable use test would have extricated the courts from the antidevelopmental consequences of the common law, these two doctrines were not interchangeable. A rule of priority had monopolistic consequences: a reasonableness doctrine based on a notion of proportional use resulted in competitive development. A contest between the two was inevitable. When the doctrinal confusion began to clear up in the second quarter of the nineteenth century, virtually all courts rejected prior appropriation, because, in the prescient language of the New York Supreme Court, it meant that "the public, whose advantage is always to be regarded, would be deprived of the benefit which always attends competition and rivalry."[43]

Prescription: A fatal enemy to modern improvements

Having rejected prior appropriation, American courts turned their attack on the monopolistic and exclusionary predilections of property law to the doctrine of prescription. The conception that prescriptive rights to certain kinds of intangible property may be acquired only after adverse use for a specified period of years did not clearly emerge in American law until the nineteenth century.[44] Before that time, a wide variety of property interests could be acquired by long, but not adverse, usage. For example, the operation of a market or collection of tolls for a ferry,[45] though invalid at the outset without a grant from the king, could eventually ripen into a right to exclude all competition.[46] The owner of a house could, after a period of years, claim a prescriptive easement for "ancient lights" and prevent his neighbor from erecting any buildings on adjoining property that would interfere with his own enjoyment of sunlight.[47] Riparian owners could acquire the right to exclude all others from injurious use of water merely by having appropriated use of the water many years before.[48] Finally, prescription legitimized a variety of restraints on trade, so that well into the eighteenth century English courts allowed damage actions against consumers who departed from longstanding business relationships to take their trade elsewhere.[49] Even in the nineteenth century, English courts entertained damage actions for competitive injury to markets by prescription.[50]

The doctrine of prescription confronted courts in a developing society with the anomaly of giving preference to old over new property merely because it was old. More fundamentally perhaps, the law of prescription had come to be associated with the monopolistic and restrictive practices of a feudal society. Thus, in his thoughtful and informative edition of Blackstone's *Commentaries* (1803), St. George Tucker of Virginia insisted that the English recognition of titles by prescription had not been imported into the common law of Virginia, primarily because Virginia had been settled too recently for prescriptive rights to accrue.[51] There is no question, moreover, that Tucker was aware of the economic and social implications of prescription.

Although there is evidence that courts were becoming resistant to prescriptive claims around the turn of the century,[52] it was only in the case of *Ingraham v. Hutchinson*[53] (1818) that the doctrine came directly under attack. While the court upheld a downstream riparian landowner's prescriptive claim to obstruct the flow of water, it faced a powerful and influential dissent by one of the most distinguished judges of the period. Judge James Gould maintained for the first time that without adverse use the lower proprietor could not claim an exclusive right to water regardless of how long before he had appropriated the flow. "The use, which the plaintiff has made of the stream has been neither a legal injury, nor an inconvenience of any kind, to the defendant. It was nothing, of which the defendant had any right to complain. He has, therefore, acquiesced in no usurpation of his rights; and has been guilty of no *neglect,* in not asserting them sooner."[54] The effect of the majority's opinion, Gould pointed out, was that the upstream defendant permanently forfeited any right to build because he did not erect his mill within the prescriptive period "whether it would have been of any use to him, or not—and whether he was in a condition to build it, or not."[55]

As Gould saw, without his conception of adverse use, the doctrine of prescription was nothing more than a modified version of the rule of prior appropriation, with the same monopolistic and exclusionary results. With Gould's requirement of implied consent by one party to an invasion of his property rights, there was no way in which a later mill owner could ever claim more than his proportionate share of water without usurping the rightful share of an already existing mill.

In one sense, Gould was simply redefining the conditions necessary for finding adversity. Under the traditional view, a mill owner fortunate enough to be left alone on a stream for the prescriptive period would have been regarded as exercising a claim adverse to the potential rights of his riparian neighbors to build mills in the future. Gould, however, would only have recognized a user adverse to existing mills. Under Gould's definition of adversity, prescription could never justify monopoly. His view was widely adopted[56] once the rule of prior appropriation had been overwhelmingly rejected in the second quarter of the nineteenth century. Legal doctrine was thus fundamentally transformed to prevent the first occupant on a stream from excluding all other mill owners, regardless of how long before he had established his works.

A still greater shock to the old law of prescription was the publication of Dane's *Abridgment* in 1824. Although Dane first acknowledged that the recent settlement of the country alone was no bar to prescription,[57] he went on to mortify the defenders of tradition by declaring that prescription was a mere "modern doctrine" that "cannot yet be considered as established law."[58] In a long and technical reply to both Dane and Judge Gould, Joseph Angell piled on old legal authorities to demonstrate that the rule of prescription could be "explained and defined with very great precision."[59] But Angell had entirely missed the point. The main issue could not be resolved by appealing to the authority of an established legal tradition from which the new doctrine represented a conscious departure. By defining the issue in terms of what had traditionally constituted property rights, Angell fell out of touch with a movement to limit sharply the power of property owners to determine the scope of economic development.

In one ironic sense, Angell displayed a degree of prescience. Angell, who could not allow himself to think that his differences with Gould could be due to anything but a misunderstanding of recognized authority, argued with a confident flourish that "in principle, there is no difference whatever" between the prescription for water and for sunlight, and that Gould's position, as applied to the latter, "conflicts with what has uniformly been regarded as established law, for many centuries."[60] Indeed, the common law rule of "ancient lights," which enabled the owner of an ancient tenement to prevent his neighbor from undertaking any building that would interfere with his own enjoyment of sunlight, rested on even more long-standing legal authority than did the doctrine of prescription in water cases. And, as in the water cases, the rule protecting ancient lights could run against a property owner who had never acted. Judge Gould, however, had recognized this very analogy and declared the doctrine of ancient lights "anomalous" in the law.

Both Angell and Gould were correct in this: under Gould's approach there could be no stopping short of transforming the whole law of prescription. And in the leading case of *Parker v. Foote*[61] (1838), the New York court, following Gould's analysis, overthrew the doctrine of ancient lights.[62] It "cannot be applied in the growing cities and villages of this country," the court declared, "without working the most mischievous consequences."[63] By the Civil War, this attitude had acquired such momentum that it challenged even the well-established doctrine that a right to support of buildings could be acquired by prescription.[64]

The most striking rejection of the common law assumption that long use was sufficient to create an exclusive property interest occurred in the great *Charles River Bridge* case.[65] The bridge proprietors, claiming an exclusive franchise as successors in interest to an ancient ferry, sued a recently chartered bridge company for damages due to competitive injury. In rejecting their prescriptive claim, Justice Morton of the Massachusetts court turned to his own purposes the common law fiction that long use was evidence of an original lost grant by noting that, since there was evidence of an actual grant, "the plaintiffs cannot now call to their aid any principles or reasoning peculiar to this kind of title, or any rules of evidence applicable to this mode of proof."[66] Although Justice Story had only recently attempted to commit American law to the position that use for twenty-one years created a conclusive presumption of title,[67] Morton and others hostile to prescriptive claims treated long use as no more than rebuttable evidence of title[68] and were thus able to circumvent a long line of English cases recognizing exclusive franchises based only on prescription.

Justice Baldwin undertook an elaborate justification of the United States Supreme Court's affirmance of the *Charles River Bridge* case, emphasizing that none of the English decisions concerning prescriptive markets and ferries could be adopted in America, since "they grow out of feudal tenures, are founded on feudal rights, and are wholly unknown in this country."[69] Even Justice McLean, whose views on the merits of the case favored the first bridge company, believed that the recent settlement of the country and "its rapid growth in population and advance in improvements" prevented most property from being acquired by long usage. "Such evidence of right," he concluded, "is found in countries where society has become more fixed, and improvements are in a great degree stationary."[70]

At its deepest level, the attack on prescription represented an effort to free American law from the restraints on economic development that had been molded by the common law's feudal conception of property. By the second quarter of the nineteenth century the common law doctrine of prescription had been considerably narrowed to prevent exclusive rights from accruing merely on the basis of long use, and American law had moved into what Francis Hilliard, writing in a different context, called an "anti-prescriptive age."[71]

THE MILL ACTS: PROPERTY AS AN INSTRUMENTAL VALUE

The various acts to encourage the construction of mills offer some of the earliest illustrations of American willingness to sacrifice the sanctity of private property in the interest of promoting economic development. The first such statute, enacted by the Massachusetts colonial legislature in 1713, envisioned a procedure for compensating landowners when a "small quantity" of their property was flooded by the raising of waters for mill dams.[72] The statutory procedure was rarely used, however, since the Massachusetts courts refused to construe the act to eliminate the traditional common law remedies for trespass or nuisance.[73] After the act was amended in 1795 and 1798, mill owners began to argue that it provided an exclusive remedy for the flooding of lands.[74] As a result, the mill acts adopted in a large number of states and territories[75] on the model of the Massachusetts law were, more than any other legal measure, crucial in dethroning landed property from the supreme position it had occupied in the eighteenth century world view, and ultimately, in transforming real estate into just another cash-valued commodity.[76] The history of the acts is a major source of information on the relationship of law to economic change. For reasons of convenience the discussion that follows concentrates on the Massachusetts experience, which, though particularly rich, is not atypical.

Under the 1795 Massachusetts statute,[77] an owner of a mill situated on any nonnavigable stream was permitted to raise a dam and flood the land of his neighbor, so long as he compensated him according to the procedures established by the act. The injured party was limited to yearly damages, instead of a lump sum payment, even if the land was permanently flooded, and the initial estimate of annual damages continued from year to year unless one of the parties came into court and showed that circumstances had changed. The act conferred extensive discretion on the jury, which, in addition to determining damages, could prescribe the height to which a dam could be raised as well as the time of year that lands could be flooded. Unlike the statutes in some states, such as Virginia,[78] the Massachusetts law authorized the mill owner to flood neighboring lands without seeking prior court permission.[79] Thus, except for the power of the jury to regulate their future actions, there was no procedure for determining in advance the utility of allowing mill owners to overflow particular lands.

The exclusive remedial procedures of the mill acts foreclosed four important alternative avenues to relief. First, they cut off the traditional action for trespass to land,[80] in which a plaintiff was not required to prove actual injury in order to recover. In a mill act proceeding a defendant could escape all liability by showing that, on balance, flooding actually benefited the plaintiff.[81] Second, the statutory damage formula removed the possibility of imposing punitive damages in trespass or nuisance. The common law view had been that unless punitive damages could be imposed, it might pay the wrongdoer to "keep it up forever; and thus one individual will be enabled to take from another his property against his consent, and detain it from him as long as he pleases."[82] A third form of relief at common law allowed an affected landowner to resort to self-help to abate a nuisance.[83] Indeed, there are a number of reported cases in which mill dams not covered by the protection of the mill acts were torn down by neighbors claiming to enforce their common law rights.[84] Finally, the acts foreclosed the possibility of permanently enjoining a mill owner for having created a nuisance.[85]

In the early nineteenth century the need to provide a doctrinal rationale for the extraordinary power that the mill acts delegated to a few individuals was acute. Not only had the use of water power vastly expanded in the century since the original Massachusetts act, but there was also a major difference between the eighteenth century grist mill, which was understood to be open to the public, and the more recently established saw, paper, and cotton mills, many of which served only the proprietor. In *Stowell v. Flagg*[86] (1814), Chief Justice Parker said, "The statute was made for the relief of mill owners from a multiplicity of suits," because the legislature found that the common law remedy would "so burthen the owner of a mill with continual lawsuits and expenses" when he overflowed his neighbor's land.[87] Parker was not troubled by the statutory mode of compensation through annual payment of damages, which, in effect, compelled the landowner to make a loan to the mill owner and thereby enabled the mill owner to amortize any permanent damage he caused. Nor did he point out that under the common law there was no private right to flood adjoining land, even upon making just compensation.

By viewing the statute as entirely remedial, Parker did not have to portray the enterprise as sufficiently public in nature to bring it within the power of eminent domain. Troubled by the enormous potential for unplanned economic change that the mills acts made possible, however, Parker did suggest that the 1798 statute was "incautiously copied from the ancient colonial and provincial acts, which were passed when the use of mills, from the scarcity of them, bore a much greater value, compared to the land used for the purpose of agriculture, than at present."[88]

Parker's language seems to imply that the only public purpose required in order to justify an extensive invasion of private rights was an increase in total utility—and that such a calculation was within the exclusive domain of the legislature. If the legitimacy of a taking was a function only of the relative values of the adjoining properties, any compelled transfer would be lawful so long as compensation was required. It is more likely, however, that Parker was merely questioning the prudence of the legislative judgment, while never doubting that mills were public in terms of whom they served.

By 1814 the significance of the growing separation between public and private enterprise was only beginning to penetrate the judicial mind. Some still conceived of mills as a form of public enterprise in which competition was impermissible.[89] Business corporations were only beginning to upset the old corporate model, in which the raison d'être of chartered associations was their service to the public. Nor is there any evidence that the increasingly private nature of mills, which was painfully evident to everyone fifteen years later, had as yet caused judges the slightest conceptual difficulty. At this time in Virginia, for example, where the old corporate model still prevailed, judges were also clear that the state's mill act could be defended on the ground that "the property of another is, as it were, seized on, or subjected to injury, to a certain extent, it being considered in fact for the public use."[90]

The dramatic growth of cotton mills after 1815 provided the greatest incentive for mill owners to flood adjoining land[91] and, in turn, brought to a head a heated controversy over the nature of property rights. The original mill dams were relatively small operations that caused some upstream flooding when proprietors held back water in order to generate power. With the growth of large integrated cotton mills, however, the flooding of more lands became necessary not only because larger dams held back greater quantities of water but also because of the need to generate power by releasing an enormous flow of water downstream. In light of this fact, the Massachusetts legislature amended the mill act in 1825 to allow the flooding of "lands . . . situated either above or below any mill dam."[92] Two years later, in the *Wolcott Woollen* case[93] (1827), the Supreme Judicial Court of Massachusetts, adopting its customary posture of resignation, observed only that "the encouragement of mills has always been a favorite object with the legislature, and though the reasons for it may have ceased, the favor of the legislature continues." It thereby extended the protection of the mill act to what were essentially manufacturing establishments, even though regulating the flow and assessing yearly damages were far more difficult than in the case of upstream flooding, where the extent of damage could easily be predicted once the allowable height of the dam was determined. Even more significant, the mill owners succeeded in inducing the court to extend the act to cover a situation that the legislature could scarcely have envisioned. Thus, a mill that had purchased land and built a new dam more than three miles upstream from its existing mill, was allowed to overflow all the land between the dam and the original mill site. In effect, the court gave mill owners virtually unlimited discretion to destroy the value of lands far in excess of any benefit they might possibly receive.

Also, in 1827,[94] the court applied another new section of the act[95] to allow a defendant to escape damages entirely by showing that the irrigation benefits the plaintiff received from having his lands overflowed more than outweighed any injury he had incurred.[96] Representing the culmination of a quarter century's experience under the mill acts, this marked the final break with the eighteenth century conception of property, which regarded the flooding of land as a fundamental invasion of right regardless of actual damage. As one court at the end of the eighteenth century observed, "It is altogether immaterial whether [flooding] may be productive of benefit or injury. No one has a right to compel another to have his property improved in a particular manner; it is as illegal to force him to receive a benefit as to submit to an injury."[97] Under the mill act, however, mere interference with another's quiet enjoyment of land no longer possessed any independent claim to compensation. The only measure of damage was the effect on the productive value of land.

Manufacturing companies did not always regard coverage under the mill act as an unmixed blessing. In fact, after *Wolcott Woollen* they often sought to extract from the legislature better treatment than they could expect to win from a jury impaneled under the act. In *Cogswell v. Essex Mill Corp.*[98] (1828), for example, there occurred a reversal of the usual roles, with the plaintiff arguing for his remedy under the statute and the manufacturing company insisting on an action at common law. In its charter, the company had been guaranteed the right to raise its dam to a specified height and to flood the adjoining lands at all times of the year. The company "would not have accepted a charter containing the mill act," it argued, for that "would subject them to the fluctuating estimates of juries"[99] to determine both the time and extent of their activity. Instead, the company preferred to make a lump sum payment of damages in a common law action in return for a guarantee of certainty in all other respects. The court held for the company on the theory that the charter provision took precedence over the mill act. Yet even in this case it is striking how far the ideology of the mill acts had influenced thinking about the rights of property. Courts had ceased questioning the right of a mill owner with legislative authorization to impair the value of adjoining land for an essentially private purpose, so long as compensation was provided. The supervisory role of the jury, which initially imparted an important measure of legitimacy to this novel undertaking, could finally be dispensed with entirely.

Extension of the mill act to manufacturing establishments brought forth a storm of bitter opposition. One theme—that manufacturing establishments were private institutions—appeared over and over again. One commentator, for example, pointed out that the original mill act had been enacted at the time when "the country had been in a state of slow progress from a wilderness to cultivation" and "lands were of comparatively little value, while the support of corn and saw mills . . . was . . . of vital necessity." Under these circumstances, he said, mills could reasonably have been regarded as "*public* easements." With the extension of the acts to manufacturing establishments, however, the essential question was changed to "the right to apply the property of any one against his consent to *private* uses."[100] The distinguished Boston lawyer, Benjamin Rand, disputed the assertion in *Stowell v. Flagg* that the mill act merely substituted one remedy for another. "Such an invasion of private property," he declared, "can only be defended in a case of great public necessity and utility."[101] Finally, those whose lands were flooded complained bitterly that there was "no remedy worth pursuing," so that "but few of those who suffer seek any relief." "Generally," they observed, "the mills and mill seats are in the hands of the active and wealthy—able to make the sufferers repent, if they resort to the law."[102]

The nearly unanimous denunciation of the mill acts soon brought forth a degree of change. From 1830, when Lemuel Shaw began his thirty year tenure as chief justice, the Massachusetts court began a marked retreat away from its earlier reluctant, but expansive, interpretation of the act. And after 1830, when the legislature eliminated the most notorious windfall to mill owners by extending to the injured party the opportunity to recover permanent damages,[103] the leading attraction of proceeding under the act was eliminated.[104]

With this evidence of diminished legislative support, the Massachusetts court consistently rejected attempts to extend the statute to novel areas. Shaw's first mill act case[105] posed the question whether the protection of the act extended to mills erected on artificial canals built for the purpose of drawing water away from natural streams. This issue presented the court with the opportunity to encourage a vast geographical expansion of industry under the protective wing of the mill act by sanctioning the flooding that would result from a network of canals on which dams would be built. If the mill act applied, "the owner of a parcel of low land may erect a mill upon it and bring water to the mill from any distant pond or reservoir, through the intermediate lands, without the permission of the owners of lands, and may rely for protection upon these statutes."[106] Although the court had seemed to sanction such a procedure only five years earlier, Shaw held that the benefits of the mill act extended only to riparian proprietors. The full import of the earlier decision was just becoming apparent as the dispute expanded beyond the fairly narrow and historically separate question of the proper means of regulating riparian owners to include the more general and necessarily controversial problem of determining the rights of landowners to use their lands.

Conceding that the mill acts "are somewhat at variance with that absolute right of dominion and enjoyment which every proprietor is supposed by law to have in his own soil,"[107] Shaw nevertheless proceeded to offer a dual justification for the acts. The legislation could be defended, he wrote, "partly upon the interest which the community at large has, in the use and employment of mills"—a theory of eminent domain—"and partly upon the nature of the property, which is often so situated, that it could not be beneficially used without the aid of this power."[108] However artificial it may have been in the existing context, Shaw argued that the mill act "was designed to provide for the most useful and beneficial occupation and enjoyment of natural streams and water-courses, where the absolute right of each proprietor, to use his own land and water privileges, at his own pleasure, cannot be fully enjoyed, and one must of necessity, in some degree, yield to the other."[109]

In spite of the difficulties inherent in viewing the mill act as a mere regulation of water rights, Shaw preferred to rest at least part of his case on that restrictive but historically approved function. Thus, he put great emphasis on the relativity of rights as between riparian owners while minimizing his reliance on a theory of eminent domain. This effort to limit the importance of the power of eminent domain appears to reflect a growing realization of the essentially private nature of the interests that were being served by the mill acts and a consequent unwillingness to allow the state to intervene solely to advance private ends.[110]

While it did help to overcome some of the difficulties in applying a theory of eminent domain to essentially private activities, the main contribution of Shaw's formulation was to force courts to see that a conception of absolute and exclusive dominion over property was incompatible with the needs of industrial development. Whether the rationale for state intervention was eminent domain or a more explicit recognition of the relativity of property rights, under the influence of the mill acts men had come to regard property as an instrumental value in the service of the paramount goal of promoting economic growth.

THE PROBLEM OF IMPROVEMENTS

From the beginning of the nineteenth century, judges, faced with the perplexing problem of adjusting the law of property to a theory of value based on the productive capacity of land, were forced at every turn to deal with two central facts about the American economy. First, there was a consistent pattern of speculation and rising land prices. Second, the value of land was often intimately related to the value of the improvements on it. In hundreds of cases, courts were faced with the question whether the common law rules, premised on a preproductive conception of land and promulgated in an economy in which the price of land had remained relatively constant, should be applied to an economy of rapidly fluctuating land prices and an increasing rate of development.

What would in England *be waste*

The English law of waste was relatively clear: any fundamental alteration by a tenant of the condition of the land constituted waste for which he was liable. Clearly, however, an economy dependent on clearing land for economic development could not enforce a rule of maintaining the existing condition of land. From the moment of independence from England, therefore, American jurists devoted their efforts to modifying or overturning the received common law doctrine.

Writing in 1801 James Sullivan acknowledged the existence of the strict English rules governing waste but maintained that "we have no reports or recollections of their being adopted here. The situation of our country is so very different, in point of agriculture, from what they are in Europe, that we hardly know how to apply the precedents there to our cases." Indeed, Sullivan continued, since leases were "generally for a short space of time" and since "the nature of the soil and the manner of improvement" were different, there were in America "few occasions for actions of waste."[111] Similarly, Zephaniah Swift of Connecticut believed that "in this country, such conversions in [land use] as are compatible with good husbandry, would not be deemed waste."[112]

Even so, there remained deep disagreements in the first quarter of the nineteenth century over how far a tenant could go in transforming land for purposes of economic improvement. In New York, for example, the Supreme Court divided sharply in an action by a landowner whose 133 acre farm was "wild and uncultivated, and covered throughout with a forest of heavy timber" when his tenant took possession. The tenant had cleared most of the farm and cut down and removed the timber. Acknowledging that "what would in *England* be waste, is not always so here," the court nevertheless held that the tenant had forfeited the estate through waste. Though he "undoubtedly had a right to fell part of the timber, so as to fit the land for cultivation," he did not have a right to "destroy *all* the timber, and thereby essentially and permanently diminish the value of the inheritance."[113] The dissenting minority wanted to go still farther in changing the law to suit the necessities of development. Justice Ambrose Spencer argued that "the doctrine of waste, as understood in England, is inapplicable to a new unsettled country. . . . Men differ very widely as to how much woodland ought to be left for the use of a farm." Without a covenant between the parties "to leave a sufficient quantity of land in wood . . . we have no right to say some waste might be committed, and other waste might not." Spencer was, in effect, denying that the law of waste was an inherent part of the right to property, since the only basis for judicial prevention of waste, in his view, was a contract between the parties.[114]

No American court was willing to go as far as Spencer would have, severing entirely the right to property from the right to prevent tenants from completely altering the estate. No American court, however, enforced the strict English common law governing waste. Some, like New York, modified it, while others rejected it wholesale.[115]

An important aspect of the American redefinition of the law of waste dealt with the common problem of improvements. It was clear at common law that a tenant who erected buildings on land had no right to remove them at the end of his term. During the eighteenth century, English courts had established an exception for the removal of fixtures that were built to carry on a trade, but they made no such concession for buildings erected solely for agricultural purposes.[116] In *Van Ness v. Pacard*[117] (1829), Justice Story expressed the already established view[118] that the common law rule had never been received in America.

The country was a wilderness, and the universal policy was to procure its cultivation and improvement. The owner of the soil as well as the public, had every motive to encourage the tenant to devote himself to agriculture, and to favour any erections which should aid this result; yet, in the comparative poverty of the country, what tenant could afford to erect fixtures of much expense or value, if he was to lose his whole interest therein by the very act of erection?[119]

Although, technically, *Van Ness* held only that the fixtures in question were in fact trade fixtures and thus still within the English exception, Story's decision was, as Chancellor Kent observed, part of "a system of judicial legislation" that had "grown up . . . so as to almost render the right of removal of fixtures a general rule, instead of being an exception."[120] Indeed, Kent maintained that extension of the rule to agricultural fixtures "may be necessary for the beneficial enjoyment of the estate," since it would encourage the tenant "to cultivate and improve" the land.[121]

Waste of the inheritance

The policy of promoting improvements was not always easy to apply. One of the most complicated departures from English law involved the right of widows to dower in unimproved land or land that was improved after the widow's husband had sold it. Most states in the nineteenth century continued to enforce some version of the common law rule that a widow was entitled to a life interest in one third of the land held by her husband during their marriage. And, since a husband could not be allowed to defeat the dower right of his wife by transferring the land, in most states she could enforce this right against a purchaser of the property from her husband.

Because the value of land was increasingly determined by its productive capacity, American courts had to decide whether to follow the common law in measuring dower by the value of land divorced from its productive capacity. This issue arose in several different forms. Did unimproved lands have any greater value for purposes of measuring dower if they could be sold to speculators for a substantial price? Would the dower right be computed on the basis of existing rents and profits if speculation had raised the land's market price above any sum that could reflect its present productivity? That is, was a widow entitled, as at common law, to a life interest in one third of the land or only to a one third interest in the profits? Finally, how would dower be computed on land that a husband had sold in his lifetime and that, by the time of his death, had increased in value, either because of a general rise in land values or because the purchaser had improved it?

The courts wrestled with each of these questions from the beginning of the nineteenth century. In 1783 an unreported Massachusetts decision[122] had turned back a challenge to the common law rule that a widow was entitled to dower in unimproved lands. The first clear change in the theory of land valuation appears in *Leonard v. Leonard*[123] (1808). Commissioners appointed by the probate judge had appraised the relevant real estate, "a considerable part of which was woodland and unproductive." The widow was awarded land valued at one third of the total, though it was agreed that her actual share "comprised the most productive parts of the . . . estate." Reversing and holding that dower should be determined not by the market value but by the rents and profits of the estate, the court declared: "This rule is adapted equally to protect widows from having an unproductive part of estates assigned to them, and to guard heirs from being left, during the life of the widow without the means of support."

This rule may have appeared to be even-handed when it was announced, but in a speculative and developing economy, with the market value of land consistently above the capitalized value of its present product, dowagers would soon become victims of the theory underlying it. In *Conner v. Shepherd*[124] (1818), the Supreme Judicial Court carried the new rule of *Leonard* to its logical conclusion. Overruling its decision of thirty-five years earlier, the court denied a widow's claim to dower in unimproved lands. Since dower is to be measured by the productive value of the land, Chief Justice Parker reasoned, there could be no dower in unimproved lands. Could not the widow, if only given her traditional share, herself make the lands productive? No, said the court, for that would constitute waste. Returning for this particular situation to the discredited common law doctrine of waste, the court declared:

According to the principles of the common law, her estate would be forfeited if she were to cut down any of the trees valuable as timber. It would seem too that the mere change of the property from wilderness to arable or pasture land . . . might be considered waste: for the alteration of the property, *even if it became thereby more valuable,* would subject the estate in dower to forfeiture: the heir having a right to the inheritance, in the same character as it was left by the ancestor.[125]

Thus, the widow lost either way. A utilitarian standard of value made worthless her dower rights in unimproved lands, while a non-utilitarian rule of waste made improvement impossible.

The contradictions go deeper still. In order not to rest the argument from waste entirely on outmoded rules of common law, Parker went on to argue that improvements by a tenant in dower "would be actually, as well as technically, waste of the inheritance." "Lands actually in a state of nature may," he said, "in a country fast increasing in its population, be more valuable than the same land would be with . . . cultivation."[126] Parker here seemed to be suggesting that legal rules should be used to discourage and delay economic improvement as long as buyers were willing to pay more for undeveloped than for developed land. If so, his theory appears to be without support in other cases, and inconsistent with the conception of property then emerging in the case law.[127] Indeed, the premise that underlay the changing law of waste was that it was preferable to encourage immediate improvement by tenants, even at the risk that development might prevent other future uses and thereby impair the transferability of land.[128]

The conclusion seems inescapable that Parker's central purpose in *Conner v. Shepherd* was to undermine the right of dower itself. "Believing that [the dower right] would operate as a clog, upon estates, designed to be the subject of transfer," the court was prepared to use any method to cut off the widow's share, including reliance on dubious assumptions about the relation between market price and productive value. Perhaps the most important effect of the holding that a widow was not entitled to dower in unimproved lands, however, was to promote further the view that the value of land was exclusively a function of its productivity.

The doctrine of *Conner v. Shepherd* was clearly an exception to a general policy of encouraging improvements by those in possession of land. That policy was most uniformly pursued in the related context of dower rights asserted against purchasers rather than heirs. For example, from the first decade of the nineteenth century, the Massachusetts courts had held that a widow could not receive any of the increase in the value of land owing to improvements made by her husband's vendee.[129] Identical results were reached in all the other states.[130] New York, moreover, held that a widow could not gain the advantage either of improvements or of a general increase in land values.[131] On the other hand, the Supreme Court of Pennsylvania, which as early as 1792 had denied that dower extended to improvements,[132] refused to deprive the widow of the benefit of general increases in land values.[133] While courts thus divided over who should have the benefit of increasing land prices, they all agreed that — regardless of common law property doctrines — the value of improvements should be left with the developer.

Caveat emptor: As if the rules of equity, justice, and convenience, were the rules of law

Perhaps the most perplexing question raised at the end of the eighteenth century by the new judicial emphasis on the productive value of land involved the proper measure of damages for breach of warranty in the sale of land. In most deeds of sale the seller warranted that he had seisin, or good title, in the land; in addition or instead, the seller might insert a covenant guaranteeing the purchaser quiet enjoyment of his land. If the purchaser, after a period of time on the land, was evicted by one proving superior title, would he be able to recover from the seller any increase in the value of the land—due either to intervening improvements or to rising prices—or was he limited to recovery of the purchase price?

Again, the rule of the English common law was clear: the purchaser could recover only the original price he had paid. While some explanations for the English rule harkened back to the mysteries of feudal land law, most American courts that departed from the English view maintained, in the words of the Connecticut Supreme Court, that "the diversity . . . between British practice and ours, is undoubtedly founded in the permanent worth of their lands, as an old country, and the increasing worth of ours as a new country." In America, the court added, "it is supposed that the purchaser goes on, improves and makes the land better till he is evicted."[134]

By 1810 six states had passed upon the measure of damages for breach of a warranty of title. In four, South Carolina,[135] Connecticut,[136] Virginia,[137] and Massachusetts,[138] the purchaser was permitted to recover the ordinarily higher value at the time of eviction. In Pennsylvania[139] and New York[140] he recovered only the purchase price plus interest.

Much of the difference of opinion was said to turn on the intent of the parties to the warranties. Judge Pendleton of South Carolina, in promulgating his state's rule in 1785, declared: "Men do not make purchases, with a view of merely having interest for their money; but they contemplate the rise in value of the thing purchased."[141] Pennsylvania and New York, by contrast, looked to the presumed intention of the seller. "I question," wrote Judge Van Ness of New York, "if one [seller] out of ten thousand enters into these covenants with the remotest belief, that he is exposing himself and his posterity to . . . ruinous consequences."[142] But inevitably, the difference of opinion reflected a difference in the choice of which interests in the population to advance. While Judge Spencer, dissenting from the New York rule allowing only the lower measure of recovery, asked: "What is to become of the industrious citizen or mechanic, who has spent his hard earnings in erecting his little house or workshop, relying on the covenant in his deed, if he can only get back his purchase-money and interest?"[143] Judge Tilghman of Pennsylvania answered that it would be unreasonable to allow the higher measure where "buildings of *magnificence* are erected to gratify the *luxury* of the *wealthy.*"[144]

The story after 1810 is of a return to common law orthodoxy, this time under the aegis of newly emerging theories of contract damages. In both Kentucky and Tennessee—the first western states to rule on the question—the courts limited the recovery of an evicted purchaser to the lower contract price plus interest. The Tennessee court did so while congratulating itself for "a laudable disposition . . . to substitute certainty for uncertainty" in measuring contract damages.[145] In Kentucky, although the result was much the same, the court first distinguished, as had many of the dower cases, between recovery for the increased value of land and recovery for improvements. Compensation for loss of increased land values, it concluded, could not have been contemplated by the parties, since it represented only "a bare possible hypothetical loss." Improvements, however, were a different matter. Nevertheless, since the legislature had already passed a statute allowing a good faith purchaser to recover the value of his improvements from one who, with superior title, had ousted him, he would not be allowed to recover against the seller as well.[146]

By 1815 North and South Carolina had also adopted the common law rule,[147] the latter having abandoned its prior decisions in favor of purchasers. Virginia reversed its position soon thereafter.[148] The result was that only in Massachusetts and Connecticut could a purchaser recover the increased value of the land at the time of eviction; in these states he could do so only under a covenant of quiet enjoyment and not under a warranty of good title.[149]

The majority damage rule for breach of a warranty of title soon became a prominent target of a growing legal literature extolling the virtues of the civil law. Thomas Cooper of South Carolina, who published an annotated translation of the civilians' bible, the *Institutes of Justinian,* correctly saw the warranty decision as part of a movement to extend the rule of caveat emptor, which he denounced as "a disgrace to the law" and a most "comfortable doctrine for the land-jobbers."[150] And in a letter to William Sampson, who was something of a patron saint among the civilians, Charles Watts of New Orleans sang the praises of Louisiana's civil law and denounced caveat emptor for depriving innocent purchasers of the value of their improvements.[151] Watts elaborated on Cooper's contention that the rule limiting sellers' liability favored the interests of land speculators, who were, after the national government, the principal vendors of unsettled lands. His emphasis on the disparity of legal and economic knowledge disfavoring buyers underlined the view that caveat emptor had political and distributional consequences that could not be dismissed with a complacent assertion that the parties were free to alter them by contract.

The contract conception of land warranties, which became dominant after 1810, had some surprising consequences. Though it is conventional to view the effect of nineteenth century rules of contract damages as placing the injured party in the position he would have occupied had the contract been performed, the result was everywhere the opposite in the case of warranties of title.[152] The rule limiting warranty damages to the purchase price seems inconsistent with a dominant theme of the period — the use of legal rules to promote improvements by assuring to developers the benefits of their investments. It may, in fact, be a prominent example of the negative influence of land speculation on the course of American legal doctrine.[153] On closer inspection, however, one finds that the policy of promoting improvements was accomplished mainly in another way, by means of "good faith possession" statutes. These statutes enabled a bona fide purchaser of land to recover the value of his improvements from one who, with superior title, ousted him from possession. While this approach had the merit of forcing the actual beneficiary of the "unjust enrichment" to disgorge the value of improvements, it also forced him to take the improvements whether he wished to or not. The value of the improvements, as Chancellor Kent lamented, "may have been very costly, and beyond the ability of the claimant to refund." And appealing to traditional conceptions of property, he emphasized that the claimant might "have a just affection for the property, and it might have answered all his wants and means in its original state, without improvements."[154]

Legislative schemes to protect the value of improvements were widely adopted during the first half of the nineteenth century.[155] In the western states, where land titles were in great disarray, statutory intervention was essential to assure to landowners the minimum security required for development. These statutes, which threw the risk of bad title entirely on the innocent nonoccupying owner, probably also had the effect of neutralizing political pressure to put the risk on land speculating sellers. The Kentucky "Occupying Claimant" law, adopted shortly after Kentucky became independent of Virginia, was said to have been enacted because "there existed . . . claims to more than three times the quantity of the lands" in the state.[156] When the United States Supreme Court invalidated the law as an interference with the obligation of contract,[157] it sent a wave of apprehension throughout the states that all these acts might be voided.

Chancellor Kent lent the authority of his *Commentaries* against these statutes as well. He conceded that "peculiar and pressing circumstances" existed in many states because "the titles to such lands had, in many cases, become exceedingly obscure and difficult to be ascertained, by reason of conflicting locations, and a course of fraudulent and desperate speculation." Yet he argued that the statutes were "strictly encroachments upon the rights of property" and concluded that "such indulgences are unnecessary and pernicious, and invite to careless intrusions upon the property of others."[158] Following Kent's views, the Supreme Court of Tennessee in 1830 declared that state's occupying claimant statute "subversive of the clearest principles of natural right in relation to property, and consequently, of the constitution, which guarantees to every man the exclusive use of his own property."[159] There was no further federal constitutional challenge to these laws, however, and by the time that Kent published the third edition of his *Commentaries* in 1836, most states had enacted some form of statutory protection for good faith improvements made by purchasers.[160]

III Subsidization of Economic Growth through the Legal System

THE SLOW EMERGENCE OF THE JUST COMPENSATION PRINCIPLE

The process of economic development in the United States necessarily involved a drastic transformation in common law doctrines, which required a willingness on the part of the judiciary to sacrifice "old" property for the benefit of the "new." The most potent legal weapon used to further this process of redistribution was the power of eminent domain—coerced "takings" of private property, usually for roads, for canals, and somewhat later, for railroads. Given the central role that eminent domain played in legal controversy during the nineteenth century, it is rather surprising to see how infrequently it arose as a legal question before that time. Until the end of the eighteenth century, it appears, development was so meager that the problem of compensation for land taken or injured by public authorities hardly played a significant role in American law.

Perhaps still more surprising is that the principle that the state should compensate individuals for property taken for public use was not widely established in America at the time of the Revolution. Only colonial Massachusetts seems rigidly to have followed the principle of just compensation in road building.[1] New York, by contrast, usually limited the right of compensation to land already improved or enclosed or else it provided that compensation should be paid by those who benefited from land taken to build private roads.[2] Despite the efforts of Thomas Jefferson to establish the principle of just compensation in postrevolutionary Virginia, no law providing compensation for land taken for roads was enacted until 1785, although the state had regularly compensated slave owners for slaves killed as a result of unlawful or rebellious activities.[3] Until the nineteenth century, Pennsylvania and New Jersey still denied compensation on the ground that the original proprietary land grants had expressly reserved a portion of real property for the building of roads.[4]

Not only was eighteenth century practice strongly weighted against compensation but so was its constitutional theory. Of the first postrevolutionary state constitutions, only those of Vermont and Massachusetts contained provisions requiring compensation. By 1800 only one additional state — Pennsylvania — constitutionally provided for compensation for takings under the power of eminent domain. Even by 1820 a majority of the original states had not yet enacted constitutional clauses providing for compensation for land taken.[5] Yet, under the influence of Blackstone's strict views about the necessity of providing compensation,[6] reinforced by the antistatist bias of prevailing natural law thinking,[7] by this time statutory provisions for compensation had become standard practice in every state except South Carolina, whose courts continued to uphold uncompensated takings of property.[8] And even without the benefit of a constitutional or statutory provision, some judges were remarkably quick to hold, in Chancellor Kent's words, that "provision for compensation — in a statute — is an indispensable attendant on the due and constitutional exercise of the power of depriving an individual of his property."[9] He established a practice of enjoining public officials from undertaking any activity for which there was no advance provision for compensation.[10]

The movement toward establishing the just compensation principle during the nineteenth century reveals a variety of important conflicts over theory and practice that the process of economic development brought to the surface. At the turn of the century, there still existed a perhaps dominant body of opinion maintaining that individuals held their property at the sufferance of the state. In 1802, for example, the Pennsylvania Supreme Court upheld an uncompensated taking of land to build an incorporated turnpike, over the objection that it violated the recently enacted just compensation provision of the state constitution.[11] There had been a long standing practice in colonial Pennsylvania, the judges noted, of including a surplus 6 percent in every land grant from the proprietor on the assumption that the state could later take property for the building of roads. This practice, the judges held, was not intended to be barred by the constitution.

It was, in fact, true that colonial Pennsylvania had usually compensated landowners only for improvements but not for unimproved land taken to build roads. Beginning in 1787, however, the legislature had often, if irregularly, included compensation provisions in road and canal building statutes.[12] Indeed after the political uproar that resulted from the Pennsylvania Supreme Court's endorsing of uncompensated takings in *M'Clenachan v. Curwin* (1802) it appears to have become "usual" for the legislature to provide for compensation in turnpike acts.

While the increasingly regularized practice of statutory compensation during the nineteenth century muted much of the conflict over the state's power to take land without payment, there continued to be a strong current in American legal thought that regarded compensation simply as a "bounty given . . . by the State" out of "kindness" and not out of justice.[13] South Carolina courts continued to uphold uncompensated takings.[14] And even where a state constitution required compensation, practical consequences flowed from this view that compensation clauses were gratuitous limitations on an otherwise inherent and unlimited sovereign power. For example, beginning in the 1830s, lawyers began to argue that even where the state was required to compensate for takings for a public purpose, there was no corresponding constitutional obligation to compensate for private takings.[15] "The provision in the state constitution authorizing the taking of private property for public uses, is not a grant, but a limitation of power," a New York railroad lawyer argued in 1831. "The states, representing the whole people are sovereign, and possess unlimited power in all cases except where they are restricted by either the federal or their own constitution. The legislature of a state, unless restricted by the state constitution, would even have power to take private property for private use."[16] The thrust of this argument was that those who claimed that railroads were not "public" — and hence not entitled to exercise the eminent domain power — would nevertheless be conceding a still more potent inherent power to take for "private" purposes *without* compensation.

Another practical consequence of the view that just compensation clauses were not grants but narrowly drawn limitations on an originally unrestrained sovereignty was that it enabled its proponents to confine the scope of the constitutional protection. Chief Justice Gibson of Pennsylvania, the foremost judicial advocate of the view that just compensation provisions were "disabling, not . . . enabling" clauses, held, as a result, that the Pennsylvania constitution did not restrain uncompensated takings for private purposes. The fact that there had "usually, perhaps always," been compensation in such cases, he declared, was "done from a sense of justice, and not of constitutional obligation."[17] He also insisted that there was no constitutional requirement that "consequential damages" be compensated, since they were not literally "takings." Even though such "indirect" injuries had "usually [been] compensated," Gibson declared, "it was of favour, not of right."[18]

There is nevertheless a clear trend throughout the first half of the nineteenth century in the direction of enacting state constitutional provisions requiring just compensation. While this trend does offer unmistakable evidence of the gradual crystalization of sentiment in favor of the abstract principle of compensation, we shall see that there was also an equally clear countertrend during the same period in the direction of limiting the scope of application of the compensation principle. This tendency toward limitation often drew upon a surprisingly widespread and powerful earlier view that all property was originally held at the sufferance of the sovereign.

The significance of the delay in establishing the constitutional principle of compensation as well as the success of the movement to narrow its scope has not been fully appreciated. The basic source of early nineteenth century resistance to the compensation principle were those entrepreneurial groups who regarded it as a threat to low cost economic development. Until around 1850, when there are unmistakable signs of a shift in opinion, these emergent groups in American society generally cast their influence on the side of limiting the compensation principle or even of justifying uncompensated takings. At the same time, there were those who sounded the alarm against all redistributions of wealth through use of the eminent domain power. "At no time has there been such a spirit of improvement pervading the country, as at present," one legal writer observed with alarm in 1829. "The vast plans, indeed, which are now in embryo in most of the States for *turnpikes, canals, railroads, bridges*, and other means to facilitate internal communication, are almost without number." He feared that "in an age and in a country, where the expediency, if not the necessity, of public improvements is constantly presenting itself to the attention of the legislative bodies . . . attempted encroachments upon private property by a State" were likely to increase.[19] By the 1850s, this apprehension that the eminent domain power might be used to bring about egalitarian redistributions of wealth finally came to unite most orthodox legal opinion around the principle of compensation. By then, however, many of the important benefits of cheap economic development had already been achieved.

THE BURDEN OF DAMAGE JUDGMENTS

Just as the principle of just compensation was becoming well established in America, a period of sustained economic growth made the problem of damage judgments central to the concerns of economic planners. As early as 1795 the directors of the Western and Northern Inland Lock Navigation Companies reported that the problem of land valuations had caused the company "serious embarrassment," apparently because of large damage judgments awarded by juries. The company "humbly entreated" the New York legislature to allow the courts to appoint appraisers, "whose decisions shall be conclusive."[20] Again, in 1798 the canal company informed the legislature that the mode of assessing damages was "injurious and expensive, and . . . justice requires some amelioration" of the law's provisions.[21] This time the main emphasis of the report was on the cost of litigation, with the company observing that in one case a jury had assessed damages at only one dollar, while the costs of the lawsuit amounted to $375. The company also reiterated its earlier plea that damage judgments be taken away from juries and suggested that the legislature include in its charter a provision similar to one recently granted to the city of Albany, which made the valuations of assessors final for land taken to establish a system of water supply.[22] While the legislature immediately acceded to the company's request,[23] the courts also continued to be solicitous of the canal's interests. After the company had paid damages for land taken, it found it necessary to overflow adjoining land still owned by farmers bordering on the canal. And even though there was not the slightest suggestion that the appraisers had taken such an eventuality into account, the Supreme Court of New York in 1807 held that the destruction of the productive value of the land had already been included in the original damage award.[24]

In some cases, even the title to property was drawn into doubt under the pressure to reduce costs of economic development. For almost two decades after the completion of the Erie Canal in 1825, New York judges continued to debate whether mill owners on the banks of various large upstate freshwater rivers were entitled to compensation for injuries resulting from diversion of water into the canal.[25] Since these rivers were not subject to the ebb and flow of the tide, they were regarded as nonnavigable under the test of the English common law. And since title to the bed of nonnavigable streams had long been held to vest in the owners of the adjoining banks, it appeared that the state was required to compensate them for injury to their water rights. On the other hand, while the English test of navigability was an accurate description of reality in that small country, it was hardly applicable in America where there were many navigable fresh water rivers in which the tide did not ebb and flow.[26] If the common law rule were abandoned, the state's title to these rivers surrounding the Erie Canal would relieve it of the obligation of providing compensation.

It was estimated by the counsel for the canal commissioners in 1826 that "probably more than $100,000 on the lines of the canals [were] involved" in that particular decision,[27] and future Supreme Court Justice Cowen calculated that for the whole state, "with its immense inland waters," the legal precedent would determine property rights amounting to "an aggregate of millions."[28] In an era in which $7,500,000 invested by the state in the Erie Canal made it, according to Chancellor Kent, "a great public object, calculated to intimidate by its novelty, its expense and its magnitude"[29] all that had preceded it, the stakes involved in the decision were enormous.

For two decades the New York courts attempted to decide the question, first following Pennsylvania[30] and South Carolina[31] in rejecting the common law rule,[32] then reversing that decision,[33] and finally ending up with an inconclusive ad hoc test of whether in particular cases there was an original intent by prior owners to grant title to the stream along with the adjoining land.[34] The result was that the state was relieved of a considerable portion, but not all, of potential damage claims.

There were similar expressions of apprehension over damages in other states. In 1807 an officer of the Schuylkill and Susquehanah Navigation company in Pennsylvania reported that the company could not complete the largest branch of its canal because, among other reasons, of "the enormous sums paid for land and water rights."[35] And in Connecticut one observer noted that in building that state's turnpikes "the purchase of the land was a very heavy charge."[36] By 1807 the proprietors of the Middlesex Canal had already spent $58,000 for impaired water rights and land values out of a total expenditure of $536,000.[37] And this expenditure would have represented but a fraction of the potential cost of damages were it not for beneficial intervention of the Massachusetts Supreme Court. Under the pressure of mounting damage claims, the court held in 1815 that landowners had no common law action for the flooding of property adjoining the canal, since the canal's charter, passed twenty-two years earlier, though admittedly "obscure, confused, and almost unintelligible in its terms," had provided its own mode of recovery for damages.[38] While the court merely purported to dismiss the common law suit in order to allow the plaintiffs to pursue their statutory remedy, it failed to mention that having neglected to sue within the one year period required by the statute, these landowners were foreclosed from recovering at all.[39]

A similar concern for the cost of damage suits was evident in the building of the Erie Canal, since abutting property owners' contributions of free land "were small and [the] value almost negligible." An 1825 report of the New York Canal Commissioners complained that "the extravagance of some of these damage claims is equalled only by the pertinacity with which they are urged." Conceding that "the owners of lands through which the canals are made, must have been necessarily incommoded in the occupancy of their farms during the time of their construction," the commission nevertheless maintained that a general increase in land values and access to markets justified refusing damages entirely.[40]

Another set of problems was raised where land was not actually appropriated but nevertheless was made less valuable because of injurious actions of the state or of private adjoining landowners. In a variety of complex and ingenious ways, courts began to establish rules which substantially limited the liability not only of the state but of private corporations chartered to undertake works of economic improvement. The effect of these decisions was to ameliorate the widespread fear, as expressed by one legal commentator, that "the government may create a franchise, and yet its grantee cannot exercise it without being subject to ruinous damages. . . . If a canal company, or a railroad company, can be required to provide for consequential damages, so as to swell the cost of their enterprise, they must be remunerated in the rate of tolls, or in some other form by the public."[41]

The cry that ruinous judgments would be visited on transportion companies became especially strong on the eve of the great boom in railroads during the two decades before the Civil War. In 1841 the New York State Assembly happily noted that because of land donations the damages resulting from construction of the New York and Erie Railroad would prove "comparatively small" where, by contrast, land damages "constitute . . . a large item in the expenditure of other roads."[42] For example, the Boston and Maine Railroad reported in 1844 that its expenditure for land and land damages in building an extension road constituted almost 50 percent of its total costs.[43] By this time railroads also had begun to fear those damage judgments that resulted from personal injuries or from fires started by sparks from locomotives. In 1844 the Western Railroad warned of "the extraordinary expenses arising from accidents which human foresight cannot always prevent, the liability to which increases in a greater ratio than the business of the road."[44] In that same year the Boston and Worcester Railroad complained of increased expenses due to "some considerable charges for damages occasioned by accidents."[45]

These concerns led to a proposal which was quite popular for a time — that government alone should be held legally responsible for damages inevitably resulting from the operation of a private franchise.[46] Much of this theory was no more than self-serving propaganda of the transportation companies, and no court ever adopted this view. However, the fact that some apparently disinterested legal scholars also arrived at the same conclusion underlines the grim legal consequences facing those who entered upon vast schemes of economic improvement.

Under traditional legal doctrine, trespasses or nuisances to land could not be justified by the social utility of the actor's conduct nor could the absence of negligence serve as a limitation on legal liability for injury to person or property.[47] And since many schemes of economic improvement had the inevitable effect of directly injuring or indirectly reducing the value of portions of neighboring land, common law doctrines appeared to present a major cost barrier to social change.[48] Nor could the issue really be avoided by transferring these costs to the government, since state governments themselves were faced with the possibility of crushing expenditures for public works.[49] Because of large damage judgments before 1830, for example, Pennsylvania was compelled to abandon various public works entirely before it finally took damage assessments away from juries.[50] In short, there existed a major incentive for courts not only to change the theory of legal liability but also to reconsider the nature of legal injury. In an underdeveloped nation with little surplus capital, elimination or reduction of damage judgments created a new source of forced investment, as landowners whose property values were impaired without compensation in effect were compelled to underwrite a portion of economic development.

The Breakdown of the Principle of Just Compensation

Under the pressure of damage judgments, American courts before the Civil War began to change legal rules in order to subsidize the activities of great works of public improvement. Looking back at the developments in the law of negligence before the Civil War, the New York Supreme Court in 1873 summarized the changes that had taken place in the conception of property:

The general rules that I may have the exclusive and undisturbed use and possession of my real estate, and that I must so use my real estate as not to injure my neighbor, are much modified by the exigencies of the social state. We must have factories, machinery, dams, canals and railroads. They are demanded by the manifold wants of mankind, and lay at the basis of all our civilization.[51]

The court's statement reflected a fundamental transformation in private law doctrines that had taken place over the previous three quarters of a century. At the beginning of the nineteenth century, the law of nuisance provided an almost exclusive remedy for indirect interferences with property rights, and courts were prepared to award damages for injury to property regardless of the social utility or absence of carelessness of the actor's conduct. By the time of the Civil War, by contrast, American courts had created a variety of legal doctrines whose primary effect was to force those injured by economic activities to bear the cost of these improvements.

Consequential damages

The earliest efforts to limit the scope of liability in American law centered not on any grand conception of the nature of legal responsibility but on the need to reduce the burden of damage judgments and to make economic planning more coherent. The first movements in this direction were characterized by an effort to redefine the scope of legal injury, as judges tended to focus on the old concepts while gradually giving them imperceptibly different meanings. In the first decade of the nineteenth century, courts began to hold that certain types of costly injuries were nevertheless too trivial to be compensable[52] or that they were previously included in calculating the compensation provided for the actual taking of other land, even in cases where, in fact, the subsequent injury could hardly have been anticipated.[53] In the next decade judges also began to hold that indirect injuries to buildings on adjoining land were not compensable where they were the consequence of economic exploitation of one's own land.[54] The result was that by the end of the first quarter of the century there already existed a body of legal doctrine which immunized proprietors from liability for certain kinds of injurious activity in the interest of promoting competitive development of land.

With the increase in economic development in the second quarter of the century, these doctrines began to be expanded to a point where they directly challenged the now almost universally acclaimed principle of just compensation. Drawing on the spirit of the earlier cases, judges began to develop a distinction between immediate and consequential injuries, so that, especially where actions of the government were involved, injurious acts that were neither direct trespasses to land nor actual appropriations for public use were often held to be noncompensable. In the leading case of *Callender v. Marsh*[55] (1823), the Massachusetts Supreme Court denied recovery for an indirect, though substantial, injury to the foundation of a person's house resulting from a city's action in regrading an adjoining street. Similarly, where the state undertook to improve navigation on public rivers through various public works, the New York Supreme Court held that both overflowing of riparian lands and obstruction of access to private docks were noncompensable injuries, even though the value of the plaintiff's property was considerably reduced.[56]

The explanation of how the exemption of consequential injuries from liability could be reconciled with the principle of just compensation was never made clear. In *Callender v. Marsh*, Chief Justice Parker seemed to suggest a theory of unjust enrichment, declaring that the risk of consequential damage was already discounted in the price originally paid for a piece of land.[57] Not only was this theory circular in ascribing a set of expectations to landowners for the purpose of deriving a legal rule that in turn would determine these expectations, but of even greater importance was the court's failure to acknowledge that the principle of just compensation had created a strong expectation of compensability. Indeed, it is difficult to see why the theory of unjust enrichment would not also lead the court to conclude that the necessity of compensation was always superfluous, since the threat of actual appropriation by the state was equally discounted in the price paid for any property. This was precisely what the Pennsylvania Supreme Court had held when it decided in 1802 that the state's 6 percent system of land grants had already compensated landowners in advance for all subsequent takings.[58] In any case, by the second quarter of the nineteenth century, most judges agreed with a similar, though more candid, statement by the New York Supreme Court in 1828 that "every great public improvement must, almost of necessity, more or less affect individual convenience and property; and where the injury sustained is remote and consequential, it is *damnum absque injuria*, and is to be borne as part of the price to be paid for the advantages of the social condition. This is founded upon the principle that the general good is to prevail over partial individual convenience."[59]

The mere suggestion that "the general good" could prevail over "partial individual convenience" without compensation was enough to convince many that the doctrine of consequential damages was nothing more than an excuse for violation of the principle of just compensation. Although both Kent and Story vigorously attacked this line of decisions on precisely this ground,[60] courts after 1825 usually continued to defend it as a vindication of the general good.

From the very beginning the exemption of consequential damages from the general principle of just compensation stood as a doctrine in search of a rationale. At the outset, it was often justified by a mechanical conception of causation borrowed from the emerging doctrine of contributory negligence. Since a person was liable only for the "natural and proximate" causes of his act, courts began to suggest that every indirect injury was also remote and hence entailed no liability.[61] At every point, however, they were confronted with the objection that, whether indirect or not, the action of the defendant was the sole cause of the resulting injury and "that justice would seem to demand that the compensation should proceed from the quarter to which the benefit flows."[62] By the middle of the century, this mechanistic explanation was buttressed by a theory which sought to explain on the same ground the immunity from consequential damages in both tort and contract. Consequential injuries resulting from breach of contract had long been regarded as noncompensable because they were not part of the risk which the contracting parties had undertaken. In a similar fashion, the attempt to extend the contract analogy to tort damages was based on the growing conviction that economic planning required that the liability of entrepreneurs be limited only to those injuries which they could have anticipated.[63]

But this theory was never really adequate either, since most consequential injuries to land were, in fact, entirely predictable, so that the question invariably returned to which party was to bear the cost of economic improvement. By 1850 the New York Supreme Court was prepared to assert that it was "a very common case, that the property of individuals suffers an indirect injury from the constructing of public works."[64] Yet, the only justification it could offer for denying compensation was Justice Parker's original theory "that when men buy and build in cities and villages, they usually take into consideration all those things which are likely to affect the value of their property."[65] And the court never did explain why this theory did not also obliterate the necessity of providing compensation even when property was actually appropriated.

The law of nuisance

While immunity from consequential injuries was usually extended only to the state,[66] another application of the same doctrine provided similar immunity to private companies. Economic development brought forth a host of nuisance actions for intangible but continuing injuries to property because of the smells and noises emitted from neighboring enterprises or because access to one's property or water privileges was impeded by various public works. While other areas of the law were changing to accommodate the growth of American industry, the law of nuisances for the longest time appeared on its face to maintain the pristine purity of a pre-industrial mentality.

With almost complete unanimity, American courts before the Civil War continued to echo Blackstone's view that even a lawful use of one's property which caused injury to the land of another could be enjoined as a nuisance, "for it is incumbent on him to find some other place to do that act, where it will be less offensive."[67] One of the rare early suggestions of the pressure for change was expressed by Judge Tapping Reeve in his 1813 lectures at the Litchfield Law School. In his discussion of "whether a stable could be erected so near a house as that the stamping of the hooves would keep the people awake," he contended that "this would depend very much on the necessity of the case — as whether the man could exercise his business in another place — for men must be allowed to carry on their business."[68] But there are few indications that such an explicit balancing of interests was actually undertaken by many other American judges.

Perhaps the most notable exception to the general unwillingness of American courts before the Civil War openly to admit the need to accommodate the law of nuisance to the demands of a developing society appeared in one of the earliest railroad nuisance cases. In *Lexington & Ohio Rail Road v. Applegate*[69] (1839), the highest court in Kentucky reversed an injunction issued by the chancellor, who had restrained the Lexington & Ohio Rail Road from running its trains through the city of Louisville on the ground that it consti-

tuted a nuisance. Expressing sentiments that rarely were acknowledged in judicial opinions of the time, the Kentucky Court of Appeals asserted that "private injury and personal damage . . . must be
expected from . . . agents of transportation in a populous and pros-
pering country."[70]

The onward spirit of the age must, to a reasonable extent, have its way.
The law is made for the times, and will be made or modified by them. The
expanded and still expanding genius of the *common law* should adapt it
here, as elsewhere, to the improved and improving conditions of our country and our countrymen. And therefore, railroads and locomotive steam-
cars—the offsprings, as they will also be the parents, of progressive improvement—should not, in themselves, be considered as *nuisances,* although, in ages that are gone, they might have been so held, because they
would have been comparatively useless, and therefore more mischievous.[71]

In balancing the social utility of railroads against the resulting
injury, the Kentucky decision stands virtually alone among pre-Civil
War cases in its candid attempt to adapt the law of nuisance to the
demands of economic development.[72] Even in England it was only
after 1865 that the courts began to acknowledge that a process of
weighing utilities and not the mere existence of injury was necessary
for deciding whether a particular use of land constituted a nuisance.[73]

In light of the nuisance doctrine that prevailed throughout most
of the nineteenth century, it seems difficult at first glance to understand how the United States could have succeeded in becoming in-
dustrialized. Where other less well-established legal categories were
more amenable to early doctrinal change, the law of nuisance long
continued to reflect the deepest eighteenth century notions of the
absolute prerogatives of private property. The abundance of undeveloped land was surely a major factor in postponing the profoundly antidevelopmental effects of the law of nuisance on the
course of industrialization. Indeed, this abundance allowed courts
to ignore what the relative scarcity of resources already forced them
to see in the area of water rights: that an absolute and exclusive conception of property inevitably retarded the emergence of competition.

Still, the prevailing nuisance doctrine immediately threatened
most transportation companies not only with the likelihood of substantial damage suits but also with injunctions.[74] If most judges did
not follow the Kentucky court in openly reshaping the law of
nuisance, they shared that court's partiality toward economic devel-

opment. As a result, while the formal doctrine appeared to change very little, judges began to establish a variety of ingenious variations in its application that eventually transformed the substantive doctrine itself. The effect of these changes was that individuals who sought damages due to injuries from great works of public improvement were frequently denied the benefits of a nuisance doctrine that, formally at least, seemed to provide the injured party with all the advantages.

One of the most heavy-handed limitations on recovery grew out of a theory that had been suggested by the New York courts as early as 1807:[75] that after an eminent domain proceeding there could be no recovery for any subsequent injury to one's land, since the original damage award had already accounted for such injury. In 1843 the New Jersey Supreme Court moved one step further and denied recovery to property owners injured by the Delaware and Raritan Canal "whether [the injuries] were clearly to be seen and easily estimated before the construction of the canal, or whether they were uncertain and doubtful results from such construction."[76] In short, the court was even prepared to deny recovery for all *unanticipated* injuries!

A more pervasive limitation on the nuisance action arose out of a new and extensive application of an old doctrine. Though a distinction between public and private nuisances had long been recognized by the common law, American courts before the Civil War succeeded in reshaping this distinction into a major barrier against individual interferences with the process of internal improvements. "The law," Blackstone had written, "gives no *private* remedy for anything but a *private* wrong." Therefore, the only means by which a public or common nuisance could be reached was through indictment instituted by public authorities "because the damage being common to *all* the king's subjects, no *one* can assign his particular proportion of it; or if he could, it would be extremely hard, if every subject in the kingdom were allowed to harass the offender with separate actions."[77] The reason assigned by Blackstone bore a close relationship to the theory of consequential damages simultaneously emerging in American law. While consequential damages were usually conceived of as causally remote injuries, at other times they were regarded as noncompensable because they were spread evenly throughout the entire community. However, the distinction between public and private nuisances did not necessarily lead to noncompensability according to Blackstone, for "where a private person suffers some extraordinary damage, beyond the rest of the

king's subjects, by a public nuisance . . . he shall have a private satisfaction by action."[78] Yet as the process of internal improvements
progressed, courts frequently expanded the public nuisance concept
into a barrier to all private suits.

Among the earliest cases in which courts used the public nuisance
doctrine to defeat private damage remedies were actions by wharf
and dock owners on navigable streams who complained that the
value of their property had been substantially reduced by the
actions of public authorities in diverting water or impairing access
to their wharves.[79] When dock owners complained that New York
State's construction of the Albany basin had rendered their property
inaccessible to ships moving up the Hudson River, the New York
Supreme Court was quick to hold that if the nuisance "operates
equally, or in the same manner, upon many individuals constituting
a particular class, though a very small portion of the community, it
is not a special damage" sufficient to allow a private recovery.[80] On
the other hand, the New York court allowed municipal authorities
to use the nuisance doctrine to eliminate a privately owned floating
storehouse situated on the Albany basin on the ground that it obstructed navigation.[81] This time the court maintained that any
member of the public had standing to abate a common nuisance,
thus freeing these officials from any obligation to provide compensation for destruction of private property.

The most significant fact about the public nuisance doctrine was
that it enabled courts to extend to private companies virtually the
same immunity from lawsuits that the state received under the
theory of consequential damages. As the transportation revolution
progressed, the most frequent beneficiaries of the public nuisance
doctrine were railroads. When the city of Boston closed down a portion of Market Street so that the Boston and Maine Railroad, as provided in its charter, could extend its tracks through that part of the
city, landowners attempted to show that this impairment of access
had rendered their property less valuable. But Chief Justice Shaw
revived again the once narrow and technical public nuisance doctrine in order to dismiss the plaintiff's claim. Even though the
plaintiff "may feel [the injury] more," Shaw wrote, "in consequence
of the proximity of his lots and buildings; still it is a damage of like
kind, and not in its nature peculiar or specific."[82] The fact that "he
suffers a damage altogether greater than one who lives at a distance" does not establish the basis for recovery, because "in its nature it is common and public." Thus, recovery for injuries to property began to turn more and more often on whether an injury was

labeled as one of degree or one of kind. And ironically, the more extensive the "indirect" injury from public improvements, the more often the public nuisance doctrine was invoked to defeat any recovery.[83] While the formal notion of what constituted a nuisance still did not admit of judicial evaluation of the usefulness of a particular project, the public nuisance concept, which purported to be only a technical remedial doctrine, nevertheless enabled courts to decide whether the utility of an undertaking was sufficient to immunize it from private damage suits or injunctions.

Statutory justification

In England, where the issue of just compensation had long ceased to be a major issue by the end of the eighteenth century, judges were far more candid than their American brothers in dealing with the problem of damages arising from social change. In two important cases at the turn of the century,[84] English courts held that no damages could be recovered against public officials who caused injuries to land inevitably resulting from public works authorized by Parliament. The effect of these decisions was to bar recovery for injury caused without negligence by public authorities, even in cases where the same injury, if caused by private parties, would have been compensable. Although a small number of American cases in the early years of the nineteenth century seemed to assume this principle,[85] it was never expressly followed as American judges appeared unwilling to adopt a rule that would openly acknowledge that acts of public and private officials were not subject to the same rule of law.[86] Others openly deplored the English cases for allowing the taking of property without compensation.[87]

The appeal of the English cases, however, revived during the second quarter of the nineteenth century as state courts were persuaded gradually to extend the theory to include private companies within the scope of public immunity from damage suits. One of the first such beneficiaries of the new theory was a Pennsylvania canal corporation[88] in which the state owned a small portion of the shares.[89] And in 1831 the Maine Supreme Court extended the immunity to injuries inflicted by a wholly private canal company. "It does not necessarily follow," the court declared, "that because a plaintiff may have sustained a serious injury in his property, consequent upon the voluntary acts of a defendant, that therefore he has a right to recover damages for that injury." There were some acts, the court observed, which might be "justified by an express provision of law." Other kinds of damage may have arisen from "acts which others might lawfully do in the enjoyment and exercise of their own rights and management of their own business." Finally, injury "may have resulted from the application of those principles by which the general good is to be consulted and promoted, though in many respects operating unfavorably to the interests of individuals in society."[90]

It was only after 1840, however, that the English cases were widely cited, as American lawyers contended that all legislatively authorized acts, whether by public officials or private holders of franchises, were equally immunized from damage actions if exercised with care. In defense of this position, one legal writer complained that under the existing law "the government may compel its grantee to carry [the franchise] into effect in such a manner that injury to private interests will be inevitable, and the courts whilst acknowledging the authority of the government, sustain an action in favor of the proprietor against whom the government authorized the alleged wrong."[91] This contention was especially well received by Theodore Sedgwick, the brilliant New York legal writer. In his influential *Treatise on Damages* (1847), Sedgwick constructed out of a handful of prior English and American cases a "general rule . . . that where the grantees have not exceeded the power conferred on them, and when they are not chargeable with want of due care, no claim can be maintained for any damage resulting from their acts."[92] Yet, very few American cases had ever expressly gone so far as to immunize private injuries to land from damages on the ground that the acts had been authorized by the legislature,[93] and some courts had expressly rejected this theory.[94] But so powerful was the impulse to reduce damage judgments that by 1863 the highest court in New York could hold that even though the street railway built through New York City was an "interference with, and injury to the use and enjoyment" of the neighboring property "to such an extent that the same would constitute a continuous private nuisance" were it not for legislative authorization, still the state could constitutionally immunize private businesses from suit without compensation to those who were injured.[95] Two years earlier, the New York court had allowed the New York Central Railroad to build a road across a stream, by which it obstructed the flow and flooded the property of an adjoining landowner. There could be no recovery without negligence, the court held, since a road built under legislative authority for a public purpose could not be a nuisance.[96]

With these cases, the public law rule of just compensation began to be limited and eroded by the still developing private law principle of negligence, which held that injuries resulting from socially useful conduct were not compensable in the absence of carelessness. Yet, it was still too early for courts, in the absence of a legislative determination, openly to decide which activities were sufficiently desirable to suspend the historic rules governing liability for nuisance.[97] The result was that the courts fashioned two different standards for recovery for nuisance: one under which the ordinary companies remained absolutely liable for injuries caused by nuisance and another for authorized works of public improvement, though privately financed, for which only negligent conduct would permit an injured property owner to recover.[98] Still more important, perhaps, were the many cases which in 1800 had been analyzed as "nuisance" cases but which by 1850 were simply conceived of as "negligence" cases, thereby imposing a less stringent standard of care. Liability of railroads for fire was a prominent example of this reclassification from nuisance to negligence.[99]

The idea of legislatively conferred immunity from nuisance actions was carried to still greater lengths by courts determined to remove restrictions on economic development. Just as statutory authorization of internal improvements came to be recognized as a defense to private suits, some courts extended the notion still further to prevent even public officers from abating public nuisances. Where the state of Pennsylvania had turned over to the officers of the Erie Canal Company the task of building a local connection with the canal, the draining of swamps became necessary in order to raise a reservoir for the waterway. Later, the state indicted the officers of the canal company after one of these reservoirs had become "stagnant, putrid, and noxious, from whence unwholesome damps and smells arise, and the air is greatly corrupted and infected." Nevertheless, the court assumed that "works of internal improvement, erected at the expense and by the officers of the state, for the benefit of the citizens at large, never could be regarded by the law as a nuisance; for the sovereign authority has expressly intended them to advance the prosperity of the community."[100]

Changes in the theory of damages

As courts were transforming the conception of legal liability, legal writers sought to adapt the theory of damages to the needs of a developing society. It was not until the nineteenth century that the measure of damages came to be regarded by orthodox legal thinkers as "a question of law." I will return to this theme in later chapters and seek to describe how the judiciary took control over rules of damages in contract and commercial cases during the nineteenth century. Not only did this development represent a major shift in power between judge and jury, as well as between commercial and anticommercial interests, but it also reflected the triumph of the precise and calculating mercantile spirit over the rather rough and informal justice characteristic of common law jury determinations.

In tort law as well the nineteenth century represented a clash between past and future. Were damage judgments to continue to reflect the ancient peace keeping and paternalistic underpinnings of tort law or were they to accommodate to the new insistence of entrepreneurial groups that certainty and predictability of legal consequences were essential for economic planning?

In an earlier period in which violations of the tort law were universally regarded as unjustified and antisocial acts, there was little moral pressure to calibrate damage judgments to the precise level of injury, since deterrence and the prevention of "unjust enrichment" were the characteristic goals of the law. Where, however, personal or property injury had begun to be thought of as an inevitable "cost of doing business," legal thinkers were faced with the paradox of imposing damages on activities that, in general, they regarded as socially beneficial. How then to adapt an earlier moralistic and penal conception of damages to a developing economy in which damage judgments were themselves crucial "costs" of development?

In later chapters, I will examine the institutional and procedural responses to these questions, which eventually resulted in bringing all damage questions under the control of judges. At this point, I would like to focus on only one somewhat narrow debate over the theory of damages that particularly concerned tort law—the status of punitive damages.

As early as 1830 Theron Metcalf, later a justice of the Massachusetts Supreme Court, argued that "neither on principle, nor by the preponderance of authority, can damages be estimated by any other standard than the actual injury received—that the extent of the injury is the legal measure of damages."[101] Metcalf attempted to demonstrate that the common law had never allowed for punitive damages in tort and that while the evidence of a malicious intent might be admitted to show the extent of actual damages, it was otherwise irrelevant in determining the measure of damages. "There would seem to be no reason why a plaintiff should receive greater damages from a defendant who has intentionally injured him, than from one who has injured him accidentally, his loss being the same in both cases." Although, "it better accords . . . with our natural feelings, that the defendant should suffer more in the one case than in the other," he added, ". . . points of mere sensibility and mere casuistry are not allowed to operate in judicial tribunals; and if they were so allowed, still it would be difficult to show that a plaintiff ought to receive a compensation beyond his injury." Nor would it be any less difficult to demonstrate "on principles of law or ethics . . . that a defendant ought to pay more than the plaintiff ought to receive. It is impracticable to make moral duties and legal obligations, or moral and legal liabilities, co-extensive."[102] Thus, Metcalf attempted to divorce tort law from its ancient function of private peace keeping. "Damages are given as a satisfaction for an injury received by the plaintiff, not by the public," he concluded.[103]

In an age when juries were suspected of partiality to small landowners whose property was damaged by the activity of large transportation companies, Metcalf's theory of damages raised the possibility of removing the subjective and thus inherently uncontrollable issue of punitive damages from a jury's consideration, thereby providing judges with a greater measure of control over damage judgments. Moreover, at a time when many injuries to land resulted from a considered engineering decision of a canal or turnpike corporation, any assurance that intentional injury would not give rise to punitive damages was of major economic importance. Yet, beyond the significant fact that this theory would have had a substantial impact on the size of damage judgments, it also expressed a widely shared need in the first half of the nineteenth century to import greater certainty and predictability into the law of damages, so that entrepreneurs could more accurately estimate the costs of economic improvements.

Metcalf's theme was taken up in 1846 by Simon Greenleaf in his impressive *Treatise on the Law of Evidence*. Damages, Greenleaf asserted, "should be precisely commensurate with the injury; neither more, nor less."[104] Yet, one year later his views were challenged by Theodore Sedgwick, who argued in his *Treatise on the Measure of Damages* that "the theory of compensation is not the theory of our law." Contending that "in cases of tort, the suit at law appears to have public as well as private ends in view," Sedgwick could see "no reason why the defendant should not, in a civil suit, be punished for his act of fraud, malice or oppression, nor why the pecuniary mulct, which constitutes that punishment, should not go into the pockets of the plaintiff, instead of the coffers of the state."[105] In his second edition, published in the next year, Greenleaf, though conceding that Sedgwick's position "may appear to be justified by the general language of some Judges," argued at length that these expressions were mere dicta.[106]

It is tempting to interpret the debate over punitive damages as reflecting a fundamental political cleavage over legal theory. The Jacksonian Sedgwick's insistence on retaining the paternalistic and regulatory functions of tort law stands in sharp contrast to the more orthodox and apolitical Greenleaf's willingness to transform tort law into an amoral system of what amounted to selling licenses to injurers. But there are difficulties with any such interpretation. Despite numerous instances in which nineteenth century legal thinkers were willing to overthrow well-established theories, there were many other occasions on which they regarded the weight of the received legal tradition as just too overwhelming to allow for innovation. And it seems clear that in this debate Sedgwick's basic loyalty was primarily to the established legal tradition itself.

At the most general level of damage theory, there was, in fact, no debate between Greenleaf and Sedgwick. Not only is Sedgwick's *Treatise on the Measure of Damages* the most brilliant and boldly innovative American antebellum legal treatise, but its publication — the first treatise on damages — itself marks a broad shift in legal theory. Sedgwick himself observed with satisfaction that it was only "at comparatively a recent period that the jury has relinquished its control over actions even of contract, and that any approach has been made to a fixed and legal measure of damages." And subject to the exception for punitive damages, he, like Greenleaf, applauded the fact that even with respect to tort damages, "by degrees, the salutary principle has been recognized . . . that the amount of compensation is to be regulated by the direction of the court, and that the jury cannot substitute their vague and arbitrary discretion for the rules which the law lays down."[107] On issues for which the received legal tradition allowed a greater measure of flexibility, then, Sedgwick was among the first to insist that the law of damages be made more certain and predictable. His discussion of consequential damages, for example, was among the earliest to suggest that in principle the measure of damages in tort and contract was the same, the effect of which was to import the precise and calculating spirit of contract law into the unpredictable process by which tort damages traditionally had been measured.[108] More than anything else, his *Treatise* stood prominently for a proposition that was just gaining general assent by the middle of the century: that the question of damages was an issue of law and that judges, if necessary, could set aside jury verdicts that were excessive.

The erosion of trial by jury

Standing beside the numerous changes in legal conceptions was an important institutional innovation that began to appear after 1830 — an increasing tendency of state legislatures to eliminate the role of the jury in assessing damages for the taking of land. It was long a commonplace that juries increased the size of damage judgments. As we have seen,[109] the New York legislature in 1798 responded to the pleas of that state's first canal company to take land valuations away from juries. In Massachusetts, juries had awarded landowners only 10 percent more than the proprietors of the Middlesex Canal had offered for land damages.[110] But that modest appreciation had occurred during the first five years of the century, when the enthusiasm for public works was fairly widespread and the proprietors were conscious of the need to offer settlements that would satisfy public opinion. Later, in Pennsylvania, however, use of the jury system for settling damage claims arising out of public works "led to excessive damage awards, especially in those counties which were dissatisfied with the works administration, and in some cases the state was actually compelled to abandon construction plans because of the expense involved."[111] When in 1830 the legislature provided for a board of public works, it "was attacked as a partisan rather than a judicial arm of the state, interested not in justice but in the lowest evaluations of damages, and in truth its harassed members had to admit that they were 'sometimes obliged to act apparently the part of an attorney for the Commonwealth,' discovering on their own initiative facts which favored state interest."[112] Although there were other early instances in which legislatures eliminated the jury's role in assessing damages,[113] it was only in connection with the building of railroads that this movement gained real force. Between 1830 and 1837 such statutes in New Jersey,[114] New York,[115] Ohio,[116] and North Carolina[117] were upheld over the objection that they violated constitutional provisions guaranteeing trial by jury.

As states turned over the task of assessing damages to commissions, many statutes also allowed these commissioners to offset estimated benefits to land against actual damages.[118] The result was that railroad companies were often allowed to take land while providing little or no compensation. Rejoicing at an Ohio court decision that overturned a commission's damage award for being too small, an Ohio journal advocating constitutional reform proclaimed that " a brighter day seems to be dawning—a day when courts will not aid [corporations] in riding rough shod over individuals. Heretofore when they have wanted the property of individuals to aid them in their splendid schemes of speculation, it had been seized and appropriated under the false and lying pretext that it was for public use, and little or nothing paid for it."[119]

THE TRIUMPH OF NEGLIGENCE

At the beginning of the nineteenth century there was a general private law presumption in favor of compensation, expressed by the oft-cited common law maxim *sic utere*.[120] For Blackstone, it was clear that even an otherwise lawful use of one's property that caused injury to the land of another would establish liability in nuisance, "for it is incumbent on him to find some other place to do that act, where it will be less offensive."[121] In 1800, therefore, virtually all injuries were still conceived of as nuisances, thereby invoking a standard of strict liability which tended to ignore the specific character of the defendant's act. By the time of the Civil War, however, many types of injuries had been reclassified under a "negligence" heading, which had the effect of substantially reducing entrepreneurial liability. Thus, the rise of the negligence principle in America overthrew basic eighteenth century private law categories and led to a radical transformation not only in the theory of legal liability but in the underlying conception of property on which it was based.

The origins of the modern negligence doctrine

One is surprised to learn how really late it was in the nineteenth century before the action for negligence became a significant factor in American law. Although judges of the eighteenth century "were familiar with the name and idea of negligence," until it was freed of its associations with bailment and contract it was still "too early to speak boldly of an action of negligence."[122] In the eighteenth century, American courts had applied the negligence concept to hold common carriers or other bailees liable on a theory of contract,[123] and to suits in which doctors or lawyers were charged with breach of a contractual relationship with their patients or clients.[124]

Another, and perhaps the most important, eighteenth century line of cases in which negligence was a factor involved both common law and statutory actions against sheriffs for taking insufficient bond[125] or for allowing imprisoned debtors to escape.[126] American judges had followed Blackstone in distinguishing between voluntary (or collusive) and negligent escapes,[127] and it is clear that by the end of the eighteenth century a distinction between some kinds of intentional and unintentional injuries was already important in America.

Blackstone regarded the duty of the sheriff as part of the "class of contracts, implied by reason and construction of law" by which "every one who undertakes any office, employment, trust or duty, contracts with those who employ or entrust him, to perform it with integrity, diligence and skill."[128] Yet, there is no indication, as is often assumed, that this distinction presupposed any limitation on strict liability. Negligence in this context simply meant the long established right of an individual to sue a public officer for failure to perform a duty imposed by law. Indeed, the same theory explains eighteenth century statutes allowing an action against towns for failure to repair roads and bridges,[129] as well as early nineteenth century cases that imposed liability for neglect on chartered turnpike and canal companies.[130]

We should inquire at the outset to what extent the notion of negligence in these public duty cases had anything in common with the modern negligence doctrine. For example, in the case of the duty of sheriffs to prevent the escape of imprisoned debtors, two important legal consequences flowed from the distinction between "voluntary" and "negligent" escapes. For a voluntary or intentional escape, the sheriff was liable whether or not he subsequently captured the prisoner; for a negligent escape, he was liable only if he failed to recapture the prisoner before trial. In addition, the sheriff was liable for the full amount of the prisoner's debt if the escape was voluntary, while for a negligent escape the jury had discretion over the amount of damages and could discount the likelihood of the debt's ever being collected.[131]

There is no suggestion that carelessness or inadvertence forms the core of the action. The primary meaning of negligence was "nonfeasance" and hence the sheriff was held strictly liable even for a negligent escape. A report of a 1795 New Jersey escape case[132] clearly underlines the prevailing conception:

The ground of defense adopted by the defendant [sheriff], was to prove that there was no want of attention on his part; that every precaution consistent with humanity, and sometimes even bordering on rigor, had been adopted with regard to [the escaped debtor]. This was made out by several witnesses, and it was proved that the escape was occasioned by circumstances not to be foreseen, and which could not be prevented by even more than ordinary exertions and caution.

Nevertheless, the Supreme Court approved the instruction of the trial court that, even though there was no evidence of a collusive or voluntary escape, "every escape not happening by the act of God, or the public enemies was, in the eye of the law, considered a negligent escape. The law admits no other excuse."

The dominant understanding of negligence at the beginning of the nineteenth century meant neglect or failure fully to perform a preexisting duty, whether imposed by contract, statute, or common law status. To be sure, actions on an implied contract against doctors or bailees often alleged carelessness or unskillfulness. Yet, even here one strongly suspects that carelessness was merely presumed from failure to perform, or, as Roscoe Pound has put it, "The negligence is established by the liability, not the liability by the negligence."[133] For example, a study of Baron Comyns' _Digest of English Law,_ one of the most influential English law texts used in America at the end of the eighteenth century, illustrates the extent to which even "misfeasance" was conceived of differently from modern negligence. Not only does Comyns' title dealing with the "Action upon the Case for Negligence" refer almost exclusively to cases of nonfeasance or nonperformance, but, surprisingly, there is hardly a trace of the concept of carelessness in the material listed under the heading "Action upon the Case for Misfeasance." Indeed, Comyns shows no conception of modern negligence when he writes that "an Action lies for Misfeasance, tho' the Damage happen by Misadventure [accident]."[134]

Even granting that there were certain kinds of eighteenth century cases in which lawyers understood negligence more or less as we do today, they surely did not see any general conflict between strict liability and negligence. For it was only in the nineteenth century that carelessness—and the associated problem of establishing a standard of care—became the central allegation in an action for negligence. A glance at the chapter on negligence in Nathan Dane's _Abridgment of American Law_ underlines the extent to which the dominant meaning of negligence, even as late as 1824, continued to be that of nonfeasance. The largest number of Dane's illustrations still deal with the failure to perform a public duty. For example, "If the commissioners of a lottery neglect to adjudge a prize to him who draws it, this action [for negligence] lies against them." Or "It is a general rule of law, that if a man neglects to do what he is bound to do, as to pay toll either by prescription or otherwise, this action lies against him." Most of these cases deal with the failure of a public officer to fulfill his duty. "In every case," Dane concluded, "in which an officer is intrusted by common law, or by statute, to perform a service, and *neglects* it, this action lies against him by the party injured."[135] Yet, although it still remains clearly a subsidiary theme, the common law strand of misfeasance already had begun to assume a larger position in the lawyer's consciousness. Not only does Dane cite English cases for "negligently and carelessly" planting trees[136] and for a surgeon's "gross ignorance and want of skill in his profession,"[137] but he also mentions a handful of American cases involving misfeasance, including a ship collision "through negligence and want of skill"[138] and a fire case in which "what is negligence or misconduct in me or my servants must depend on all the circumstances of the case."[139] In short, the story of the rise of negligence in America involves a process of analyzing the almost imperceptible changes in emphasis by which an older status-oriented conception of failure to perform a duty is gradually laid aside and the distinctively modern emphasis on careless performance begins to take its place.

In America, as in England, "the prime factor in the ultimate transformation of negligence from a principle of liability in Case to an independent tort was the luxuriant crop of 'running down' actions reaped from the commercial prosperity of the late eighteenth and early nineteenth centuries."[140] The earliest running down cases were lawsuits involving collisions between ships or horse-drawn carriages that began to appear at the turn of the century.[141] What was new was that as courts began regularly to enforce legal duties between strangers, they were compelled to see that a theory of liability for negligence could no longer be derived from either a common law status or from a contractual agreement between the parties.[142] Perhaps of still greater importance, the collision cases were the first to involve joint actors, a factor that inevitably led judges and juries beyond the simple inquiry into whether an injury had been committed in order to determine which party had "caused" the injury. The inquiry into causation shifted the attention of courts to the question of carelessness. And as liability for carelessness began to be understood as deriving from an independent social obligation imposed by the state, the search for a standard of care encouraged judges to regard themselves as social engineers and legislators, whose decisions to impose liability were influenced by broader considerations of social policy.

Yet it was some time after the advent of the negligence action that judges realized its potential for affecting the course of social change. It was one thing for courts to recognize an independent action for negligence; it was quite another for them to develop habits of thought which would undermine the basic presumption of compensation for injury that had been erected over several centuries of the common law. And although American judges talked the language of negligence from the beginning of the nineteenth century, it was quite some time before they used the negligence concept in order to mount a general attack on the prevailing standard of strict liability.

The influence of the forms of action

It has become quite commonplace to assume that one of the most important factors restraining the emergence of an independent action for negligence was the elaborate system of common law forms of action. According to conventional wisdom, the importance of the distinction between the common law writs of trespass and trespass on the case (or "case") was that the former was based on strict liability while the latter required an allegation of negligence. Since the action for trespass was limited to direct injuries, so this argument goes, there could be no independent action for negligence until the strict liability—negligence distinction between trespass and case was obliterated. This is regarded as the significance of Chief Justice Shaw's decision in the famous case of Brown v. Kendall[143] (1850). But can this version of the development of negligence stand once it is realized that there was no well-developed conception of negligence even for an action on the case before the nineteenth century?

As long as the action of trespass on the case was largely based on an implied contract theory in the eighteenth century, judges had little conceptual difficulty in distinguishing between trespass and case. Given the upsurge of collision cases involving strangers, however, judges were forced to rethink the underlying theory of the actions, and from the end of the eighteenth century English courts began to distinguish between actions in trespass and case on the basis of whether an injury was direct and immediate or indirect and consequential.[144] But there is no evidence that American judges who adopted this pleading distinction ever assumed that an action in trespass was based on strict liability while an action on the case was grounded in an allegation of negligence. Until some judges later found in the writs the modern distinction between intentional and negligent injuries, the basic legal consequence that flowed from the forms of action related solely to the formalities of good pleading: for an indirect injury the proper plea was in case.

At the end of the eighteenth century, when the conception of negligence revolved around nonfeasance, liability in trespass and case was equally strict and there was little inducement to distinguish between them. Nuisance, which dominated tort actions for injuries, was itself a strict liability doctrine and was, in fact, pleaded in case. When in the course of the first half of the century the idea of misfeasance begins to prevail, it transforms not only the action on the case but that of trespass as well. The result was that some judges came to require that negligence be proven even in trespass, while others distinguished between the actions on the basis of whether the injury was intentional or negligent. But whatever the period studied, there is no indication that American judges, unlike their English brethren,[145] ever regarded the substantive law governing the two writs as turning on a distinction between strict liability and negligence.[146]

In light of this analysis, it should be seen that an exaggerated significance has been assigned to the opinion of Chief Justice Shaw in the case of *Brown v. Kendall*.[147] In that case, the Chief Justice held that in the absence of negligence the defendant was not liable even in trespass for unintentionally hitting another person with a stick swung while trying to part two fighting dogs. Ever since Holmes celebrated Shaw's opinion as a bold and virtually unprecedented effort to subject the trespass action to the test of negligence,[148] we have been told by commentators that the decision "marked [a] departure from the past," and that Shaw "gets most of the credit for the establishment of a consistent theory of liability for unintentionally caused harm."[149] Unfortunately, these commentators have not studied with equal care the dramatic and simultaneous transformation of the action on the case.[150]

As early as 1814 the Massachusetts Supreme Court had indicated impatience with a defendant who attempted to take advantage of the always elusive difference between direct and indirect injuries,[151] and in that same year, the Massachusetts court allowed a plaintiff to amend his declaration from case to trespass, leading Nathan Dane to conclude, that "the distinction" between the actions "is not so important."[152] In 1823 the court upheld a trespass action for "negligently driving" a carriage,[153] but it was not until 1833 that the Massachusetts court held for the first time that as a matter of law there could be no recovery in an action on the case for failure to prove negligence.[154] Moreover, even by 1846 there were only three or four such reported cases in Massachusetts.[155] Therefore, the decision to subject the trespass action to the test of negligence in *Brown v. Kendall* was merely part of a very recent flowering of the negligence action having nothing to do with any earlier recognition of a substantive distinction between writs.[156]

The developments in Massachusetts merely repeated a process of change that first had taken place in New York and Pennsylvania somewhat earlier. In the 1843 case of *Harvey v. Dunlop*,[157] the New York Supreme Court held that negligence had to be proved in a trespass action, but this decision merely represented the culmination of a uniform course of New York decisions which since 1820 had assumed that carelessness had to be shown in both trespass and case.[158] Indeed, much the same process occurred in Pennsylvania, where in 1839, only six years after the first decision denying recovery for failure to prove negligence,[159] a lower court held that negligence needed to be shown in trespass as well.[160]

There was one other distinction between trespass and case that has misled students into assuming a difference in substantive law governed the two writs. In eighteenth century England, actions against a master for injuries caused by his servant were usually brought in case, since the theory of the action was contractual. Later commentators have been puzzled by Blackstone's statement that "if a smith's servant lames a horse while he is shoeing him, an action lies against the master, and not against the servant."[161] Because we have come to regard the duty of care as an obligation owed to the world,[162] we have been confused by the fact that Blackstone could have regarded the servant as immune from liability for negligence. Yet, it is clear that he conceived of the sole source of legal obligation as arising from the blacksmith's status in relation to the customer, and consequently, in the absence of any implied contractual relationship, the servant himself was not liable for his own negligence. American lawyers at the beginning of the nineteenth century typically continued to regard both the liability of masters and the allegation of the servant's negligence in similarly contractual terms,[163] so that the allegation of negligence was viewed as performing the limited function of proving a breach of an implied contract.

In 1795, however, we find Zephaniah Swift of Connecticut stating that when a trespass is committed by the servant with the express or implied command of the master "the master shall be deemed guilty of it, as well as the servant."[164] Though Swift had moved beyond a status-oriented theory of implied contract, there was still no thought here that negligence in the servant was essential to establish the master's liability to a stranger.[165] However, as strangers started to sue for injuries due to collisions, the allegation of negligence began to appear outside its ancient status-oriented setting. Yet, even here it appears initially to have performed another, quite limited, function in the master-servant context. At the end of the eighteenth century, we find the English courts holding that all master-servant actions must be brought in case, since even a direct injury committed by a servant could not be regarded as the master's trespass.[166] Moreover, English judges held that the master was not liable at all when the act of the servant was willful, for the theory behind the master's liability had gradually shifted to an inquiry into whether he had expressly or impliedly commanded his servant to perform the injurious act.[167] Allegation of the servant's negligence, therefore, merely served the limited purpose of showing that his action was unintentional, which otherwise would have placed it beyond the master's command.[168] As a result, American judges at the beginning of the nineteenth century were still prepared to impose liability on masters in an action on the case even though the injurious act of the servant was "merely fortuitous and accidental."[169] While the negligence principle, when it finally did emerge, was a means of limiting defendants' liability, negligence in the master-servant context originally served the entirely different purpose of negating willfulness and thereby establishing the master's liability. For just as in the sheriff cases, negligence was often equated merely with unintentional behavior. Eventually, however, by almost imperceptible changes in emphasis, the function of negligence in the master-servant area merged into the general stream of negligence liability that was developing in the nineteenth century.

The earliest reported American master-servant case in which failure to prove carelessness in the servant becomes the basis for denying the master's liability appears in a decision of the South Carolina high court in *Snee v. Trice*[170] (1802). The plaintiff sued after a fire set on adjoining land by the defendant's slaves spread and destroyed the plaintiff's house and cornfield. Although the bulk of the court's opinion deals with the question of whether the English law of master and servant was applicable to slaves, the case undoubtedly introduces a modern strand of negligence. Troubled by the fact that slaves represented "a headstrong, stubborn race of people, who had a volition of their own, and the physical power of doing great injuries to their neighbours and others, without the possibility of their masters having any control over them," the court was tempted to restrict the operation of an English rule that "was by no means applicable to the local situation and circumstances of Carolina, where almost the whole of our servants are slaves."[171] But since the case represented "the first of the kind ever known to have been brought and discussed in this country,"[172] the court hesitated to change the English master-servant doctrine.[173] Instead, it held that the cause of the injury "had more of the appearance of accident than negligence," since "the morning was still, and the fire had burnt down, but towards the middle of the day, the wind arose, and blew up the sleeping embers which communicated the fire to the building."[174]

With the help of an ambiguous passage in Blackstone, the South Carolina court thus seemed to reject the common law doctrine imposing strict liability for fire. The allegation of "negligence" appears in English fire cases as early as 1401,[175] even though there always was strict liability for fire at common law[176] until the rule was changed by statute in England at the beginning of the eighteenth century.[177] In these early cases, negligence meant no more than "neglect" or "failure" to keep the fire on one's own land, having no modern overtones of fault or carelessness.[178] Blackstone also wrote of a master's liability for a servant who "kept his master's fire negligently,"[179] but there is likewise no reason to suppose that he meant anything but this old-fashioned notion of neglect.[180] Nevertheless, by the time the South Carolina court was called upon to infuse these words with meaning in *Snee v. Trice,* there was a strong incentive to limit the scope of liability for the acts of slaves. As a result, we find the first unambiguous recognition in American law of a legally imposed standard of care not arising out of contract.[181]

There is no indication that the South Carolina decision had any influence on the course of American law. Outside South Carolina, the decision was ignored, probably having been regarded as limited to the peculiar problem of slavery. And, in fact, when the next slave case came before the South Carolina court in 1825, the judges suppressed whatever creative impulses they earlier had shown by misreading *Snee v. Trice* to hold that a master was not liable for the negligence of his slave.[182] The result was that the law of negligence remained dormant thereafter in South Carolina for quite some time.

Another problem with the usual interpretation that assumes substantive differences for actions in trespass and case is that it does not explain why plaintiffs should have been so indifferent to the forms of action if trespass represented the better writ because it was based on strict liability. There are many early instances of direct injuries caused by a principal in which the plaintiff would be expected to sue in trespass and yet brings the action in case.[183] Moreover, after 1817, when the New York Supreme Court allowed plaintiffs to choose between the actions in suits involving direct injuries caused by negligence, they sued in case as often as in trespass.[184] And even in those states which did not permit such an election, it is difficult to suppose that plaintiffs would have so often made the mistake of suing in case if they were aware of the substantive advantages that trespass is supposed to have conferred. After 1833 the English courts also allowed an election between trespass and case,[185] yet the seeming indifference of plaintiffs to the forms of action equally puzzled two distinguished English legal historians, who concluded that there were many secondary advantages that still induced English plaintiffs to sue in case.[186] Whatever may have been so in England, however, it appears that most of the secondary advantages in America also lay with trespass.[187] In short, the failure of plaintiffs consistently to sue in trespass for direct injuries confirms what the statements of American judges and lawyers show directly: that there were never substantive advantages to suing in trespass. Rather, both trespass and case simultaneously responded to the rise of negligence in the nineteenth century.

In New York, Massachusetts, and Pennsylvania the rise of the negligence action was accompanied by the assumption that even for direct injuries the plaintiff must prove negligence. In other states, however, the attempt to preserve the distinction between the actions caused judges to ignore the express language of the English decisions[188] and to attempt to put the difference between trespass and case in terms of whether the defendant had brought about the injury through design or through negligence.[189] A decision of the New Hampshire court in 1826 allowing an election between actions for a direct injury caused by negligence[190] preserves this formulation and marks the beginning of the modern distinction between intentional and unintentional torts. Not only does this case underline the disintegration of the direct-indirect distinction, but, more important, it reveals the inability of American courts to sanction the English doctrine of strict liability in trespass after the negligence doctrine had begun to emerge.

Negligence as a matter of law

There are three quite different stages in the rise of the negligence doctrine in nineteenth century America. In the first, beginning early in the century, the important element is the shift in emphasis from an implied contract theory of nonfeasance to a tort conception of misfeasance. In the second stage, after 1820, judges realize, under the influence of the collision cases, that proof of injury and proof of carelessness constitute separate inquiries, and they develop the idea of duties owed to noncontracting strangers with sufficient clarity to deny recovery as a matter of law where there is no proof of carelessness. In the third stage, beginning around 1840, the negligence doctrine breaks out of its rigid confinement to highway and ship collision cases and begins directly to challenge the presumption of compensation for injury in settled areas of the law.

If the collision cases changed the understanding of negligence in an action on the case, they equally challenged the assumption of strict liability in a trespass action. For even where courts continued to insist that trespass was the only proper plea for direct injuries, it was logically impossible, even in trespass, to avoid considering the question of carelessness in a collision case.[191] Whereas the presumption of strict liability had flourished in a society in which the typical tort action involved the suit of an inactive plaintiff who had been injured by an active defendant, the collision cases, which involved joint actors, forced to the surface for the first time the question of who, in fact, had caused the injury. Thus, modern negligence made its first limited appearance as a question of causation or, in contemporary terminology, of contributory negligence.

Even so, there are no American cases in which an independent defense of contributory negligence is made a question of law until after 1823,[192] and even more surprisingly, only one of these cases before 1838 involved a collision.[193] Instead, in this period the typical contributory negligence defense arose under a Massachusetts statute holding towns liable in double damages for neglecting to repair highways under their control.[194] Since the courts regarded the statute as imposing a crushing burden on the towns, judges were willing to allow them to argue that they were not liable because the plaintiff had failed to use due care to guard against the dangerous obstruction. But any assumption that the collision cases failed to leave a clear mark on the developing law of negligence because they did not give rise to a formal contributory negligence doctrine is surely misleading. Since an injury from a carriage or ship collision always involved the simultaneous action of both parties, juries inevitably had to decide who caused the injury by determining which party was at fault. As a result, the problem of negligence was concealed in jury deliberations about who, in fact, was responsible for the injury. The really lasting influence of the collision cases on the development of American law, therefore, was to obscure the common law assumption that individuals were presumed liable for all injury resulting from risk-creating activity and to direct the law's focus exclusively to the immediate question of which party's carelessness had brought the injury about. While the collision cases were the first consistently to introduce carelessness into the analysis of negligence cases, contributory negligence only arose as an independent question of law when courts were faced with the later highway repair cases. In these cases, since the actions of plaintiff and defendant were separate in time, judges were finally forced to see that the question of causation was not merely a matter of failure to repair the road but also involved the temporally separate issue of whether the plaintiff had driven his vehicle with due care.[195]

Nevertheless, even after contributory negligence arose as an independent question of law, it still was not understood by contemporaries as a fundamental departure from the basic system of strict liability. In most states until 1840 cases involving joint actors remained sufficiently rare to be treated as an isolated category in the law. Courts conceived of the issue of contributory negligence as merely a threshold inquiry about who, in fact, caused an injury, and once a court was satisfied that the plaintiff was not the proximate cause of the injury, it often returned to the traditional language of strict liability.[196] For example, until as late as the middle of the century courts allowed contributory negligence to defeat recovery in trespass or nuisance actions, where, but for the plaintiff's own negligence, they were still prepared to impose liability on the defendant without inquiring whether he had been careless.[197] Conversely, in New York, which had a thriving system of negligence by 1830, a formal doctrine of contributory negligence played no significant role until much later.[198]

It is important to see precisely where the negligence doctrine stood in America as late as 1830. Only in New York was it clear both that negligence would be determined on the basis of a standard of care and that defendants were not liable as a matter of law without proof of carelessness. Massachusetts and Vermont also recognized contributory negligence as a defense to legal liability, but the question was still regarded as a limited and preliminary inquiry into causation. Moreover, as long as the negligence doctrine was associated with causation, its potential for expansion into areas involving inactive plaintiffs, in which causation was undisputed, remained minimal. And it was precisely in this latter area, which often involved railroads, where the great burst of negligence jurisprudence after 1840 took place. In short, there were no cases outside New York before 1833 in which failure to prove negligence became the basis for denying recovery as a matter of law.[199]

In that year, however, the courts of Massachusetts and Pennsylvania rendered decisions recognizing modern negligence principles. In *Sproul v. Hemmingway*[200] Massachusetts Chief Justice Shaw held that the defendant in a ship collision case not involving contributory negligence was immune from liability in the absence of negligence,[201] but it was still more than a decade later before the law of negligence played any important role in that state's jurisprudence.[202] In Pennsylvania the transition was far more dramatic, since there was no existing doctrine of contributory negligence in the commonwealth. In an action for destruction of a bridge caused by flooding due to a break in a dam, Chief Justice Gibson boldly announced that in the absence of carelessness the defendant was not liable.[203] It was only the fourth time[204] — and the first outside New York — that an American court had used the negligence standard as a bar to recovery by an inactive plaintiff. With this case, the courts of both New York and Pennsylvania has already clearly decided that, regardless of causation, injury brought about by risk-producing activity was itself no ground for imposing legal liability. The test of negligence had begun to be regarded as a general limitation on legal liability.

The general triumph of negligence

It is important to emphasize that at the beginning of the nineteenth century the principle of strict liability for injury to property was regarded as just another application of the growing general presumption in favor of compensation for the taking of property. Nor was it crucial that constitutional provisions for just compensation applied only to takings by the state. Chartered transportation companies that committed trespasses or nuisances were equally regarded as agents of the state,[205] and most of the cases involving injuries to person or property after 1840 were brought about by the activity of canals or railroads. Nevertheless, even a charter was not necessarily crucial in imposing a legal obligation in favor of compensation. The mill acts, which suspended common law nuisance remedies for flooding of land in favor of a statutory scheme of compensation, were analyzed on the assumption that any rule of law which denied all forms of compensation to injured property owners would amount to an unconstitutional taking.[206]

The clash between the principle of compensation and the emerging doctrine of negligence did not become apparent until courts were prepared to extend the negligence standard to injuries in which the nuisance standard of strict liability had once dominated. Indeed, the negligence system began to expand beyond the relatively recent and narrow field of collisions at the same time as the abstract principle of just compensation finally triumphed in America. Not only were the expansion of the just compensation principle and the rise of negligence chronologically related, but, in fact, the rise of the negligence doctrine was seen by contemporaries as an attempt to escape from the moral imperative of just compensation.

One of the most dramatic clashes between the old and the new systems of analysis occurred in the Connecticut case of *Hooker v. New-Haven & Northampton Co.*[207] (1841). In a 3-2 decision, the Supreme Court held that the defendant canal company could not justify injury to adjoining land on the ground that it was not caused by negligence. The minority had relied on an earlier case that appeared to hold that in the absence of carelessness a chartered company could escape liability if there were a statutory authorization for its activity. But the majority still referred the legal question to the test of the old morality, holding that the only issue was just compensation.

An individual may use his own property, without intent to injure his neighbour; but if in so doing, he does him a damage, he must be answerable. For all civil actions, the law doth not so much regard the intent of the actor as the loss and damage of the party suffering; and although a man does a lawful thing, yet if damage do hereby befal another, he shall answer it, if he could have avoided it. [Citing cases] and the reason of all these cases, is, because he that is damaged, ought to be recompensed.[208]

Within the two decades before the Civil War, however, the negligence standard began to invade the general law governing injury to persons and property. One of the most important routes was the idea that a chartered company, being authorized by the legislature, did not exceed its jurisdiction unless it acted carelessly. Another important area concerned the spread of fire. In the first generally recognized American case in which a court denied recovery in the absence of negligence, the New York Supreme Court held in 1811 that a defendant was not liable for an injury resulting from a fire that spread from his land "because it was lawful for him to keep his fire there."[209] The decision not only represented a dramatic departure from the common law rule of strict liability for fire,[210] but since it denied recovery to an inactive plaintiff who had been injured by

an active defendant, it was the first decision in American law completely to separate the question of negligence from the issue of causation. Nevertheless, there are few other reported fire cases before
1840[211] and none that refuses to impose liability for failure to show
negligence. The question of spreading fire, however, became generally important in the two decades before the Civil War as railroad
locomotives commonly sent sparks onto neighboring land. Indeed,
the first reported noncollision negligence case in many states raised
this question of liability for fire[212] as courts began to extend the negligence standard to all actions in tort.

The subversion of the expanding public law principle of just compensation by the increasingly ruthless application of the private law
negligence principle must be seen as a phenomenon of industrialization. It is not surprising, therefore, that the negligence standard
rose earliest in New York, Pennsylvania, and Massachusetts, to challenge the assumption of strict liability,[213] for there was a relatively
advanced level of economic development in these states. Indeed, the
rise of the negligence principle was only part of a more general attempt to limit the scope of application of the principle of just compensation. This effort in turn was intimately associated with the
need to reduce the crushing burden of damage judgments that a system of strict liability (or just compensation) entailed.

As it developed in the course of the nineteenth century, one legal
commentator later observed, the American attitude toward legal
liability was based on the assumption that the "quiet citizen must
keep out of the way of the exuberantly active one."[214] Indeed, the
law of negligence became a leading means by which the dynamic
and growing forces in American society were able to challenge and
eventually overwhelm the weak and relatively powerless segments of
the American economy. After 1840 the principle that one could not
be held liable for socially useful activity exercised with due care
became a commonplace of American law. In the process, the conception of property gradually changed from the eighteenth century
view that dominion over land above all conferred the power to prevent others from interfering with one's quiet enjoyment of property
to the nineteenth century assumption that the essential attribute of
property ownership was the power to develop one's property regardless of the injurious consequences to others.

THE CONSEQUENCES OF LEGAL SUBSIDIZATION

One of the most striking aspects of legal change during the antebellum period is the extent to which common law doctrines were trans-

formed to create immunities from legal liability and thereby to provide substantial subsidies for those who undertook schemes of economic development. This pattern of subsidization raises several important questions concerning the relationship between legal and economic change in the period. Was legal subsidization socially efficient? Did it encourage investment in areas in which, as the welfare economist would put it, social benefits exceeded social costs even though private costs were greater than private benefits?[215] Or did it in fact promote overinvestment in technology, which might be inferred from the strikingly low contribution to the gross national product that Robert Fogel has claimed railroads, for example, actually made?[216] Because of the difficulties in accurately determining the indirect benefits that flowed from a particular technology, any conclusions about the social efficiency of subsidization must be advanced quite hesitantly.

Perhaps a more basic and manageable question for the historian is the need to explain why there developed so clear a pattern of subsidization through the use not of the tax system but of the legal system. One of the most consistent features of state economic policy during the first several decades of the nineteenth century was the pattern of extraordinarily low state budgetary expenditures. In Massachusetts, for example, where the state budget between 1795 and 1820 remained constant at roughly $133,000, the Handlins have identified a clear pattern of state use of legal instruments such as monopolies and franchises as an alternative to cash outlays.[217] Even states like New York, which amassed an enormous debt in building its canal system, regarded these financial arrangements as involving profitable investments and not as cash subsidies out of the tax system. Indeed, despite a geometrical increase in its debt during the 1820s and 1830s, New York did not impose a general property tax until 1843. "Taxation played a very unimportant role during the first fifty years of the state's existence."[218] Pennsylvania also was "unwilling to embark upon effective taxation" until 1842, when "a vigorous tax program was finally initiated." Investment in canals had simply "supplemented [a] strong anti-tax bias" with "the idea of a positive profit-making state—a state in which taxes were abolished."[219] In short, every bit as significant as overt forms of direct legislative financial encouragement of enterprise were the enormous, but hidden, legal subsidies and resulting redistributions of wealth brought about through changes in common law doctrines.

What factors led antebellum statesmen generally to turn to subsidization through the legal, rather than through the tax, system? One explanation seems fairly clear. Change brought about through technical legal doctrine can more easily disguise underlying political choices. Subsidy through the tax system, by contrast, inevitably involves greater dangers of political conflict. Beyond these general observations about the consequences of choosing between seemingly nonpolitical as opposed to overtly political forms of subsidization, however, can we in addition say anything about the specific redistributive consequences of the one as opposed to the other? Until we know much more about the potential redistributive effects of state tax systems in this period, it would be dangerous to make any firm comparisons. Nevertheless, it does seem fairly clear that the tendency of subsidy through legal change during this period was dramatically to throw the burden of economic development on the weakest and least active elements in the population. By contrast, it seems plausible to suppose that in a period when the property tax provided the major share of potential state revenue, the burdens of subsidy through taxation would have fallen disproportionately on the wealthier segments of the population.

There is reason to suppose, therefore, that the choice of subsidization through the legal system was not simply an abstract effort to avoid political contention but that it entailed more conscious decisions about who would bear the burdens of economic growth. It does seem likely, moreover, that regardless of the actual distributional effects of resorting to the existing tax system, a more general fear of the redistributional potential of taxation played an important role in determining the view that encouragement of economic growth should occur not through the tax system, but through the legal system. It is, after all, quite striking that a dramatic upsurge of explicit laissez-faire ideology in America can be correlated with a dramatic increase in state taxation during the 1840s.[220] One clear result of this ideological change was that state judges during the 1840s and 1850s began to restrict the scope of redistributive legal doctrines like eminent domain, which formerly had been aggressively used to promote economic development.[221] Thus, whether or not legal subsidies to enterprise were optimally efficient or instead encouraged overinvestment in technology, it does seem quite likely that they did contribute to an increase in inequality by throwing a disproportionate share of the burdens of economic growth on the weakest and least organized groups in American society.

THE INFLUENCE OF INDIVIDUALISM

There were essentially three stages in the development of American law relating to conflicting uses of property. In the first stage, which continued until roughly 1825, the dominant theme was expressed by the maxim *sic utere*. Dominion over land was defined primarily as the right to prevent others from using their property in an injurious manner, regardless of the social utility of a particular course of conduct. This system began to break down in the second quarter of the nineteenth century as it became clear not only that common law doctrines led to anticompetitive results but that the burdens on economic growth under such a system might prove overwhelming. Through limitations imposed on the scope of the nuisance doctrine, the emergence of the negligence principle, and the riparian doctrine of reasonable use, courts began to strike a balance between competing land uses, freeing many economically desirable but injurious activities from legal liability if they were exercised with due care. Thus, in a second stage which crystalized by the middle of the nineteenth century, property law had come largely to be based on a set of reciprocal rights and duties whose enforcement required courts to perform the social engineering function of balancing the utility of economically productive activity against the harm that would accrue.[222]

In the two decades before the Civil War, however, one detects an increasing tendency by judges to apply the balancing test in such a way as to presume that any productive activity was reasonable regardless of the harm that resulted. And out of this intellectual climate, a third stage began to emerge, which self-confidently announced that there were no legal restraints at all on certain kinds of injurious activities. In a number of new and economically important areas, courts began to hold that there were no reciprocal duties between property owners; that courts would not even attempt to strike a balance between the harm and the utility of particular courses of conduct. While this trend only reached its culmination after the Civil War, its roots were deep in an antebellum change in the conception of property. Dominion over land began to be regarded as an absolute right to engage in any conduct on one's property regardless of its economic value. Judges began to withdraw to some extent from their role of regulating the type and degree of economic activity that could be undertaken. And the mercantilist character of American property law was diluted by an emerging laissez-faire ideology.

From a fairly early period in the nineteenth century, American cases dealing with the right of adjoining landowners to lateral support reflected a strong tendency to encourage competitive improvement of land. As far back as the seventeenth century, English courts had decided that although adjoining landowners owed a duty of lateral support to land in its natural state, there was no similar duty of support for buildings erected on the land.[223] But when the subject first became of general economic importance in the nineteenth century, the English courts seemed to undermine the vitality of the common law rule by holding that an adjoining landowner was liable for injury to both land and buildings, if the weight of the plaintiff's buildings did not contribute to the injury.[224] By contrast, in a series of cases early in the century, American courts first established the general principle that a landowner owed no duty at all to support buildings on adjoining land.

In the widely influential case of *Thurston v. Hancock*[225] (1815), Massachusetts Chief Justice Parker announced the general principle that "the proprietor of land, unless restrained by convenant or custom, has the entire dominion, not only of the soil, but of the space above and below the surface, to any extent he may choose to occupy it." There was no way "of accounting for the common law principle which gives one neighbour an action against another, for making the same use of his property which he has made of his own," Parker reasoned, unless it were based on the narrow "qualification" of a right acquired by prescription.[226] Without such a right gained by long use, he concluded, there was no basis in law for restraining one's neighbor from doing anything on his own land which one originally could have done oneself.

The decision reflected the same social and economic considerations that were leading to a rejection of the principle of prior appropriation in the area of water rights, as courts began to see the anti-developmental consequences of allowing the first occupant to determine his neighbor's future course of conduct.[227] Moreover, in rejecting the claim of the first builder to control his neighbor's subsequent development of land, Parker dramatically shifted the idea of dominion over land from its traditional emphasis on the limitations that others could impose on land use.

Thurston v. Hancock, nevertheless, expressed an interesting tension between old and new principles. Although it appeared to contradict his principle of absolute dominion over property, Parker reaffirmed the common law duty of lateral support owed to land in its natural state. While the plaintiff thus could not recover for injury to his buildings, he could still be compensated for the nominal injury to land. Attempting to explain this difference in treatment by a strikingly modern theory of property rights based on fulfillment of individual expectations, Parker maintained that the plaintiff could not recover for damage to his building because "he who first built his house should have taken care to stipulate with his neighbor, or to foresee the accident and provide against it."[228] On the other hand, he held that land damage was recoverable because "the defendants should have anticipated the consequences of digging so near the line."[229] However circular and contradictory this distinction was, its practical result seemed a throwback to an eighteenth century common law dichotomy between "natural" and "artificial" activity which reflected profoundly agrarian preferences. In English water cases, for example, a similar distinction between "artificial" and "natural" uses of water became an important ideological tool for preferring agrarian over industrial uses of water.[230] As it applied to the rights of landowners, however, this bias ironically had the opposite effect of freeing economic development from the legal restrictions that a rule of priority imposed on competitive land use. For as long as buildings were considered an unworthy artificial use of land, courts were prepared to withhold the extensive protection that rules of priority accorded to land itself.[231]

In spite of its dual outlook, *Thurston v. Hancock* nevertheless marks a radical break with common law tradition and reveals the early impact of individualism on the development of American law. The decision was the first in the nineteenth century to hold that in certain kinds of economic activity there existed no correlative rights and obligations between adjoining landowners. Some of the cases that purported to follow the Massachusetts decision were fitted into a more traditional mold, holding that there was no duty of lateral support for buildings only if the injurious activity was reasonable.[232] But *Thurston v. Hancock* went much further by absolutely refusing to examine the social utility of the actor's conduct.[233]

By the middle of the century, under the influence of Parker's opinion, there were those who were willing to go still further and reject even the last vestige of conventional doctrine in that opinion. Even the duty of lateral support of land, the New York Supreme Court declared in 1850, "would often deprive men of the whole beneficial use of their property. An unimproved lot in the city of Brooklyn would be worth little or nothing to the owner, unless he were allowed to dig for the purpose of building." If a landowner "may not dig because it will remove the natural support of his neighbor's soil," the court concluded, "he has but a nominal right to his property. . . . A city could never be built under such a doctrine."[234] By the middle of the nineteenth century, in short, the meaning of property rights was in the midst of a major transformation—from the notion that dominion over land entailed the power to prevent all injury by others to the view that many externally imposed limitations on land use were themselves violations of absolute property rights.

One of the most dramatic reversals of legal premises was illustrated by the virtually unanimous refusal of American courts after 1840 to extend the recently developed riparian doctrine of correlative rights to the law governing waters percolating in subterranean channels. Although several early English and American cases had assumed that the riparian doctrine applied equally to subsurface waters,[235] the question only came to the fore after technological innovations created a "minerals-dominant economy" after 1840.[236] In the leading American case of *Greenleaf v. Francis*[237] (1836), the Massachusetts Supreme Court, neglecting even to mention the common law doctrine relating to surface streams, held that every landowner had the right to appropriate the entire flow of a subterranean stream. "Every one has the liberty of doing in his own ground whatever he pleases," the court declared, "even although it should occasion to his neighbor some other sort of inconvenience." The victorious defendant, of course, cited *Thurston v. Hancock*.

Though there were rational distinctions to be drawn between surface and subsurface streams, based on whether landowners could foresee the injurious consequences of their acts,[238] American courts usually preferred to rest their decisions on general conceptions of the nature of property or on a policy of encouraging economic development. In the Pennsylvania case of *Wheatley v. Baugh*[239] (1855), for example, the court insisted that to apply the riparian doctrine of reciprocal rights to percolating streams "would amount to a total abrogation of the right of property." And in *Frazier v. Brown*[240] (1861), the Ohio Supreme Court put a large part of its decision on the ground that an extension of the riparian doctrine to underground percolating waters "would interfere, to the material detriment of the commonwealth, with drainage and agriculture, mining, the construction of highways and railroads, with sanitary regulations, building and the general progress of improvement in works of embellishment and utility." Indeed, the difference in treatment accorded underground and surface waters can largely be explained by the fact that the first cases directly involving the former arose only after laissez-faire assumptions firmly took hold of the imaginations of American judges.

Most of the reasons advanced for not applying correlative rights to underground streams were equally applicable to surface streams.

If reciprocal rights to subsurface streams were recognized, the New York Supreme Court declared in 1855, "no one will be safe in purchasing land adjoining or near a stream of water, as he may be restrained forever from making some valuable, and frequently, from the progressiveness of the age, necessary improvements." Any restraint, it concluded, would contradict "the rule that a man has a right to the free and absolute use of his property" unless he caused direct injury.[241] In fact, as the New Jersey Supreme Court saw many years later, the rule of no correlative rights could not easily be defended simply on grounds of encouraging economic development.

It is sometimes said that unless the English rule be adopted, landowners will be hampered in the development of their land because of the uncertainty that would thus be thrown about their rights. It seems to us that this reasoning is wholly faulty. If the English rule is to obtain, a man may discover upon his own lands springs of great value for medicinal purposes or for use in special forms of manufacture, and may invest large sums of money upon their development; yet he is subject at any time to have the normal supply of such springs wholly cut off by a neighboring landowner, who may, with impunity, sink deeper wells and employ more powerful machinery, and thus wholly drain the sub-surface water from the land of the first discoverer.[242]

Under the orthodox rule, the court added, "*might* literally makes *right*, and we are remitted to the simple plan, that they should take who have the power, and they should keep who can."

As the court thus pierced the abstract claims to equality of the noncorrelative rights doctrine, it also saw that the real explanation of its origins was one of power. For whatever the theoretical equality of both large and small landowners to upset the expectations of the first discoverer, the opportunity to do so, the court saw, depended upon the ability to "sink deeper wells and employ more powerful machinery." Mining companies, in short, were the natural beneficiaries of the rule.

Contemporaries actually perceived that the movement to distinguish between the rules governing surface and subsurface streams was essentially an effort to free the law from the regulatory premises of riparian doctrines. Courts, one legal commentator protested, were shaping the law to give preference to enterprises according to "the difference in the magnitude of the interests [involved]. . . . The simple fact, that a man may carry on a very profitable business, by only doing his neighbor a little injury, is no sufficient excuse for the injury done. If there be a great difference in the interests, the greater may well afford to pay for the damage done to the less."[243]

One result of the increasing tendency to refuse to impose reciprocal rights and obligations on business enterprise was to remove the judiciary entirely from certain forms of economic regulation. Courts, for example, began to refuse even to examine whether particular uses of property were undertaken solely for the motive of injuring a neighboring landowner, an inquiry which, after all, was often indispensable for determining the social utility of harm-producing conduct.[244] To inquire into motives, a New York court declared, "would be highly dangerous to the security of the enjoyment of real property."[245] This tendency, of course, was part of a more general view that had begun to gain ascendancy before the Civil War—that economic development could best be promoted by giving free rein to individual property owners to develop their land.[246] Nevertheless, as courts attempted to overcome the regulatory framework of the common law, the notion of unrestrained dominion over land was often entirely dissociated from its economic foundations to become a functionally autonomous dogma of its own.

As in the case of underground percolating waters, most courts held that the right to rain waters was also the absolute property of the owners of land on which it fell. Unlike subsurface streams, however, surface rain waters were almost never employed for economically useful purposes, and legal controversies thus invariably involved the right of one landowner to drain these waters onto his neighbor's property.[247] Though some courts applied the "reasonable use" test in deciding the extent to which such drainage was allowable,[248] the large majority of American courts, proceeding from a conception of absolute property rights, adopted the "common enemy" rule to hold that each landowner could do all he could to keep the water off his land, even if it harmed his neighbor.[249] In a series of cases before the Civil War, the Massachusetts Supreme Court developed the prevailing doctrine,[250] and Chief Justice Bigelow, in a case immediately following the war,[251] summed up these developments. "*Cujus est solem, ejus est usque ad coelum* [He who possesses land possesses also that which is above it] is a general rule, applicable to the use and enjoyment of property," he wrote, "and the right of a party to the free and unfettered control of his own land above, upon and beneath the surface cannot be interfered with or restrained by any consideration of injury to others."

In the two decades before the Civil War, the ideologies of laissez-faire and rugged individualism had finally established a prominent beachhead in American property doctrine. Though never entirely able to overthrow the regulatory assumptions behind the earlier law, these new doctrines nevertheless underlined a deep tendency in the application of even conventional doctrine to favor the active and powerful elements in American society.

IV Competition and Economic Development

A common law policy in favor of competition emerged from a changing conception of property relations as well as from a fundamental transformation in the relationship between the state and business enterprise. By the end of the eighteenth century, the new American states had become involved in the process of promoting economic development by granting corporate charters and franchises to private investors. Though this pattern has often been portrayed as economically inevitable, it actually seems to have arisen out of conscious considerations of policy. In every state after 1790 a political decision to avoid promoting economic growth primarily through the taxing system seems to have crystallized. While economic disarray and postrevolutionary suspicion of government partially explains this turn of events, historians have never really attempted to understand the effects on the distribution of wealth that an alternative taxing policy might have entailed as compared with the system of private financing that actually came into being.

In fact, there is some slight evidence during the 1780s that states were considering use of the taxing power for economic development. In 1781, for example, the Pennsylvania legislature repealed a 1761 statute "to enable the owners of Schuylkill Point Meadow Land . . . to keep the bank, dams, sluices and floodgates in repair, and to raise a fund to defray the expenses thereof." The statute had provided for a scheme of decentralized improvements, whereby each abutting landowner would assume voluntary responsibility for improving the banks. In 1781 the legislature, finding "by experience that the mode of supporting the bank by individual allotments is attended with diverse inconveniences," provided instead for a more coercive scheme of taxation. It created a new company with the independent power to tax each of the owners for his share of the improvement.[1] By the early nineteenth century, however, little more is heard of such independent taxing authorities. In Massachusetts, in fact, the Supreme Judicial Court in 1807 flatly refused to uphold the taxing power of a similar legislatively created corporation chartered to improve roads.[2]

By 1800 a pattern of private ownership of banks, insurance companies, and transportation facilities had become dominant in America, although existing legal theory continued to enable judges and jurists to regard those enterprises as arms of the state. Eighteenth century English legal theory, in fact, concealed two fundamentally different conceptions of the legal status of corporations and franchises. A much older medieval definition of this relationship treated it as just another species of property, sharing many of the monopolistic and exclusionary characteristics that medieval law had generally assigned to property interests. A second, much later, conception of franchises and corporations tended to explain the grant of monopolistic privileges by a more restricted theory that underlined their public character. Blackstone presented both of these legal strands, though he clearly emphasized the latter. Later, in nineteenth century American law these two strands began to diverge, reflecting the beginnings of a sharp distinction between public and private law. As a result, the question of whether the power to exclude competition was inherent in the general right of property or instead represented a more limited grant of a public privilege soon became a central issue in American law.

Until the second quarter of the nineteenth century, however, there was little cause directly to unravel these contradictory eighteenth century assumptions; they were often therefore simply ignored. In general, a policy of encouraging the investment of scarce private capital in an underdeveloped society coincided with a legal system whose doctrines guaranteed a measure of economic certainty. Private law property conceptions also continued early in the nineteenth century to promote monopolistic development by enforcing the exclusionary privileges of first entrants. And those private law property ideas were simply incorporated into the law governing charters and franchises, so that both public and private law equally applied anticompetitive rules of priority to the new forms of property thus created.

The principle of competition emerged in an entirely different intellectual and economic climate. By the second quarter of the century, it was becoming clear in a number of areas of the law that a conception of property as monopolistic and exclusionary placed an unmanageable burden on continued economic growth. At the same time the growing recognition of a separation between public and private interests induced courts to question whether eighteenth century anticompetitive doctrines were still consistent with economic realities and objectives. In the process, a presumption in favor of competition unambiguously emerged in the common law for the first time, as courts established a general policy that competitive injury was noncompensable.

There is a clear relationship between competition and economic development in nineteenth century America. In an underdeveloped society, with little available private capital, a policy of encouraging development required that the legal system provide legal arrangements that guaranteed private investors certainty and predictability of economic consequences. Perhaps the most important of these guarantees was protection against freedom from competitive injury. To accommodate this policy, courts promulgated rules reflecting a view of property as essentially exclusive and monopolistic, so that every attempt to draw business away from an existing enterprise was usually treated as an injury to property itself.

As it became bound up with a state policy of promoting development, private investment was also regarded as an extension of state efforts to further economic growth. The legal system thus rarely distinguished between public and private forms of investment. But as development proceeded, the early monopolistic strategy for encouraging economic growth soon became a legal barrier to further growth. Certainty and predictability of legal arrangements became incompatible with sustained economic development. Previous state concessions to private interests thus had come to represent obstacles to continued growth, and for the first time state efforts to encourage economic growth began to diverge from private efforts to preserve existing legal expectations. Under the continuing pressure to encourage further investment, the legal system gradually began to distinguish between public and private interests. A monopolistic and exclusionary conception of property was replaced by legal rules that allowed various uncompensated injuries to property. Eventually, out of this sweeping redefinition of property, the legal presumption in favor of competition emerged full blown.

THE SEPARATION OF PUBLIC AND PRIVATE INTERESTS

The change in the conception of the corporation marks one of the fundamental transitions from the legal assumptions of the eighteenth century to those of the nineteenth. The archetypal American corporation of the eighteenth century is the municipality, a public body charged with carrying out public functions; in the nineteenth century it is the modern business corporation, organized to pursue private ends for individual gain. By the time of the decision in the *Dartmouth College* case[3] (1819), holding that a corporate charter was a contract, the conception of the corporation as a public body had been on the decline for almost a generation, although the implications of this trend were only beginning to be explored. The distinction between public and private corporations, so prominent in Justice Story's concurring opinion, was above all a response to the dramatic growth in the number of business corporations during the previous generation. As late as 1780, colonial legislatures had conferred charters on only seven business corporations, and a decade later the number had increased to but forty. However, in the last ten years of the eighteenth century 295 additional corporate charters were granted.[4]

During the early years of the nineteenth century both the weight of the old corporate model and the pressure for change are everywhere in evidence. In 1809 the Virginia Supreme Court upheld a legislative act amending the charter of an insurance corporation. "With respect to acts of incorporation," Judge Spencer Roane observed, "they ought never to be passed, but in consideration of services to be rendered to the public. . . . It may be often convenient for a set of associated individuals, to have the privileges of a corporation bestowed upon them; but if their object is merely *private* or selfish; if it is detrimental to, or not promotive of, the public good, they have no adequate claim upon the legislature for the privileges."[5] Yet, three years earlier the Massachusetts Supreme Court for the first time recognized the private nature of a recently organized turnpike corporation. Probably intending to put to one side the long history of state regulation of the affairs of municipal corporations, it observed that "the rights legally vested in this, or in any corporation, cannot be controlled or destroyed by any subsequent statute, unless a power for that purpose be reserved to the legislature in the act of incorporation."[6]

The Virginia insurance case involved an action by a member of the corporation objecting to an increase in his insurance premium resulting from legislative amendment of the corporate charter. The court was able to avoid deciding whether the amendment could be adopted over opposition of the corporation by holding that the company had accepted the change. And since "decision by a majority is a fundamental law of corporations in this country and in England,"[7] the majority could bind a corporate member without his consent. The model the court was applying was still the eighteenth century conception of a municipal corporation in which an individual could be bound by a majority vote.

Proceeding also from this earlier conception, the Massachusetts legislature in 1804 had created a corporation for the purpose of laying a street, designating as members the owners of land over which the street would pass. The corporation was authorized to assess its members for the costs of the project, and after one of the prospective members refused membership he was sued for his share of the assessment. Now, however, the Massachusetts Supreme Court held that the corporation could not assess persons who did not assent to membership. "If it were a public act," Justice Parker wrote, "predicated upon a view of the general good, the question would be more difficult. If it be a private act, obtained at the solicitation of individuals, for their private emolument, or for the improvement of their estates, it must be construed, as to its effects and operation, like a grant. We are all of the opinion that this was a grant or charter to the individuals who prayed for it, and those who should associate with them; and all incorporations to make turnpikes, canals, and bridges, must be so considered."[8] The legislature, in short, could not create corporations modeled on municipalities with coercive powers of taxation over its members.

As corporations thus were classified from public to private bodies, they inevitably invoked emerging constitutional restrictions on the state's discretion to deal with vested property rights. So successfully was the conception of the corporation detached from its eighteenth century foundations that while members of a municipality were personally responsible for corporate liabilities,[9] courts gradually assumed in the nineteenth century that only limited liability existed for members of a business corporation.[10] On the other hand, private power could not necessarily be controlled by the legal instruments traditionally used to keep public bodies in check. When a canal company in 1810 attempted to take advantage of the eighteenth century rule that a private party could not maintain an action for breach of a corporate duty, the Massachusetts Supreme Court again invoked the new distinction between public and private corporations to permit the company to be sued.[11]

For a time, the corporation continued to occupy a twilight zone in the eyes of the law, sometimes conceived of as a public instrumentality, at other times regarded as a private entity. While they sought to emphasize their recently acknowledged private nature when claiming constitutional protection of corporate property, corporations continued to underline their public service functions in order to claim both the power of eminent domain and freedom from competition. Attempting to take advantage of the eighteenth century notion that public instrumentalities were protected from competition, corporations continued to argue both that their charters were grants of exclusive property interests and that economic rivalry was, in effect, a private law nuisance to property. In turn, this claim forced courts not only to reconsider the legal position of the corporation but also to reexamine the general theory of competition in America.

THE THEORY OF COMPETITION

As the conception of the corporation was in the process of transformation, a startling variety of complicated economic changes began to unfold, which further undermined eighteenth century legal assumptions. Common roads were displaced by turnpikes, and ferries were threatened by the erection of bridges. Soon turnpikes were in turn challenged by canals, and canals by railroads, with each new step creating a complex of legal questions for which the past supplied only the dimmest of guidance.

Despite a variety of judicial attacks on monopoly during the struggles against prerogative government in seventeenth century England, the spirit of competition was only partially recognized by the common law. Reflecting earlier feudal or mercantilist influences, English courts throughout the eighteenth century continued to relieve entrepreneurs from the effects of injurious competition. It is true that Blackstone, repeating a long-standing premise of English law, pointed out that an owner of a mill had no right to sue a competitor for economic injury, though he could bring an action for interference with the flow of water.[12] And he also observed that one had the right "to set up any trade, or a school, in a neighborhood or rivalship with another: for by such emulation the public are like to be gainers; and, if the new mill or school occasion a damage to the old one, it is *damnum absque injuria.*"[13]

Nevertheless, for hundreds of years English courts had enforced anticompetitive doctrines under a somewhat different theory. Even as late as 1779, in an opinion by Lord Mansfield, the English judges entertained an action for damages against a tenant who, departing from custom, had ground corn at a mill owned by his landlord's competitor. And it was not until 1824 that a procompetitive policy was finally introduced by the English courts.[14] True to his apologetic nature, Blackstone, in fact, also supported these and other restraints on trade on the theory that the businesses in question were essentially public enterprises. A mill is erected, he noted, "by the ancestors of the owner for the convenience of the inhabitants, on condition, that when erected, they should all grind their corn there only." And he emphasized that there were a number of other business enterprises such as bake houses and malt houses where the same anticompetitive theory prevailed.[15] Indeed, English decisions of the seventeenth and eighteenth centuries allowed a mill owner to sue or enjoin any inhabitant of a town who established a competitive mill, or even, in certain cases, to destroy the mill. One declaration drawn up around 1735 brings an action against a town's inhabitant for setting up a mill and "contriving and intending to deprive [the plaintiff] of the Profitts of the Toll of his . . . Mills."[16]

Thus, even as the principle of competition was dimly recognized in English law, the public nature of an enterprise continued to be acknowledged as a virtually all-inclusive shield of immunity from competition. Even at the beginning of the nineteenth century English courts still enforced the rule of the medieval common law, that no market or fair could be established within seven miles of another.[17] Just as markets were regarded as franchises, since they were originally established by grants from the king, so too were ferries, and the same exclusionary policy prevailed. In a passage that was cited and recited by American lawyers, Blackstone pointed out that even an ancient ferry, one whose owner's rights had been established by ancient usage, was also entitled to an exclusive franchise to carry passengers.[18] Ancient usage had long been regarded by English judges to be evidence of a lost grant from the king, though this presumption was understood to be a legal fiction established for other purposes. Still, English courts were prepared to hold that even without an express grant conferring an exclusive right, ancient usage was evidence of such a right. Thus, even after the rise of parliamentary supremacy, the anomaly persisted in England of regarding custom as legitimizing those long-standing monopolies which the king himself no longer had the power to create without a parliamentary act.[19] Virtually every modern English case in which exclusive ferry or market privileges were asserted involved an ancient franchise for which the owner could produce no grant because such grants had long fallen into disuse.[20]

The lesson that Americans derived from these cases was that the monopolistic privilege emanated not from any express act of the sovereign but from the nature of property itself. While Blackstone encouraged this conclusion, he justified it on policy grounds, regarding the ferry as a quasi-public agency. Since the law required that every ferry owner keep his vehicle in repair and readiness on pain of being fined, he explained, "it would be therefore extremely hard, if a new ferry were suffered to share his profits, which does not also share his burthen."[21]

While the English precedents thus established an unchallenged tradition that the grant of common carrier privileges necessarily impied the right to exclude all competition, they were not consistently followed in America. When in 1794 a ferry owner sued for lost profits due to competition from a recently erected bridge over the Merrimack River, the Massachusetts Court of Common Pleas accepted the bridge company's defense that, since the plaintiff had "no legal right to a property in the . . . ferry," the ferry owner had sustained "such damages as without any injury or wrong done to him."[22] On the other hand, in 1798 the Massachusetts Supreme Judicial Court accepted a referee's report allowing damages against a bridge company that had drawn away some of the business of an eighty year old ferry. Though the owner could produce no grant of ferry privileges, the court accepted his claim of a property interest by prescription and allowed the action.[23]

FRANCHISES AND COMPETITION IN AMERICA

It is not always appreciated that the law governing franchises in America at the beginning of the nineteenth century reflected a curious and confusing mixture of different policies. One important reason for securing a legislative grant was to immunize the proprietor from an action for nuisance. In the absence of state approval, a mill or bridge built on a navigable stream could be abated as a nuisance or its owner subjected to criminal prosecution for interference with the passage of boats or fish.[24] Likewise, a turnpike toll gate located on a free highway was regarded as a nuisance unless authorized by the state.[25] A second objective in securing a franchise was, in the case of turnpikes, and later, canals and railroads, to enable a group of entrepreneurs to share in the state's power to take land for public purposes, although as the growing separation between public and private enterprises became clearer, this power was challenged at every step of the way.[26]

Before the nineteenth century, these two objectives were regarded by many as the only legitimate reasons for granting a corporate charter. Writing in 1792, Massachusetts Attorney General James Sullivan urged repeal of the charter of the Massachusetts bank, since it was a grant of an *"exclusive privilege"* insofar as it allowed the bank to take a higher rate of interest than private lenders. However, he distinguished the charters of bridge and canal companies. They did not convey exclusive privileges, he maintained, since "in the first case, the government grants a property in the river, which belongs to the state; and in the last it is only a grant of power to use the property and soil which they have bought, or may buy of others."[27] In short, Sullivan refused to recognize that a charter inherently conveyed a property right to exclude competition.

The legal standing of corporate charters was thrown into further confusion, however, once it was acknowledged not only that the state had the power to make exclusive grants but that a corporate charter extended the benefit of limited liability to individual entrepreneurs. As a consequence it was often impossible to know whether a corporate charter was sought only as a grant of an exclusive franchise. Even as late as 1812, when one would have supposed that policy already had settled in favor of state-licensed competitive banking, Chancellor Kent continued to maintain that all bank charters as well as turnpike, canal, bridge, and ferry charters were grants of exclusive franchises.[28] Moreover, there were those legislative grants whose express purpose was to confer a monopoly on the beneficiaries as did the Fulton-Livingston Steamboat grant in New York,[29] and for a time it was unclear whether even these franchises gave the recipients any private property interest.

Because of the different purposes which underlay legislative grants, it was not always easy to distinguish those enterprises in which competition was the rule from those in which it was impermissible. The eighteenth century theory, borrowed from Blackstone, assumed that enterprises which were arms of the state were to be free from injurious competition. The earliest banks in America, for example, were modeled on the great public banks of Europe, and "though banks grew numerous, there persisted a strong conviction that a charter was a covenant which the grant of other charters violated."[30]

Later, as the separation between public and private enterprise became clearer, there were those, such as Chancellor Kent, who assumed that the grant of monopoly privileges was mainly a *quid pro quo* for undertaking a large and risky investment. If competition were allowed to flourish, Kent declared, "all our statute privileges of the like kind, on which millions have been expended, would be rendered of little value, and the moneys have been laid out in vain."[31] Justice Story in his *Charles River Bridge* dissent also protested that there was "no surer plan to arrest all public improvements, founded on private capital and enterprise, than to make the outlay of that capital uncertain, and questionable both as to security, and as to productiveness."[32] Story, nevertheless, put forth a new theory which modified the old Blackstonian public enterprise theory in light of the recently acknowledged separation between public and private interests. Story first emphasized that franchises were private grants in order to extend to them the benefits of constitutional protections of property. It was therefore no longer possible, Story conceded, to maintain that "every grant to a corporation becomes, *ipso facto,* a monopoly or exclusive privilege" or that "the grant of a bank, or of an insurance company, or of a manufacturing company, becomes a monopoly; and excludes all injurious competition." But what then was left of the eighteenth century theory? What remained, he argued, was a distinction between "such grants as involve public duties and public matters for the common benefit of the people, and such as are for mere private benefit, involving no such consideration."[33] In short, the grant of exclusive privileges was in consideration for the imposition of public duties on private enterprise.

Whatever the theory, there seemed to be an almost infinite potential for removing most business enterprises from the test of competition. For example, there had been a long history in the colonies of regarding mills as the leading species of public service enterprise.[34] Mill sites were often appropriated from common lands, and mills were frequently exempted from taxation by towns.[35] Towns or legislatures often regulated the rates charged by mill owners, and there were instances in the eighteenth century when municipal authorities had prevented owners of grist mills from putting the mills to other uses on the ground that the proprietors originally had obtained water privileges from the town.[36] Although as a practical matter mill owners usually were able to avoid raising the question of liability for competitive injury by suing rivals for interference with water rights, they occasionally also attempted to recover directly for losses resulting from competition, as their English counterparts often had done. In the Massachusetts case of *Clark v. Billings*[37] (1783), the plaintiffs not only complained of injury to their water privileges but also alleged that the defendants had erected their mills "intending to deprive the pltfs. of the profits of their cornmill . . . and great quantities of corn did grind to the detriment of pltfs'. old corn mill." While the defendants argued that there could be no liability for competitive injury, the Massachusetts Supreme Court allowed the action, since "the diversion of the water was a sufficient ground for an action although the bare loss of custom might not be."

Despite this expression of doubt about whether the plaintiff could sue on a pure theory of competitive injury, in the next year the Massachusetts Supreme Court submitted to a jury a two-count claim for injury from a rival mill.[38] Again, the first count alleged interference with water rights, while the second count put forth a theory of competitive injury similar to those that had appeared in *Clark v. Billings* and in many English cases throughout the eighteenth century. In his second count, the plaintiff alleged that he "ought to receive and Grind at the said Mill all the Grain there brought by people living in the . . . Town of Mendon and the Towns round about" and that the defendant erected his mill "contriving and fraudulently intending . . . unjustly to oppress [the plaintiff] and of the toll and profits of his said Mill to deprive and hinder." While the jury found for the defendant, there is no indication that the court regarded the plaintiff's allegation of competitive injury as legally insufficient.

Despite other occasional dicta implying that there could be no action by mill owners for loss of trade from competition,[39] even as late as 1813, on the eve of the great boom in American manufacturing, Judge Tapping Reeve, in his influential lectures at the Litchfield Law School, suggested that it was still an open question whether competition between mills was permissible. "In the first settlement of this state land was granted for mills," he was quoted as saying, "on condition of having a good one kept — hence has arisen a question, whether other people have a right to build so near as to lessen his profits. No decisions appear on this subject."[40] Though no decisions ever did appear, as the nineteenth century manufacturing establishment severed whatever resemblance it bore to the eighteenth century gristmill, there was no logical reason why Reeve's and Story's public benefit theory could not have applied.[41]

However often Blackstone's theory of franchises was cited after 1815, it is remarkable how little attention was actually paid to his views earlier in the nineteenth century. For a time in these early years, American courts seemed unwilling to accept the monopolistic implications of Blackstone's views even in the face of a growing number of state-promoted monopolies. The early assumption that banks were monopolies "gave way, slowly and obstinately, and what overcame it was not logic but self-interest, and corruption."[42] Though he had suggested the contrary only five years earlier, by 1817 even Chancellor Kent was prepared to acknowledge that unless restrained by the legislature the right to enter the banking business was "a common law right belonging to individuals, and to be exercised at their pleasure."[43] Nor did American courts ever adopt the English doctrine of exclusive markets, although there was a well-established line of English decisions treating markets and fairs as public enterprises.[44] Attempts in the early part of the century to induce courts to imply new and questionable areas of monopoly were usually rebuffed. A private effort to establish a claim to exclusive fishing privileges in the Delaware River was denounced by the New Jersey Supreme Court in 1812 as "a title set up against the liberties and privileges of the people" and "a restraint on their natural rights."[45] Two years earlier, in order to prevent riparian owners from claiming exclusive fishing rights in the large rivers of the commonwealth, the Supreme Court of Pennsylvania had rejected the common law rule designating rivers in which the tide did not ebb and flow as nonnavigable.[46] The court also denied that an exclusive right could be acquired by prescription. In this early period judges were willing to sacrifice even exclusive property rights in order to prevent monopoly. Thus, in 1811 the Connecticut Supreme Court extended the common right to fish so far that they allowed an individual to gather shellfish lodged in the river's bed, even though in so doing he was concededly trespassing on another's land.[47]

On the other hand, many legislative grants expressly conferred exclusive privileges.[48] In the Connecticut case of *Perrin v. Sikes*[49] (1802) the plaintiff, who had been granted the exclusive right to carry passengers by stage from Hartford to Boston, sued an unlicensed competitor, as provided in his grant, for the forfeiture of his stage and horses or double their value in cash. The court upheld the penalty, even though the defendant had shown that his stage had transported passengers not to Boston but to Albany, since "the spirit of the grant was, that [the plaintiff] should enjoy the exclusive right to transport passengers to the Massachusetts line."[50]

Such generous constructions of legislative grants were not often indulged by courts in this early period. In 1815 the Connecticut Supreme Court itself substantially restricted the scope of its earlier decision, holding that an exclusive stagecoach franchise was a personal right that could not be inherited, nor could such a right be acquired by prescription.[51] And perhaps the most representative, but in light of later developments, the most surprising, decisions of this early period occurred in two New York cases, which revealed great resistance to any conception of an individual property right in an exclusive franchise. In *Donelly v. Vandenbergh*[52] (1808) two holders of an exclusive stagecoach franchise sued a competitor for a $500 penalty provided by the legislature. The defendant showed that, in all, six persons had been jointly granted the franchise and that two of them, not the plaintiffs, had licensed him to run the stage. The plaintiffs replied that the six grantees had agreed to divide the state into separate exclusive routes and that the defendant could not receive a license to compete from one of the parties to the agreement. By a 2-1 vote, with the future Chancellor Kent in the majority, the Supreme Court held that the plaintiffs could not recover the penalty. In view of the fact that "public accommodation and convenience were the objects the legislature had in view" in granting the franchise, the court refused to regard the grant as a property interest that could be partitioned "so as to give exclusive and independent rights, in distinct parcels of the road." And since the defendant had received his license from one of the grantees, he could not be sued for the statutory penalty.

Even the dissent of Justice Spencer reveals the great distance that the law was yet to travel in defining the nature of a franchise. Though he believed that the defendant was liable because the grant constituted a divisible property interest, he regarded the franchise as exclusive only because the legislature had expressly made it so. The act of the legislature, he declared, was for "the public . . . interest and convenience" because without the statutory penalty anyone was free to set up a stagecoach, since it was "a right, before enjoyed by all."

One year later, these views were reinforced in the case of *Almy v. Harris*[53] (1809), where the Supreme Court, again with the concurrence of James Kent, reversed a judgment for damages in favor of a ferry owner against an unlicensed rival, holding that his only remedy lay under an act which prohibited unlicensed ferries on pain of a five dollar penalty. If the plaintiff, the court wrote, "had possessed a right, at the common law, to the exclusive enjoyment of this ferry," then he would be authorized to sue in an action for damages, notwithstanding the statute. But since he "had no exclusive right at the common law, nor any right but what he derived from the statute," he was limited to recovery of the statutory penalty despite a grant of exclusive ferry privileges. In thus denying any common law right in the exclusive grantee, the court not only continued the view expressed in *Donelly v. Vandenbergh* that an exclusive franchise gave no property interest that the common law was bound to protect, but it also adopted the opinion of Justice Spencer that the only purpose of a franchise was to enforce the state's interest in regulating the number of ferries on a river. In short, until judges became convinced of the state's interest in using exclusive franchises in order to promote economic development, American courts continued to resist the English view that a franchise was a property right which a grantee could enforce for the purpose of preventing injurious competition.

THE STATE'S POWER TO CREATE MONOPOLIES

Though courts often attempted to limit the scope of exclusive grants, the power of the state to confer such privileges appears never to have been challenged judicially until the controversy over the Livingston-Fulton steamboat monopoly in New York.[54] In 1787 the New York legislature granted John Fitch "the sole and exclusive right and privilege of constructing, making, using, employing and navigating, all and every species or kinds of boats, or water craft, which might be urged or impelled through water, by the force of fire or steam, in all creeks, rivers, bays and waters whatsoever, within the territory and jurisdiction of this state" for a term of fourteen years. Eleven years later, Fitch having failed to develop a steamboat, the legislature revoked its prior grant and conferred a new twenty year franchise on Robert Livingston, upon the condition that he should give proof within one year of having built a boat meeting certain specifications.[55] Although success continued to elude Livingston, the legislature in 1803 extended the franchise for another two decades provided that a suitable boat could be developed within the next two years. Having again failed to meet the deadline, Livingston and his partner, Robert Fulton, nevertheless succeeded once more in inducing the legislature in 1807 to renew the twenty year grant and to allow them two more years in which to satisfy the state of their success. Finally, on the heels of their triumph, the legislature in 1808 provided that the franchise would be extended five more years, up to a total of thirty years, for every additional boat they built.

In 1811 Livingston and Fulton appealed to New York Chancellor John Lansing for an injunction against a rival steamboat that had begun to navigate the waters of the Hudson River between New York and Albany. Although the defendants did not challenge the legislature's inherent power to make an exclusive grant, except as it conflicted with congressional power under the Commerce and Patent Clauses, Lansing, an avid Jeffersonian who, two decades earlier, had fought against ratification of the Constitution, put his refusal of an injunction on the far broader ground of a probable conflict with natural right. He declared that he could find no instance in England of a grant of exclusive rights to navigation "and it would seem that it was considered contrary to the *jus publicum,* that such a grant should be made."[56] Moreover, if this grant were valid, "where is the line of distinction to be drawn between what has been granted, and what is unsusceptible of grant?" Could the legislature go still further and restrict the right to navigate sailboats or rowboats? "If carried to this extent, would it not be an abridgement of common rights? Could it comport with the constitutional provision, that the citizens of all states are to have like privileges and immunities with the citizens of the several states?"[57]

Having purported to find merely a degree of doubt about the legitimacy of the Livingston-Fulton grant, Lansing was able to invoke the traditional discretion of a court of equity to deny an injunction without specifically holding the grant invalid. Even so, he was unanimously reversed by the Court for the Correction of Errors, which, modeled on the English House of Lords, comprised the justices of the Supreme Court, including Chief Justice Kent, as well as the entire membership of the New York Senate. It is notable that in rearguing the case on appeal the defendants continued to ignore the argument from natural rights in spite of the favorable opinion below, which suggests that they too regarded Lansing's opinion as an aberration. Undoubtedly, they estimated correctly. It was simply too late in the day to mount a general attack upon the legislative power to confer exclusive grants. "To deny to the legislature this right," Justice Thompson observed, "would be at once striking from our statute books grants, almost innumerable, of a similar nature." It would draw into question "all our turnpike roads, toll-bridges, canals, ferries, and the like."[58] Though the defendants had never really attacked the general legislative power to issue grants, the reversal of Lansing's decision was regarded as establishing that power, so that by the time of the second challenge to the steamboat monopoly in 1820 the Court of Errors regarded any attack on the legislative power "as no longer open for discussion here."[59] And it was not until Roger Taney's 1833 private legal opinion as attorney general, about which more will be said later,[60] that there was any renewed legal challenge to the state's power to grant monopolies.

Besides arguing that the Constitution gave Congress exclusive power over commerce as well as patents, the *Livingston* defendants' main contention dealt with the appropriateness of issuing an injunction. Only three years earlier the New York Supreme Court had held in *Almy v. Harris* that there was no private right at common law to recover damages for infringement of an exclusive ferry franchise and that the only available remedy was recovery of a five dollar statutory penalty. Without a recognized property interest, it seemed to follow that there was also no right to an injunction by private individuals. Nevertheless, the court ignored that decision and ordered that an injunction issue. It was the first recognition in the New York courts that an exclusive grant constituted a vested property interest.

There was still another problem concerning the issuance of an injunction. The English chancery courts, reflecting a longstanding hostility to monopolies, had refused until well into the eighteenth century to enjoin violations of the patent law until the plaintiff first established his rights at law.[61] And in America there was still formidable rhetorical power in the word "monopoly," as evidenced by Lansing's opinion. Moreover, "on the issue of whether their rights were monopolistic, Fulton and Livingston, as Jeffersonians, were bound to be sensitive. So sensitive were they . . . that it was basic to their defense to contend that their right was, properly speaking, no monopoly."[62] "*Monopoly* is a technical term," their counsel argued. "It is a *prerogative* grant, in hostility to the public good. Who ever heard of a monopoly erected by act of parliament? This legislative grant was intended to compensate genius for introducing, extending, and perfecting, the invention of others. A public benefit was contemplated."[63]

From a historical viewpoint, this position was not without merit. It appears that nobody had ever doubted the power of parliament to issue exclusive grants; opposition to the power to create monopolies had arisen entirely in the context of challenges to the royal prerogative.[64] Nevertheless, the decision to issue an injunction confirmed not only the already well-recognized power of the legislature to make these grants, but it also advanced the far more uncertain proposition that a franchise was just another form of property which, apart from any statutory remedies, the common law was bound to protect. Moreover, the expansive use of the injunctive remedy prevented the defendants from having their rights tried before a jury, where, one suspects, popular prejudice against monopoly would have tempered the enthusiasm with which chancellors who were free from immediate popular control subsequently extended the protections accorded the holders of franchises.

FRANCHISES AND IMPLIED MONOPOLIES

In 1814, two years after the decision in *Livingston* v. *Van Ingen*, James Kent became Chancellor of New York. Looking back two decades later at the dramatic growth of equity jurisprudence in America, Justice Story noted that before the Revolution it "had no existence at all, or a very imperfect and irregular administration." Even since the Revolution, he observed, "it has been of slow growth and cultivation" and there were some states that had refused entirely to adopt a system of equity. "It did not attain its full maturity and masculine vigor," he concluded, "until Mr. Chancellor Kent brought to it the fullness of his own extraordinary learning, unconquerable diligence, and brilliant talents."[65] Kent himself noted in 1828, with some exaggeration, that, upon becoming Chancellor, "I took the court as if it had been a new institution, never before known to the U.S. I had nothing to guide me, was left at liberty to assume all such English chancery powers and jurisdiction as I thought applicable under our constitution. This gave me great scope . . . I might once & a while be embarrassed by a technical rule, *but I most always found principles suited to my views of the case.*"[66] And it was in part due to these earlier efforts of Kent that Story was later able to declare with some satisfaction that "injunctions are now more liberally granted than in former times."[67]

One year after he became chancellor, Kent was asked to enjoin a group of individuals who had purchased land and built a toll-free private road in competition with the recently established Croton Turnpike.[68] Nothing in the general law governing turnpike corporations[69] or in the acts specifically incorporating the turnpike company had said anything about the grant of an exclusive privilege to take tolls, though the rate of tolls was regulated. In fact, one of these acts had declared that nothing in the legislation "shall be construed to prohibit" persons "from obtaining from the present or any future legislature, a law to incorporate a turnpike company" in competition with the Croton Turnpike.[70] Nevertheless, Kent held, without so much as reference to the specific terms of the statute, that the rival road was "a material and mischievous disturbance of the plaintiffs in the enjoyment of their statute privilege" and issued the injunction. "If such a contrivance as this case presents, is to be tolerated," he declared, "all our statute privileges of the like kind, on which millions have been expended, would be rendered of little value and the moneys have been laid out in vain."[71]

The decision in the *Croton Turnpike* case represented the second important turning point in the underlying assumptions about franchises in America. For there was a vast difference between this case, in which Kent merely assumed that a franchise to collect tolls was necessarily exclusive, and *Livingston v. Van Ingen,* where there was no doubt that the legislature had expressly conferred a monopoly on the plaintiffs. It was the first clear instance in America of judicial enforcement of an implied exclusive monopoly.[72] Thus, within six years the New York courts had moved from the decision in *Almy v. Harris* that even an exclusive franchise did not confer a property interest on the grantee to the position in the *Croton Turnpike* case that such a property interest was granted even without an express legislative grant.

In *Newburgh Turnpike Co. v. Miller*[73] (1821) Kent again enjoined a private group at the instance of a turnpike corporation, this time for building a private bridge that took away some of the company's toll. The bridge, he wrote, "creates a competition most injurious to the statute franchise, and becomes . . . a nuisance." Citing Blackstone's examples of exclusive ferries and markets by prescription, he declared that "no rival road, bridge, ferry, or other establishment of a similar kind, and for like purposes, can be tolerated so near to the other as materially to affect or take away its custom. It operates as a fraud upon the grant, and goes to defeat it."[74]

The assumption that a franchise was inherently exclusive was the culmination of economic policies that had prevailed in America since the end of the eighteenth century. The process of encouraging the investment of scarce private capital in works of public improvement had led state legislatures to grant a variety of monopolistic privileges to groups of individuals who were prepared to take high risks for the chance of substantial rewards. Even when it was not expressly acknowledged, it was usually understood by all parties that these grants would be exclusive. "The consideration by which individuals are invited to expend money upon great, and expensive, and hazardous public works, as roads and bridges, and to become bound to keep them in constant and good repair," Kent contended, "is the grant of a right to an exclusive toll." Whether this common understanding could be converted into an inflexible legal right, however, was something else again. There were other good reasons for securing legislative grants, which involved privileges other than monopoly. Turnpike proprietors needed legislative permission in order to receive the benefits of incorporation as well as to share in the state's power to take land for highway purposes. Bridge corporations, in addition, required state approval before they could obstruct navigation on a river. Yet, once these other purposes were ignored, it seemed to follow inexorably that the only conceivable reason for conferring a franchise was to create a monopoly and that any subsequent charter by the state involved a breach of faith, or, in Kent's words, "a fraud upon the grant."

THE STATE'S POWER TO CHARTER COMPETING FRANCHISES

The presumption of monopoly emerged out of an economic climate in which the encouragement of investment seemed to require the grant of exclusive privileges. Consequently, during the first quarter of the nineteenth century the interest of the state in promoting economic development rarely conflicted with the interests of private entrepreneurs in enforcing exclusive grants. Indeed, the conclusion that franchises conveyed exclusive property interests was developed in a context of purely private lawsuits. For example, all the decisions in which Kent had participated arose out of suits by licensed grantees seeking to exclude competitors who had received no authority from the state. Nevertheless, by the second quarter of the century, when the question of whether the state could authorize competing public works first came to the fore, it seemed that these cases, together with the parallel line of decisions in *Fletcher v. Peck* and the *Dartmouth College* case, had already decided the issue in the negative. If a franchise automatically created a monopoly, then any subsequent grant by the state was an infringement of that exclusive privilege.

Before the great *Charles River Bridge* case was decided by the Massachusetts Supreme Court in 1830, however, there were few direct considerations of the state's power to authorize competing public works. In the Haverhill Bridge controversy in Massachusetts in 1798, a technicality prevented the court from considering the bridge company's defense that it had won approval of the state.[75] In 1793 the New Hampshire legislature refused to approve a bridge charter until the existing ferry proprietor, who held under an exclusive grant from the colonial governor, gave his consent.[76] But it is by no means clear that this was anything but a contest of opposing political interests. And although in 1797 the North Carolina Supreme Court held over a vigorous dissent that the state had no power to establish a competing ferry,[77] the existence of such power was never doubted when the court next considered the question in 1815.[78] In South Carolina, which alone among the states had continued to hold that compensation was not required for the taking of property for public use, the Supreme Court in 1818 allowed the uncompensated transfer of a ferry franchise by the legislature.[79] In Massachusetts, it was claimed that of the fourteen instances in which bridges had impaired the profitability of ferries between 1780 and 1811, in only four had the legislature seen fit to consider the question of compensation.[80] Similarly, in 1811 the New Jersey legislature chartered a competing water supply company over the protests of the established company. The sole ground of opposition apparently was economic and not legal, with the existing company contending alternatively that high risks deserved high profits and that there was insufficient demand to support more than one company.[81] Even in the case of *Livingston v. Van Ingen,* where the sanctity of private property was much discussed, Livingston and Fulton breathed a sign of relief when the New York high court ignored the opposing counsel's suggestion that legislative revocation of an earlier grant of the steamboat monopoly might adversely affect their claims.[82]

It appears therefore that before the possibilities in the *Dartmouth College* decision had begun to be fully explored, no one doubted that the state had the power to set up rival public works. Without the aid of American public law theory, English private law monopoly theory did not appear to restrict the state.

The first nineteenth century case even to suggest that a competing franchise would be unconstitutional really seemed to establish the opposite principle. In *Enfield Toll Bridge Co. v. Connecticut River Co.*[83] (1828) the central question was whether the legislature could charter a second company to construct locks around the same falls. There was "no doubt," the court declared, that the grant was exclusive. "It is not like cases put at the bar of grants of turnpike companies, of ferries, of banks, etc. A grant to a company to construct locks around the bars at the upper falls, exhausts the power of the legislature of locking those falls. A grant to another company to lock the same falls, would be a palpable infringement of the first grant."[84] Far from announcing any great constitutional principle, the court appeared to base the first company's exclusive right on the physical impossibility of making two grants to construct the same set of locks, treating the second grant as in actuality a revocation of the first.[85] And for the same reason, the court refused to maintain that the typical franchises of this period—for turnpikes, ferries, and banks— were exclusive.

In 1823, the year in which Chancellor Kent was forced to retire because of age, a new tone had already begun to enter judicial discussion of the franchise question, not entirely uninfluenced by the political climate then existing in New York. In the preceding year a constitutional convention had extended original equity jurisdiction to all lower court judges, reflecting the view, as one prochancery observer remarked, "that this high Court of Chancery is not only expensive and tedious, but very aristocratic." He noted that "the current of popular opinion sets strongly in favour of the boldest and most extravagant plans of reformation. Pains have been taken to create and diffuse prejudices against the higher Courts of the State; and we regret to say that this temper was openly manifested on the very floor of the convention."[86]

In the midst of this reformist atmosphere, the New York Supreme Court was asked to interpret the charter of the Cayuga Bridge Company, which expressly conferred the exclusive privilege of collecting tolls within a three mile limit, even from private travelers who crossed the lake over which the bridge was erected. The company brought suit against a person who had traveled across the frozen lake on foot, beginning his passage outside the three mile limit but completing his crossing within it. Holding that there could be no recovery, the Supreme Court declared that the act should be strictly construed. "The act confers upon the company certain privileges, and restrains the right of the citizen. . . . In the construction of statutes made in favor of corporations or particular persons, and in derogation of common right, care should be taken not to extend them beyond their express words or their clear import."[87] When this principle was again applied in 1830, in the same year as the *Charles River Bridge* case was decided in Massachusetts, the Cayuga Bridge Company lost another, even more disastrous, legal battle involving the state's power to authorize a competing bridge.[88] In 1797 the company was granted exclusive privileges within three miles of its bridge, which was destroyed in 1809. Having failed to rebuild by 1812, the local inhabitants applied to the legislature for the right to erect another bridge. After pressure from the legislature resulted in a compromise, the company was authorized to construct a new bridge on the old site as well as to build another bridge a few miles away. In addition, the legislature conferred on the company all the rights and privileges which were granted by the original act. Nevertheless, in 1825 the legislature authorized the local citizens to erect a free bridge more than three miles from the original site but only one mile from the more recent bridge. In dissolving an injunction issued by his predecessor against the free bridge, Chancellor Walworth held that the legislature had not extended an exclusive three mile limit to the second bridge. "Such injudicious grants of exclusive privileges," he declared "should not be farther extended by construction. . . . So far, therefore, as the last two acts are in derogation of common right, they must be construed strictly against the company according to the principles of the common law."[89]

THE *Charles River Bridge Case:* THE CHANGING STATE ROLE IN
ECONOMIC DEVELOPMENT

With the *Charles River Bridge* case state courts were finally forced
to confront the implications of an important shift in the conception
of the proper role of the state in encouraging economic development.[90] Established judicial doctrines concerning exclusive franchises had become fixed just as canals and bridges were beginning to
displace turnpikes and ferries on a large scale. Though earlier
grants of monopoly privileges may have been necessary in an underdeveloped society in order to promote private investment, the restrictive consequences of these grants were becoming apparent by
the second quarter of the nineteenth century. Not only did the originally chartered facilities often prove inadequate to handle expanding commercial activity, but the existing and entrenched network of
transportation was also becoming rapidly outmoded.

When the Charles River Bridge was built in 1785, connecting
Boston with Charlestown, the population of Boston amounted to
17,000 people and that of Charlestown to only 1,200.[91] By 1827,
when the Massachusetts legislature authorized the erection of the
neighboring Warren Bridge, Boston's population had increased to
more than 60,000 and Charlestown's to 8,000, with the surrounding
suburbs enjoying a like expansion in population.[92] As a result of
these changes, Chief Justice Parker pointed out, "The profits of the
bridge have been great beyond the example of any similar institution in this country."[93] On an initial investment which the company
claimed amounted to $51,000, plus a subsequent outlay alleged at
$19,000, the bridge was collecting tolls of $30,000 per year by the
time the Warren Bridge was chartered.[94]

The rival bridge was built only 260 feet from one end of the
Charles River Bridge and 915 feet from the other,[95] and its charter
provided that the proprietors could charge a toll until they received
a specified return but that, in any event, passage over the bridge
would become toll free when it reverted to the state within six years.[96]
Thus, by the time the case was argued in the Supreme Judicial Court
of Massachusetts in 1829, less than a year after the Warren Bridge
was opened, the Charles River Bridge Corporation already had suffered a decline in revenues of between one half and two thirds,[97] and
by the time the case was decided by the United States Supreme
Court in 1837, the earnings of the corporation had been virtually
eliminated.[98]

Not only was the state's role in promoting economic development beginning to change, but in other areas of the law settled expectations that had grown up in a static and underdeveloped society were giving way to a conviction that ownership of property in a dynamic and changing environment necessarily entailed many risks and uncertainties for which the law could offer no protection. From the beginning of the century, legislatures had frequently compromised traditional property rights in order to promote economic development. Writing in 1801, James Sullivan, the leading Jeffersonian statesman in Massachusetts, argued for "the economy and advantage" of a state law which drastically reduced from twenty years to three the period in which the owner of mortgaged premises could redeem the equity in his property. The rule was established, he asserted, so "that the improvement of real estates thus taken might not be delayed or lost to the mortgagee."[99] Courts also were coming to modify even conventional conceptions of property rights in order to allow for economic change. In the economically important area of water rights, judges had begun to reject the common law doctrine prohibiting exploitation of water for mills when such use injured another proprietor on the same stream.[100] They had also rejected the doctrine of prior appropriation, denying the claim of the first proprietor on a stream to the exclusive use of water for his mill.[101] As early as 1805 Justice Livingston perceived that the rule of prior appropriation sanctioned a system of monopoly in which "the public, whose advantage is always to be regarded, would be deprived of the benefit which always attends competition and rivalry."[102] The goal of economic development had also led courts to modify the English common law doctrine of waste, thereby allowing life tenants greater leeway to improve land at the expense of remaindermen. "The law of waste, in its application *here*," a judge of the Virginia high court declared, "varies and accommodates itself to the situation of our new and unsettled country."[103]

In a similar vein, American courts modified the absolute conception of property which underlay the common law rule concerning ownership of fixtures. In *Van Ness v. Pacard*[104] (1829), Justice Story rejected the rule that the landlord owned all fixtures which the tenant had annexed to the land. "The country was a wilderness," he noted, "and the universal policy was to procure its cultivation and improvement. The owner of the soil as well as the public, had every motive to encourage the tenant to devote himself to agriculture, and to favour any erection which should aid this result; yet, in the comparative poverty of the country, what tenant could afford to erect fixtures of much expense or value, if he was to lose his whole interest therein by the very act of erection?" One year later, the authors of *A Treatise on the Law of Fixtures* denounced the same common law rule as "both inequitable in its principles, and injurious in its effects to the spirit of improvement."[105]

The tendency to modify property relationships in order to encourage development was not always greeted with enthusiasm. Chancellor Kent, for example, viewed with suspicion analogous laws which allowed a good faith possessor of land to receive the value of his improvements after being ousted by one with a superior title.[106] Most of these laws, Kent conceded, were passed in Western states where "titles to such lands had, in many cases, become exceedingly obscure and difficult to be ascertained." Yet, he declared, "They are strictly encroachments upon the rights of property, as known and recognized by the common law of the land," and concluded that "in the ordinary state of things . . ., such indulgences are unnecessary and pernicious, and invite to careless intrusions upon the property of others."[107]

Other conventional notions of property rights had begun to give way under the pressures of economic development as judges began to see that a conception of property as absolute and exclusive gave a single landowner the power to prevent all sorts of economically desirable, but injurious, modes of activity on neighboring land. In a series of cases beginning in 1815, courts had begun to restrict this power of individual landowners to determine the course of economic development, likening the damage caused by conflicting land use to competitive injury for which there was no legal remedy.[108] They had also established a parallel line of decisions under which actions by the state that impaired the value of a person's land were regarded as "consequential" injuries that were noncompensable if they did not amount to physical invasions of a person's property.[109]

The opinions of the judges of the Massachusetts Supreme Court in the *Charles River Bridge* case above all reflected these changing conceptions of the nature of property. While the court was equally divided over whether the Charles River Bridge franchise constituted an implied exclusive monopoly, even the opinion of Chief Justice Parker, who concluded that the franchise was exclusive, revealed the extent to which the spirit of development was undermining traditional notions of property. "The whole history and policy of this country from its first settlement," Parker began, "furnish instances of changes and improvements, the effect of which has been to transfer the adscititious value of real estate in one town, resulting from its favorable position for trade, to another, which, by alteration of roads, erection of bridges, or more recent interior settlements, had taken its place as a thoroughfare, or as a place of transit or deposit for articles of merchandise. Losses of this kind never have been, and probably never will be compensated." Moreover, compensation could not be provided for businesses which "find their profits and emoluments diminished and sometimes destroyed by the change of fashion, or by new inventions for carrying on the same branch of business in a cheaper and more acceptable way." He concluded that "it is necessarily one of the contingencies on which property is acquired and held, that it is liable to be impaired by future events of this kind . . . so that property is in fact held upon a tenure which admits of its deterioration in value from causes of this kind."[110]

Parker's opinion reads as if it were originally written in support of the conclusion that the first bridge franchise did not confer a monopoly. He recognized, as did all of the judges, that the grant of a franchise represented a contract which the state could not revoke. He also maintained "that if the ferry right is restricted to the ferry ways, it is of no value, for the value of the franchise depends upon tolls, and these, upon passengers, and if another ferry could be set up along side of theirs, or a bridge more especially, their franchise is destroyed."[111] But he confessed that he did not see why his views about the risks which property ownership entailed "do not apply to property in ferries and bridges to a considerable, if not to the whole extent."[112]

For Parker's brethren, Justices Morton and Wilde, the way out of this conceptual dilemma was by reference to experience, which revealed a host of exceptions to any claim that a consistent principle of monopoly existed. Though "it was once contended by learned and respectable men," Justice Morton wrote, "that the first bank charter gave this extensive exclusive right," this view "is now generally exploded."[113] Moreover, "in many cases the legislature have established new turnpikes, some nearly parallel with and diverting travel to the injury and sometimes ruin of former ones."[114] Since there were so many instances of diversion of traffic and diminution of profits resulting from rival facilities, Morton contended, the consequence of Parker's position would mean that later entrepreneurs "have been acting and expending their money on the faith of void charters, and all their acts have been unauthorized." In any event, the judiciary was in no position to decide "how near . . . must the new one be, and how much of the travel must it divert" before it infringed the prior grant. To recognize an exclusive right over the entire line of travel, he concluded, "would amount to a stipulation, that the channels of communication and course of business, and in fact the state of society and of the country itself, should remain stationary."[115]

The *Charles River Bridge* case represented the last great contest in America between two different models of economic development. For Justice Putnam of the Massachusetts court and for Justice Story of the Supreme Court, the essential elements for economic progress were certainty of expectations and predictability of legal consequences. With these conditions satisfied, Putnam predicted, "men of capital and energy would embark their funds in enterprises of a public character, in the hope that their own fortunes might be advanced with the public prosperity. . . . But let the reverse of this be suspected, and public credit will be paralyzed."[116] Story could "conceive of no surer plan to arrest all public improvements, founded on private capital and enterprise, than to make the outlay of that capital uncertain, and questionable both as to security, and as to productiveness."[117] Justice Morton, on the other hand, though conceding that "exclusive rights for short periods sometimes encourage enterprise, of public usefulness," believed nevertheless that "generally their tendency is to impede the march of public improvement, and to interrupt that fair and equal competition which it has ever been the policy of our country to encourage."[118]

The problem with the Story-Putnam position was that there was no longer any necessary connection between high risk and sizable investment and the quasi-public nature of a business enterprise. There now existed manufacturing enterprises for which the level of risk and the extent of investment far exceeded those of many bridge corporations, not to speak of ferries. Yet, even Story and Putnam agreed that in these new cases investors were not entitled to a special measure of legal protection against competition. As public service corporations were transformed into private, profit-making organizations, the historic legal categories were no longer practical guides for determining the course of public policy. The decision, Morton concluded, was best left to the legislature.

All the judges on the Massachusetts court conceded that the legislature had the power to make grants of exclusive franchises, however undesirable they might be. For them, the only question in the case was whether the legislature should be bound in the absence of an express commitment. But as the case was winding its way up to the United States Supreme Court, where Chief Justice Taney would uphold the state's freedom to grant a competing franchise, there appeared in 1833 the first authoritative legal attack since the contest over the Livingston Steamboat franchise twenty-two years earlier on the very power of the state to confer such monopolies. The author of the challenge was President Jackson's Attorney General Roger B. Taney.[119]

In 1832 the New Jersey legislature granted the Camden and Amboy Railroad the exclusive right to railroad transportation between New York and Philadelphia in exchange for shares of the company's stock, which promised enough revenues to cover the entire costs of state government. Shortly thereafter, the Philadelphia and Trenton Railroad acquired a controlling interest in the Trenton and New Brunswick Turnpike Corporation for the purpose of inducing the legislature to allow it to lay tracks over the turnpike.[120] The new owners of the turnpike company then sought to bolster their case by soliciting paid opinions from leaders of the bar such as Kent and Webster as well as Taney on the legal issues involved.[121] All three favored the position of the turnpike company but on strikingly different grounds. Kent and Webster, as would be expected, argued the orthodox view that the grant to the Camden and Amboy Railroad was void because it infringed the prior franchise to the turnpike company. Taney, however, contended that the legislature possessed no power at all to make any grant of exclusive privileges.[122]

Though he began by conceding that the view that a charter was an irrevocable contract was "now too well settled to be disputed," Taney's opinion came as close as any in American law to directly challenging the doctrine of the *Dartmouth College* case. Since the legislature exercises only delegated power, he argued, "it is quite clear that they cannot bind the state by contract or otherwise, beyond the scope of the authority granted to them by their constituents." The legislature is the agent of the sovereign power, the people, "and when it steps beyond the limits of its authority, its acts are void" and do not bind subsequent legislatures. "The question is, have the people of New Jersey delegated to the legislative body, the power to make such contracts and to deprive them for the time specified of the power of prosecuting such works of internal improvements as they may deem necessary to advance their interests and promote the prosperity of the state?"

Though he acknowledged that there was "high authority for regarding this power as an incident to the power of legislation," Taney concluded that "a legislative body, holding a limited authority under a written constitution," could not "by contract or otherwise, limit the legislative power of their successors." Echoing Chancellor Lansing's opinion in the *Livingston* case he attempted the final *re-*

ductio ad absurdum. "If they can deprive their successors of the power of chartering companies of a particular description or in particular places, it is obvious that upon the same principle, they might deprive them of the power of chartering any corporations for any purposes whatever; and if they might by contract, or otherwise, deprive their successors of this legislative power, they could surrender any other legislative power whatever in the same manner, and bind the state forever to submit to it."

Nevertheless, Taney did retreat from the course to which his logic had seemed to lead him: that any future commitment by the state was void. He conceded that "there are cases no doubt in which the acts of the legislature irrevocably bind the state. . . . This happens in all cases of delegated power where the agent is acting within the scope of his authority." Thus, where the state borrows money, or grants lands, or establishes a corporation, these acts are binding because they "are within the admitted scope of legislative authority; and being contracts made by the authorized agents of the people, they are necessarily binding on their constituents; and cannot be altered without the consent of the other party to the contract." Why the long contested power of the state to charter a corporation was "within the admitted scope of legislative authority," Taney did not say. In fact, his opinion concluded with the very argument that had become the familiar stock in trade of opponents of corporations.[123] "The power of surrendering attributes of sovereignty," he declared, "which are essential to the well being of the state, cannot be presumed to reside in the representative, unless expressly granted by the constituent." Though he had begun his analysis with the radical assumption that any restriction on the future action of the legislature was in principle a derogation of sovereignty, Taney merely concluded with the now discredited Jeffersonian argument which turned on whether the legislature had received express authority to bind itself, an argument that had been used by many to question even the state's power to charter corporations.[124]

Though his sentiments as attorney general undoubtedly influenced his decision as Chief Justice, Taney's opinion in the *Charles River Bridge* case was entirely conventional by comparison. His rule that "in grants by the public, nothing passes by implication"[125] was merely the culmination of a course of change that had been unfolding for more than a generation. This recognition of the private nature of business corporations goes as far back as the *Dartmouth College* decision and reflects the changing nature of economic relationships. No longer primarily representing an association between state and private interests for public purposes, the corporate form had developed into a convenient legal device for limiting risks and promoting continuity in the pursuit of private advantage. The general incorporation laws which had begun to crop up in the thirties expressed this tendency to regard the corporation as just another form of business association.[126] Yet, even after the private nature of corporations was recognized for the purpose of limiting the state's power over vested property rights, the anomaly persisted of regarding them as essentially public bodies entitled to share in the historic privileges conferred on arms of the state. In this sense, the *Charles River Bridge* case represents another chapter in the process of discarding the old habit of conceiving of corporations as the recipients of special state-conferred favors and of adjusting the law to treat the act of incorporation as nothing more than a mere license to exist.

Perhaps the most decisive influence on the courts was the advent of the railroad. Both Massachusetts and New York had chartered railway corporations in 1826. Maryland authorized the Baltimore and Ohio in 1827, and four years later New Jersey followed suit. In 1829 there were only two railroad corporations in New York state. Within the next five years, the legislature had chartered forty-eight additional corporations. Likewise, by 1829 Massachusetts had chartered only two insignificant railroads, but in the next six years it authorized fifteen more railroad lines, many of which, like the Boston and Lowell, were substantial undertakings.[127] It became clear almost immediately that the entrenched networks of transportation —bridges, turnpikes, and canals—would use every available legal as well as political weapon to stop the growth of railroads.[128] In turn, the railroad became the leading inducement for courts to adopt a new position, since the doctrine of implied monopoly threatened to saddle every railroad with an unmanageable burden of costly damage actions as well as injunctions.[129] "Let it once be understood," Chief Justice Taney wrote, "that such charters carry with them these implied contracts, . . . and you will soon find the old turnpike corporations awakening from their sleep, and calling upon this court to put down the improvements which have taken their place. The millions of property which have been invested in railroads and canals, upon lines of travel which had been before occupied by turnpike corporations, will be put in jeopardy."[130]

Despite the seemingly great distance between the opposing sides in the Charles River Bridge controversy, it is remarkable how easily, given the universal enthusiasm for economic development, these differences ultimately dissolved into rather narrow questions of legal technique. The conflict between railroad and turnpike in New Jersey led the parties to bolster their legal positions by seeking the opinions of Kent, Webster, and Taney. In a similar situation, Chancellor Kent, long retired, reported that in 1831 he also had been paid a fee for a legal opinion requested by the Mohawk and Hudson Railroad. Having realized that he was a stockholder in the rival Albany and Schenectady Turnpike Company, which had also sought his opinion, he "instantly returned the case and fee to the gentleman who handed them to me, as I did not incline to give any opinion to the adverse party, in a case in which I was essentially interested."[131] In his opinion favoring the turnpike, Kent pressed the traditional view that the act of 1826, chartering the railroad, was illegal as "a disturbance of a lawful franchise." Moreover, he defended the lawfulness of an 1830 act of the New York legislature, which permitted the turnpike company to build a railroad over its existing road, as no violation of the railroad's franchise, since the legislature originally had reserved to itself the power to alter the charter.

Although his view that franchises embodied implied monopolies had not changed, an unexpected note of hostility to exclusive grants emerged for the first time in Kent's opinion. He enthusiastically defended the power of the legislature to stipulate in advance that it could alter or modify the terms of a charter. "It is vain to talk of immoveable vested rights and privileges," he wrote, "when they are accepted under the express condition, that the same sovereign who granted them may, at his pleasure, alter and modify them." Even Kent apparently had come to realize that the grant of exclusive privileges was proving a clog on economic growth, and he now was prepared to endorse legislative reservations that, effectively, gave the state unfettered power over corporate charters.

Considering that corporations and privileges are multiplying upon us in every direction, and upon all possible subjects with astonishing fertility, the reservation of a power to *alter and modify* becomes most important to the safety and prosperity of the state. The reservation ought to be liberally construed. Grants are rapidly and heedlessly made. The 18th section of the railway act is a striking illustration of it. It seems to imply, that without this section, the hands of the legislature would be tied up from granting any other rail-way any where in any part of the state! It clearly does imply, that the legislature cannot grant any other rail-way from *the city of Albany to any other place*. And is that possible?[132]

In short, by the time the *Charles River Bridge* case was decided by the Supreme Court, recognition of the need to prevent future exclusive monopolies for transportation was no longer confined to a portion of the political spectrum. The only substantial disagreement turned on the proper means for avoiding the consequences of earlier monopolies. No longer did Kent write of the need to provide legal guarantees to encourage the investment of private capital. Nor did he emphasize the inherently public nature of transportation companies on which public duties were imposed in return for protection against competition. Within less than a generation, judges and jurists had come to agree that a policy in favor of competition was a *sine qua non* for further economic development.

V The Relation between the Bar and Commercial Interests

O NE of the phenomena that has puzzled historians is the extraordinary change in the position of the postrevolutionary
American bar — "the amazing rise," Perry Miller called it, "within
three or four decades, of the legal profession from its chaotic condition of around 1790 to a position of political and intellectual domination."[1] In the period between 1790 and 1820 we see the development of an important new set of relationships that made this position of domination possible: the forging of an alliance between legal
and commercial interests. It is during this period that the mercantile
classes shed a virulent antilegalism often manifested during the colonial period by a resort to extralegal forms of dispute settlement.
During this same period, the Bar first becomes active in overthrowing eighteenth century anticommercial legal doctrines.

Full time legal practitioners specializing in great commercial litigation involving lucrative fees appear for the first time after 1790.
In Massachusetts, for example, the young Theophilus Parsons
"ma[de] himself master of the law of prize and admiralty, of which
few lawyers then knew anything" on the assumption "that this
branch of business would probably be made profitable by the events
of war. In fact," his son noted, "he had almost the monopoly of it;
and it was very profitable."

My mother used to speak of the "prize times," . . . as the most profitable, in
her view, which she had ever known. The clients got their money easily,
and spent it as easily.[2]

The leaders of the Bar in the period after 1790 are not the land
conveyancers or debt collectors of the earlier period, but for the first

time, the commercial lawyers. Alexander Hamilton's New York practice before he entered the national government in 1789 was, Julius Goebel informs us, "basically of a bread-and-butter variety raising no exhilarating issues. . . . When Hamilton returned to the law in 1795 and to a very changed business environment, the dimensions of his commercial practice increased, as did the importance of his retainers." One of Hamilton's first cases terminated in an astronomical damage judgment of $120,000.[3]

By 1800 marine insurance litigation between merchants and insurance companies became a mainstay of commercial practice. Horace Binney, the leader of the Philadelphia bar, recalled that "insurance cases were probably never so numerous or important as in Philadelphia from 1801 to 1817. That city was the first commercial port in the United States, and her insurers were as active as her merchants."[4] In Boston and New York as well, the leaders of the Bar prospered beyond the most active imagining of the preceeding generation. Reflecting these developments, in 1808 John Hall of Baltimore established the country's first legal periodical, *The American Law Journal*, for the purpose of reporting on commercial law decisions. "Every merchant," he wrote, "must sensibly feel the inconvenience and perplexity which result from his ignorance of those laws by which his dearest rights and most important privileges are regulated."[5]

The active involvement of lawyers in commercial affairs marks a major transformation in the relationship between legal and mercantile interests. By 1822 Daniel Webster "took the liberty" of informing Justice Story that commercial interests disapproved of a case he had recently decided. "The merchants are hard pressed," he wrote, "to understand why there should be so much good law, on one side, & the decision on the other."[6] Nor was Story inattentive to the desires of merchants. After he extended the federal admiralty jurisdiction to marine insurance cases in *De Lovio v. Boit*[7] (1815), he noted that "to my surprise . . . the opinion is rather popular among merchants. They declare that in mercantile causes, they are not fond of juries; and, in particular, the underwriters in Boston have expressed great satisfaction at the decision."[8]

It should have come as no surprise to Story that in most cases "merchants were not fond of juries." For one of the leading measures of the growing alliance between bench and bar on the one hand and commercial interests on the other is the swiftness with which the power of the jury is curtailed after 1790.

Three parallel procedural devices were used to restrict the scope of juries. First, during the last years of the eighteenth century American lawyers vastly expanded the "special case" or "case reserved," a device designed to submit points of law to the judges while avoiding the effective intervention of a jury. In England, Lord Mansfield had used a similar procedure to bring about an alliance between common lawyers and mercantile interests.[9] "This procedure was used with great effect in marine insurance litigation, and . . . converted what had been largely a matter of merchant custom into a recognizable body of common law precedent."[10] "During the years of Hamilton's professional activity in New York," Goebel has noted, "an ingenious enlargement" of this procedure "had come nearly to supplant traditional practice."[11] In Massachusetts, as well, virtually all cases submitted to the Supreme Judicial Court during the last two decades of the eighteenth century proceeded from an agreed statement of facts.[12]

A second crucial procedural change—the award of a new trial for verdicts "contrary to the weight of the evidence"—triumphed with spectacular rapidity in some American courts at the turn of the century. The award of new trials for any reason had been regarded with profound suspicion by the revolutionary generation. "The practice of granting new trials," a Virginia judge noted in 1786, "was not a favourite with the courts of England" until the elevation to the bench of Lord Mansfield, "whose habit of controling juries does not accord with the free institutions of this country; and ought not to be adopted for slight causes."[13] Yet, not only had the new trial become a standard weapon in the judicial arsenal by the first decade of the nineteenth century; it was also expanded to allow reversal of jury verdicts contrary to the weight of the evidence, despite the protest that "not one instance . . . is to be met with" where courts had previously reevaluated a jury's assessment of conflicting testimony.[14] In both New York[15] and South Carolina[16] this abrupt change of policy was first adopted in order to overturn jury verdicts against marine insurers. In Pennsylvania[17] too the earliest grant of a new trial on the weight of evidence occurs in a commercial case.

These two important restrictions on the power of juries were part of a third more fundamental procedural change that began to be asserted at the turn of the century. The view that even in civil cases "the jury [are] the proper judges not only of the fact but of the law that [is] necessarily involved"[18] was widely held even by conservative jurists at the end of the eighteenth century. "The jury may in all cases, where law and fact are blended together, take upon themselves the knowledge of the law . . . ," William Wyche wrote in his 1794 treatise on New York practice.[19]

During the first decade of the nineteenth century, however, the Bar rapidly promoted the view that there existed a sharp distinction between law and fact and a correspondingly clear separation of function between judge and jury. For example, until 1807 the practice of Connecticut judges was simply to submit both law and facts to the jury, without expressing any opinion or giving them any direction on how to find their verdict. In that year, the Supreme Court of Errors enacted a rule requiring the presiding trial judge, in charging the jury, to give his opinion on every point of law involved.[20] This institutional change ripened quickly into an elaborate procedural system for control of juries. In Massachusetts, as well, the first decade of the nineteenth century marked the beginnings of the decline of the jury's power. Legislation in 1804 and 1805 eliminated the colonial system of trial before the entire Supreme Court bench together with one of its bizarre consequences: multiple and often conflicting instructions to the jury. Single judge trials were substituted, and in 1808 the Supreme Judicial Court required for the first time that trial judges instruct the jury on every material point at issue.[21] Finally, between 1805 and 1810, the high court began regularly to order new trials for errors in the proceeding below.

By 1810, it was clear that the instructions of the court, originally advisory, had become mandatory and therefore juries no longer possessed the power to determine the law. Courts and litigants quickly perceived the transformation that had occurred and soon began to articulate a new principle — that "point[s] of law . . . should . . . be . . . decided by the Court," while points of fact ought to be decided by the jury.[22]

These procedural changes made possible a vast ideological transformation in the attitude of American jurists toward commercial law. The subjugation of juries was necessary not only to control particular verdicts but also to develop a uniform and predictable body of judge-made commercial rules. We shall soon see the process by which American courts after 1790 overthrow particular anticommercial doctrines of the eighteenth century. At this point, however, I wish to focus on a more general jurisprudential transformation which permitted the Bar after 1790 to identify commercial law with "natural reason" and "universal law."

The influence of Lord Mansfield's conception of a "general jurisprudence" of commercial law became overwhelming in America after 1790.[23] Many of Supreme Court Justice James Wilson's "Law Lectures" were devoted to singing the praises of Mansfield and of his efforts to reduce mercantile law to "rational and solid principles."[24] Wilson called for the establishment of special "courts of commerce" to handle mercantile affairs[25] and reiterated Mansfield's self-serving identification of commercial law with the law of nations. Commercial and maritime law, Wilson repeated, were "not the law of a particular country, but the general law of nations."[26]

The identification of commercial law with a universal law of nations served several important functions. In both England and America, it allowed procommercial judges to go outside the existing legal system to import novel and congenial rules of law. It was also a profoundly antilegislative conception of the nature and source of law. Since commercial rules were part of "the general law of nations," James Sullivan observed in 1801, judges were obliged to "depend" on the law of nations for "their origin and their expositions," rather than on any municipal regulations of particular countries. This meant that "the most important interests of mankind cannot be secured, directed, and governed by the special acts of legislation in a country . . . ," but, rather, by judicial pronouncements on commercial law.[27]

Thus, the intellectual foundation was laid for an alliance between common lawyers and commercial interests. And when in 1826 Chancellor Kent wrote to Peter DuPonceau about the arrangement of his forthcoming *Commentaries*, he underlined the extent to which he would pay attention only to decisions of the courts of commercial states. "My object," he wrote,

will be to discuss the law . . . as known and received at Boston, New York, Philadelphia, Baltimore, Charleston, etc. and as proved by the judicial decisions in those respective states. I shall not much care what the law is in Vermont or Delaware or Rhode Island, or many other states. Cannot we assume American common law to be what is declared in the federal courts and in the courts of the states I have mentioned and in some others, without troubling ourselves with every local peculiarity? I shall *assume* what I have to say to be the law of every state where an exception is not shown, because I mean to deal in *general Principles* and those positive regulations, legislative and judicial, which constitute the basis of all American jurisprudence.[28]

The search for "general Principles" and the extinction of "every local peculiarity" in the treatise tradition which Kent inaugurated was grounded on a legal ideology that was no more than a generation in the making. The law of "Boston, New York, Philadelphia, Baltimore, Charleston" would have had virtually no meaning to an American lawyer as late as 1790. But within little more than a generation, an alliance between the mercantile classes and the legal profession that came to serve it had been successfully forged. The leaders of the Bar lost interest in the law of "Vermont or Delaware or Rhode Island." And questing after "general Principles," they completed their "amazing rise, within three or four decades . . . to a position of political and intellectual domination."

As the Bar was molding legal doctrine to accommodate commercial interests, a far-reaching and seemingly paradoxical institutional change was also underway. At precisely the same time as the formal legal system was becoming more receptive to mercantile concerns around the turn of the century, the availability of extralegal forms of dispute settlement by the commercial classes was being sharply curtailed. The mercantile interest for the first time was required to recognize the legal primacy of the Bar. In turn, the legal profession was becoming ever more willing to destroy those precommercial doctrines which were ultimately responsible for the antilegalism of the commercial classes during the colonial period.

Ever since Lord Coke's decision in *Vynior's Case*[29] (1609), refusing to enforce an agreement to arbitrate, the common law had not been protective of arbitration agreements. This seemed to matter little in Coke's time, when it was common to use penal bonds to reinforce promises to arbitrate. Only after this self-policing mechanism broke down with the statutory abolition of penalties in 1697 did judicial failure to enforce agreements to arbitrate become serious. One important result was the passage of the first Arbitration Act in England in 1698.[30]

From this date, arbitration in England is derived almost exclusively from legislative authorization. "A price, however, was exacted for this [statutory] recognition, because the Act marks the initiation of a period in which arbitrations came under close scrutiny and review by the courts. . . . The process of judicial intervention into arbitration can be seen growing throughout the eighteenth century, as the functions of the law courts and the practice of the mercantile community coalesce into a coherent system."[31]

In America, use of extrajudicial means of settlement through arbitration and reference was very widespread among commercial interests during the colonial period but remained essentially unregulated by courts.[32] Arbitration was popular for two important reasons. First, it allowed for great informality, and hence was faster; it was thus substantially cheaper. For example, in 1751 a New York newspaper printed a letter from one English friend to another urging him not to take his case to a lawsuit.

You will not consent perhaps *now* to submit the Matter in Dispute to Reference; but let me tell you that after you have expended large Sums of

Money, and squander'd away a deal of Time & Attendance on your lawyers, and Preparations for Hearings one Term after another, you will probably be of another Mind, and be glad *Seven Years* hence to leave it to that
Arbitration which you now refuse.[33]

Again in 1762 the New York merchant, John Watts, was inclined to
accept a proposal from his insurers that their dispute be submitted
to arbitration "as the most speedy and just determination, rather
than be put to the expence of two or three lingering Law suits, that
may be spun out for Years in the way the Law is here." Two years later, in another maritime matter, Watts defended his decision to
submit the dispute to arbitration to avoid "throw[ing] ourselves into
Expensive endless Law."[34]

The second and perhaps most important inducement for extrajudicial settlement of disputes was that it enabled colonial merchants
to avoid a precommercial consciousness which, they frequently complained, was prevalent among "Officers of the Courts and Lawyers,
who never Trade."[35] This conflict between merchants and a legal
profession not yet attuned to commercial interests was a general
theme in colonial America. It has been overshadowed in the historical literature by a quite different, though entirely compatible,
theme of lower class hatred of lawyers as "hired guns" of oppressive
creditors and mortgagors. But this more vocal antilawyer sentiment
directed at the traditional functions of the Bar should not obscure
the strong strand of middle class antilegalism that developed among
merchants during the colonial period. In Virginia, for example,
there was a strong "animadversion toward lawyers as a professional
class . . . far into the eighteenth century." While "much of the litigation that went on in colonial Virginia during this period was confined to commercial matters . . . the common lawyers . . . were pre-
dominantly interested in the law as Coke had understood it."[36] In
New York "the ruling class of merchants and patroons" displayed
marked "aversion" to lawyers in general,[37] while it was still commonplace to "identify . . . lawyers with conservative and landed interests."[38] "The Gentlemen of the Law," wrote Cadwallader Colden
of New York in 1765, "both the Judges and the principal Practitioners at the Bar are either, Owners, Heirs or strongly connected in
family Interest with the Proprietors."[39] In fact, "the most compelling
and persistent grievance against the landed elite concerned the
power it exercised over the New York judiciary."[40] A 1768 New York
election campaign was fought under the slogan of "No lawyer in the
Assembly." "Venomous broadsides called the attention of the voters

to the fact that New York was a commercial city owing its prosperity to merchants and not to men of law. Debates were held on the question of 'Whether a Lawyer could possibly be an honest man,' and too many were of the opinion that he could not." A bill to curtail the jurisdiction of the New York Supreme Court after the election was interpreted by a contemporary "as an effect of almost universal clamor against the law and its practisers. You cannot conceive the violence of the people's prejudices." The campaign against lawyers was still alive two years later. In a public controversy between a merchant and a lawyer, the former called upon the general public to choose between them, arguing that "the Word of a reputable Merchant is in all Places as good, and in this City will go much further than that of a quibbling Lawyer." "That a New York merchant could utter these words in the public journals in 1770 without fear of retribution," a historian of the controversy writes, "serves to illustrate that the vigorous campaign against the lawyers of the previous year had not subsided."[41]

Colonial merchants were ever ready to complain that "a Number of Pettifoggers are allways ready to . . . puzzell the Laws, which are far from being explicit with respect to Commerce."[42] And lawyers were regularly accused of turning litigation into "an endless Piece of Business" in order to increase fees. Only on the day of trial, after many delays, it was contended, would a lawyer submit a commercial matter to arbitration; "for truly the Jurors are incompetent Judges of so intricate a Business, and Referees will admit as Evidence, what the Court, for want of strict legality, may reject."[43]

Likewise, before Mansfield ascended the English bench, Lord Campbell maintained, "mercantile questions were so ignorantly treated when they came into Westminster Hall, that they were usually settled by private arbitration among the merchants themselves." As in England, American colonial courts were "governed by precedents of agricultural rather than mercantile origin," which "were ill adapted for the settlement of merchants' disputes."[44] Even as late as 1793 English merchants felt it necessary to establish a system of arbitration in the town of Newcastle-upon-Tyne. In addition to the fact that legal decisions "were often grievously expensive," they were "frequently different from what sea-faring persons conceived to be just" because of "the ignorance of lawyers in maritime affairs." Over opposition "principally from the gentlemen of the law, . . . a number of gentlemen, respectable for their knowledge in mercantile and maritime affairs" established an "association for general arbitration."[45]

Similarly, amid the campaign against lawyers in New York, the Chamber of Commerce in 1768 established a Committee on Arbitration, which surviving records indicate handled over two hundred cases between 1779 and 1792. The largest number of disputes concerned maritime affairs and virtually all, of course, involved commercial matters.[46] The rules of the Boston Chamber of Commerce in 1794 also provided for arbitration of matters "respecting Commerce, Insurance, or Maritime Affairs."[47]

This theme of avoidance of common law settlement of mercantile matters was a precise recapitulation of the English experience in the one hundred and fifty years before Mansfield ascended the King's Bench in 1756.[48] In fact, the precommercial attitude of American lawyers was still a significant cause of the hostility to lawyers even after the Revolution. Most articulate of the lawyer haters was the Massachusetts merchant, Benjamin Austin, who bluntly proposed in 1786 that the legal profession be abolished and that most judicial proceedings be replaced by a system of informal, binding arbitration. "Seven-eighths of the causes which are now in their hands might have been settled by impartial *referees*," he wrote. "Why cannot the disputes of the merchant, etc. be adjusted by reference, rather than by a long tedious Court process? Or why should we engage lawyers who are wholly unacquainted with all mercantile concerns?"[49]

In his "Law Lectures," delivered in 1792, United States Supreme Court Justice James Wilson also emphasized the unfamiliarity of American lawyers with commercial transactions. "Before the revolution," he wrote, "we were strangers, in a great measure, to what is properly called foreign commerce." American lawyers had none of the "advantages of use and habit to form precedents for their [foreign] transactions. . . . The rules, therefore, of admitting foreign testimony, and of authenticating foreign transactions, have been but lately the objects of much consideration." Thus, it was necessary for Americans, "in order to improve the opportunities, with which they are favoured, and to avail themselves, as they ought, of the happy situation, in which they are placed," to "encourage commerce by a liberal system of mercantile jurisprudence." Deploring "the numerous embarrassments, which arise from the want of a proper commercial forum . . . well known and severely felt both by the gentlemen of the bar, and by the gentlemen of the exchange," he applauded the use of arbitration and advocated the establishment of a chancery as well as a commercial court to resolve mercantile disputes.[50]

Two decades later, Benjamin Austin's attack on legalism and the legal profession was taken up in a pamphlet entitled *Sampson against the Philistines,* ascribed to the Philadelphia reformer Jesse Higgins. Higgins argued that the only means of breaking the hold of the legal profession over settlement of disputes was to establish compulsory arbitration. It would not only assure "cheap justice [and] speedy justice"; it was also "most convenient among a free, enlightened, and commercial people."

Higgins admired "the merchants of New York" who, he wrote, "at a very early period, established a chamber of commerce for the adjustment of differences on commercial regulations generally." Most important, he noted, they established a system of arbitration of commercial disputes. "This hint from New York has been taken by the merchants of this and other cities," he concluded.[51]

In 1791 New York for the first time passed a general arbitration statute for the benefit of "merchants and traders and others," which extended legislative support of private dispute settlement well beyond any previous enactment.[52] Prior statutes had narrowly limited arbitration to disputed mercantile accounts, although there was an additional common law power to refer disputes while they were pending before the courts.[53] The 1791 statute extended arbitration to cases where no action had yet been brought and enabled the parties to seek a rule of court, backed by the contempt power.

From the limited evidence we have, however, it appears that this great legislative advance in encouraging extrajudicial dispute resolution had little effect. While it is clear that in the eighteenth century arbitration was a significant forum by which the commercial classes sought to avoid common law decisions, it appears that resort to arbitration had begun to decline in the last years of the eighteenth century. Despite the vastly increased opportunity for arbitration made possible by the New York statute of 1791, Julius Goebel has compiled statistics which show that during the years 1784-95, more than twice as great a proportion of cases were sent to referees by action of the New York courts than in the 1796-1807 period.[54] Even more surprising, from another study of arbitration cases brought to the New York courts beginning in 1800 it is clear that "most of these cases [did] not involve commercial disputes among merchants."[55] Though, because of the state of judicial records, we do not have equivalent statistics concerning arbitrations for the pre-1800 period, or indeed for the great bulk of arbitrations informally entered into, one still has a strong sense that resort to arbitration among New York merchants had begun to decline after 1795.

Before we turn to other states, where no equivalent statistics have been assembled, several other observations about New York arbitrations should be mentioned. First, from a careful reading of New York arbitration cases beginning with the published reports in 1799, one has the strong impression of a new willingness of the judges to reverse arbitration awards for technical deficiencies. Indeed, one is struck by the fact that the traditional common law deference to the reports of arbitrators had begun rapidly to dissipate.[56] In a case decided at the turn of the century, the New York Supreme Court reversed a referee's award solely because "facts in this case are intricate, and there exists so much doubt and obscurity on the subject, that there is reason to apprehend that the referees did not possess all the lights which may now be afforded them, and which may lead to a more satisfactory result."[57] The mood behind this and similar decisions was underlined two years later with the publication of the New York lawyer George Caines' *Lex Mercatoria,* the first American treatise on commercial law. Even an arbitration agreement in an insurance contract, Caines wrote, cannot deprive the parties of a right to sue at law "because the tribunals of the land are not, by the contracts of individuals, to be thus ousted of their jurisdiction."[58] These trends became rather clear when, in *De Hart v. Covenhaven*[59] (1801), the New York Supreme Court deprived merchants of their most important inducement to seek arbitration—the arbitrator's greater familiarity with commercial law—by holding that it would not refer any matter to arbitration where it appeared that a question of law would arise.[60] With this decision, the judges had finally established their position as the sole and authoritative expositors of New York law. Thereafter, up through the time of the Civil War, arbitration in New York was confined to fact finding, and referees were required to follow judicially established rules.[61] In short, during the first few decades of the nineteenth century, both juries and arbitrators were deprived of their prerevolutionary shares of lawmaking power.

From the increasing frequency of appeals from arbitration awards, one can see that the New York courts at the beginning of the nineteenth century had succeeded in reversing the colonial pattern of merchants' deference to these awards. By contrast, a New York merchant writing in 1764 declared that he "could not in decency appeal from their [the arbitrators'] Judgment as it is contrary to all Practice." Besides, he observed, "We should have appeard with an ill Grace, having the Award of people in Commerce against us, to be offerd to a Jury of the same profession, for which reason it is invariably lookd upon as a point of Justice and propriety to submit to the referrees, or why leave it to them at all, the Looser is seldom content or satisfyd."[62]

A similar pattern of judicial hostility to arbitration appeared in Massachusetts decisions in the early nineteenth century. Under the pressure of radicals like Benjamin Austin, who coupled their attacks on lawyers with proposals for a statutory scheme of arbitration, the Massachusetts legislature in 1786 enacted a Referee Act.[63] Though Massachusetts lawyers and judges acknowledged that in passing the act "the legislature intended to provide for a speedy and equitable mode of deciding civil causes" and that its "object . . . was to provide for justice speedily and cheaply," they did their best to undermine its purposes.[64] From the outset, the Supreme Judicial Court multiplied the technical grounds for reversing arbitration awards on the theory that "this, being a special jurisdiction, in derogation of the common law, [it] must be strictly pursued."[65] In another case the court reiterated that the terms of the arbitration statute must be strictly construed. "When parties leave the common law for these peculiar remedies, they cannot expect the Court to show them particular favor," the judges declared.[66] Applying this premise in still another case, the court held that the Referee Act did not allow arbitration of questions concerning real estate, though, as a commentator pointed out, this result was required neither by the words of the act or the common law.[67]

Under the influence of Quaker antilegalism, Pennsylvania had been at the forefront of the colonies in providing for arbitration. Quaker doctrine urged that even though "among the best men differences would unavoidably arise from their intercourse in business and other causes," disputes "should be settled in a Christian manner. Therefore . . . no member should appeal to law; but . . . he should refer his difference to arbitration by persons of exemplary character in the Society."[68] English Quakers, for example, established "a concise statement of . . . rules recommended by the Society in the case of arbitrations," including the imposition of discipline on "any member going to law with another, without having previously tried to accommodate matters between them, according to the rules of the Society."[69] Seventeenth century Pennsylvania Quakers introduced a system of arbitration boards, consisting of three or four laymen known as "common peacemakers," to settle all disputes between citizens.[70] And even as late as 1790 James Dallas reported that under "the present practice" "referees" were "entrusted" with "a very great share of the administration of justice" in Pennsylvania.[71]

Nevertheless, in postrevolutionary Pennsylvania, there is strong evidence, similar to that in other states, of increased judicial resistance to extrajudicial resolution of disputes. In 1788 the Pennsylvania Supreme Court observed that since the Revolution it had intervened more often to correct findings of facts and conclusions of law in referees' reports.[72] One result was that referees sought to protect their awards from reversal by requesting advisory opinions on the law from the courts or by rendering special verdicts subject to the judges' decisions on particular points of law.[73] Another result was that lower court judges acknowledged that they would enter "further into the merits of the case, than they usually do on reports of Referees," even though "they do not think that they depart from the spirit of former decisions."[74]

By 1803 the Pennsylvania Supreme Court announced that it would set aside referees' reports for "a clear, plain, evident mistake, either in law or fact, which affects the justice and honesty of the case." Though it continued to claim that it would be liberal in judging reports, the court insisted that "the nature and settled forms of actions must be preserved by [referees] equally as by juries" and that reports would be given only "the same effect as verdicts."[75] Before these decisions, one judge complained, arbitrators' awards in Pennsylvania "ha[d] been considered as a kind of judgment, given by private courts" and were thus regarded as "equally binding as a contract of the parties, or a judgment of a court of competent jurisdiction."[76]

These restrictions on arbitrations clearly represented a newly emerging pattern of hostility to extrajudicial settlement of disputes as well as an effort to consolidate all law-declaring functions in the judiciary. In *Sampson against the Philistines*, Jesse Higgins repeatedly observed that the Pennsylvania legal profession, in its united opposition to arbitration, had gone so far as to assert that it violated the right to jury trials.[77] In fact, Supreme Court Justice Hugh Henry Brackenridge, "acknowledge[d]" in 1814 that he "ha[d] been at all times, more friendly to an increase of the jurisdiction of the [local] justices, than to the system of unmanageable, and desultory arbitrations." He proposed that the appellate courts should discourage arbitration by threatening the parties with loss of the right to appeal from any trial court that issued a statutory rule of reference.[78]

The consequence of these and similar opinions was to so circumscribe the independent power and discretion of arbitrators that by 1833 a Pennsylvania legal writer could maintain that "a cause decided by arbitrators, and taken into court by appeal, comes up for trial under circumstances not in the slightest degree more favourable to him whom the arbitrators have thought right, than if the matter had never been investigated [before]." He added:

When a case comes before a court and jury, on appeal, it is tried precisely as if it had never been referred . . . and the award of arbitrators weighs not a feather. An advocate endeavoring to strengthen his case by citing even the unanimous award, in his favour, of the largest number of arbitrators provided for by the act of assembly, would be laughed at by his brethren, and silenced by the bench.

Along with most members of the Bar, the writer was overjoyed with this judicial attack upon the independent power of arbitrators, for, he argued, they had often "misinterpreted . . . their obligations and their powers."

The words of the [arbitrator's] oath—"justly and equitably"—are often construed as if intended to invest arbitrators with discretionary powers, far exceeding those of courts and juries; to decide according to principles adopted by themselves, in each particular case, in opposition to the established rules of the law, and to recur, for the ascertainment of facts, to other sources than strictly legal evidence.

"The belief has prevailed so extensively," he concluded, "that arbitrators may disregard the strict rules of law. . . . The arbitration law was designed, by those who framed it, to lessen the expenses of litigation, not to change the principles which regulate the decision of controversies. When men, chosen under its provisions, to settle disputes, disregard the rules governing the tribunals empowered to revise and reverse their judgments, they but double the cost and inconvenience of what is, at best, sufficiently vexatious and expensive."[79]

A new Pennsylvania arbitration statute passed in 1836 codified the newly orthodox view that arbitrators possessed no independent power to develop commercial rules. Though at common law "a mistake on the part of the arbitrators, either in matter of fact, or matter of law . . . [was] insufficient to set the award aside," the statute empowered the courts to reverse an arbitration award "for a plain mistake committed by the arbitrators, either in matter of fact, or in matter of law." While nonstatutory arbitration continued to be theoretically available to parties, those who took this course lost the chief benefit of arbitration under the statute—the power to levy execution on the arbitration award. And even this theoretical alternative was barred to litigants, as Pennsylvania courts went out of their way to assume that the parties intended to seek arbitration under the statute.[80]

South Carolina also began to restrain the discretion of arbitrators sometime during the early part of the nineteenth century. Departing from the strict rules against intervention that prevailed in England, where, it was acknowledged, referees' awards were "conclusive, unless some corruption, or other misbehavior of arbitrators, is proved," jurists instead had followed Pennsylvania practice so that "courts will hear allegations against [arbitrators' awards], either on the ground of an evident mistake in matter of fact, or error in matter of law, as well as for corruption, etc." The result, a commentator noted in 1839, at a time when the authority of jury verdicts had reached low ebb, was that in South Carolina "the sacredness of awards ought not to be extended beyond that of [jury] verdicts. . . . They have the same operation *here* as verdicts."[81]

The result was that an increasingly self-conscious legal profession had succeeded in suffocating alternative forms of dispute settlement. By 1836 David Hoffman observed that the arbitration process had been almost totally displaced as a central forum for establishing commercial rules and "unhappily" had become "merely preliminary to ultimate proceedings in a court of justice." The cause of this "growing evil" he ascribed to the Bar's desire for "augmentation of professional emolument" which had led it to encourage "expensive and dilatory judicial litigation."[82]

Thus, it appears that several major changes in the attitude of judges and merchants toward commercial arbitration had begun to emerge at the beginning of the nineteenth century. First, an increasingly organized and self-conscious legal profession had become determined to oppose the antilegalism among merchants which, during the colonial period, had taken the form of resort to extralegal settlement of disputes. Second, the mercantile classes, which had found the colonial legal rules hostile to their interests began, at the end of the eighteenth century, to find that common law judges themselves were prepared to overturn anticommercial legal conceptions. Third, the development of a split in the commercial interest, first manifested in the field of marine insurance, converted a largely self-regulating merchant group into one that was made dependent on formal legal machinery. Thus, one might loosely describe the process as one of accommodation by which merchants were induced to submit to formal legal regulation in return for a major transformation of substantive legal rules governing commercial disputes. The judges' unwillingness any longer to recognize competing lawmakers is a product of an increasingly instrumental vision of law.

Law is no longer merely an agency for resolving disputes; it is an active, dynamic means of social control and change. Under such conditions, there must be one undisputed and authoritative source of rules for regulating commercial life. Both the hostility of judges to arbitration and the willingness of merchants to forgo extrajudicial settlement spring from a common source: the increasingly active and solicitous attitude of courts to commercial interests.

A similar pattern of judicial resistance to competing sources of law and of increasing control over commercial litigation can be seen in the rapid demise of a favorite institution of colonial merchants — the "struck" or special jury. Juries of merchants had been familar to eighteenth century English common law practice well before Lord Mansfield, during the second half of the century, made them a staple of judicial settlement of commercial disputes.[83] In 1741 New York adopted a 1730 English statute, which regulated selection of these juries "in such Manner as special Juries have by Law heretofore been struck, for Trials at Bar."[84] By 1764 New York merchants were unwilling to appeal from an arbitration award "of people in Commerce" when it would only have resulted in trial by "a Jury of the same profession."[85] Thus, despite the dominance of anticommercial legal rules in the colonial legal system, the struck jury served widely to establish an alternative forum for settlement of disputes among mercantile interests.

Again, after the Revolution, the New York legislature authorized any party to a lawsuit to demand a struck jury and established procedures for jury selection.[86] But in 1801 the legislature for the first time placed struck juries within the judges' control, permitting them only where the court "may deem it necessary, by reason of the importance or intricacy of the case."[87]

Under New York practice, a struck jury was assembled after each party, in turn, "struck" twelve names from a panel of forty-eight jurors assembled by the clerk. The trial jury was then chosen in the ordinary manner from among the remaining twenty-four candidates.[88] The primary purpose of this procedure was to assemble jurors familiar with the intricacies of commercial practice.

The continued widespread use of merchant juries in postrevolutionary New York seems to have been part of an established system of mercantile control over the rules of commercial law. In 1797, for example, Brockholst Livingston expressed complete surprise at the fact "that a Jury of twelve respectable men [all of whom were merchants] should in a mercantile transaction shew so much deference for the opinion of the Court."[89]

It is only after the statute of 1801 deprived parties of these juries as a matter of right that cases challenging struck juries began to appear in the New York reports. Between 1803 and 1807 many such cases involving commercial litigation are brought before the state Supreme Court. But a dramatic reversal occurs thereafter. While the statute permitting struck juries continues on the books throughout the century, the published reports for the next fifty years show no commercial case submitted to a struck jury after 1807.

Before 1807 the largest number of cases involved litigation over marine insurance. They provide an extraordinary picture of the rapid disintegration of that homogeneous New York commercial community, which seems originally to have brought the system of struck juries into being. In the last decade of the eighteenth century, the New York legislature began to incorporate marine insurance companies, which in a short period of time transformed the entire structure of New York's commercial relationships. Before that time, marine insurance had been underwritten by an informal and interlocking network of merchants, each of whom would turn to the group for insurance on their own commercial undertakings. Insurance companies destroyed these circulating relationships, creating different classes of underwriters and policy holders with competing commercial interests. One result was that previously accepted rules of insurance law began to come under attack by merchants. Another was that the institution of the struck jury, previously an agreed upon procedure for selecting commercially sophisticated jurors, no longer uniformly served the interests of the commercial class. As merchants suing on marine insurance policies found that the anticorporate bias of the regular jury was more to their advantage than a special jury of merchants, they began to oppose the company's call for a struck jury. Indeed, as the interests of the corporate insurer and the general body of merchants began to diverge, the former's call for a mercantile jury might prove self-defeating as well.

Both these trends emerged in New York during the first decade of the nineteenth century. They seem to account both for the 1801 statutory limitation on the previously unrestricted right to struck juries and for the abrupt and unannounced end of requests for special juries in commercial cases after 1807. They can be seen as well in several New York cases between 1803 and 1807.

In *Barnewall v. Church*[90] (1803) we see the first challenge to the authority of merchant juries as well as the beginnings of a split over previously accepted commercial rules. The plaintiff sued to recover for a total loss of his ship under his marine insurance policy. "A struck jury was obtained, and they, from their skill in navigation, aided by their general knowledge in mercantile transactions,"[91] awarded him the large sum of $8,000. The defendants, who appear still to have been an old-style group of unincorporated underwriters, had maintained that the ship's loss was due to its unseaworthiness at the outset of the voyage, a good defense under established law. Having lost before the merchant jury, the insurers urged the court to reverse the jury's finding that the ship was seaworthy, arguing that "the bias . . . of juries, in subjects of this sort, cannot be unknown to the court." "It is now become almost a maxim for juries never to find a verdict for a defendant," they insisted, "when unseaworthiness or usury are relied on in defence."[92]

Sharply retreating from what appears to be a longstanding pattern of judicial deference to findings of merchant juries, a majority of the court reversed the verdict and granted a new trial on the issue of seaworthiness.[93] Only Judge Thompson was willing to uphold the verdict on the traditionally authoritative ground that "these points have been decided by a respectable jury of merchants."[94] But as part of the majority voting to override the jury, Judge Radcliff was ratifying a new pattern of commercial relationships that had already begun to emerge. "The circumstance that here was a struck jury," he declared, "is not of decisive weight in favor of the verdict, especially as it is founded on a point against which, as a ground of defence, it is known considerable prejudice exists."[95] Merchant juries, in short, would no longer be permitted to modify existing legal rules under the pretext of fact finding.

From the time that *Barnewall v. Church* was decided in 1803, a judicial trend against struck juries had set in. During the next four years, insurance companies moved in the Supreme Court for struck juries in five cases[96] and were turned down in all but one. Significantly, in four of these the policy holders opposed the request for a special jury, thus underlining their newly found confidence in lay jurors. In 1804 the court asserted the primacy of its lawmaking function by refusing a struck jury, even though the plaintiff did not oppose it, declaring that it needed to be independently satisfied that the case was "intricate or important." The only exception to judicial opposition to requests for struck juries is an 1806 case, in which, over the plaintiff's objection, the court granted a request after the company showed that an insurance policy valued at the large sum of $10,000 was involved. But even this case seemed to be ignored the next year, when the court turned down a motion involving an insurance claim of $27,500.

After 1807 we hear no more of merchant juries in New York commercial cases of any sort. The statute providing for struck juries remained on the books, and indeed special juries continued to be used in libel cases.[97] But the idea that a homogeneous mercantile class could provide uniform rules of commercial practice had died. Commercial law in New York would thereafter be exclusively within the province of the judiciary.

Another state that used struck juries until almost the turn of the century was South Carolina. Under a colonial statute of 1769[98] special juries could be had by the consent of both parties in cases "concerning trade, and disputes with merchants" or where the value in dispute was at least 50 pounds sterling; either party could also insist on a struck jury if he paid costs. In addition, under a 1791 statute[99] the trial court could, on its own motion, empanel a struck jury.

Even more than in New York, merchant juries seem to have exerted a powerful influence over the course of development of post-revolutionary South Carolina commercial law. A large number of complex commercial cases between 1789 and 1796 were decided by juries of merchants.[100] In a 1790 case involving the measure of damages for breach of a contract to deliver speculative government bonds, the court declared that "it is fortunate, that so respectable a jury are convened for the purpose of fixing a standard for future decisions."[101] In another commercial contract dispute in 1792, the plaintiff's counsel sought to reduce the impact of an earlier decision of a special jury, arguing that it "was not the opinion of the court, only the fluctuating opinion of a jury, which might be contradicted by another jury." However, the court followed the earlier verdict, since it "had been determined by a special jury, on very just and legal principles, and all the other cases . . . have confirmed the doctrine ever since."[102]

The power of merchant juries can best be seen in a 1795 case[103] involving the measure of damages for nonacceptance in South Carolina of a bill of exchange drawn in Boston. At the first trial, the court submitted to a lay jury the purely legal issue of which jurisdiction's law of damages should govern. Nevertheless, after the verdict, it ordered a new trial. "As the parties . . . expressed a dissatisfaction at this verdict," the court explained, "and as it was *a new case*" it "directed a new trial, before a special jury of merchants, who were now finally to settle this point." It was "therefore left . . . again to the jury . . . either to make our law on the subject their rule" or to adopt that of Massachusetts, "as they thought most agreeable to the law of merchants." After the jury decided that Massachusetts law should govern, the court concluded: "This, therefore, may be considered as establishing the law on this point, and making the *lex loci,* or law of the place where the bill is drawn, the rule."

For all this total subservience to the rule of merchants, after 1796 the reign of special juries in South Carolina appears, like New York, to have ended abruptly. Citing the use of special juries for "the purposes of delay and chicanery," the South Carolina legislature in 1797 abolished struck juries unless both parties consented. With this statutory change came the end of use of these juries by the court's own motion. Indeed, there are no reported cases thereafter in which both parties request that a jury be struck.

The end of the use of special juries in commercial litigation expresses more general trends in early nineteenth century law. Along with judicial restriction on the use of commercial arbitration, it marks the decline of the colonial merchant's strategy of submitting legal disputes to extrajudicial authorities. This trend not only reflects the breakdown in the homogeneity of mercantile interests; it is also part of a clear pattern of judicial hostility to competing sources of legal authority.

VI The Triumph of Contract

M ODERN contract law is fundamentally a creature of the nineteenth century. It arose in both England and America as a reaction to and criticism of the medieval tradition of substantive justice that, surprisingly, had remained a vital part of eighteenth century legal thought, especially in America. Only in the nineteenth
century did judges and jurists finally reject the longstanding belief
that the justification of contractual obligation is derived from the
inherent justice or fairness of an exchange. In its place, they asserted
for the first time that the source of the obligation of contract is the
convergence of the wills of the contracting parties.

Beginning with the first English treatise on contract, Powell's *Essay Upon the Law of Contracts and Agreements* (1790), a major feature of contract writing has been its denunciation of equitable conceptions of substantive justice as undermining the "rule of law." "It
is absolutely necessary for the advantage of the public at large,"
Powell wrote, "that the rights of the subject should . . . depend upon
certain and fixed principles of law, and not upon rules and constructions of equity, which when applied . . . , must be arbitrary and
uncertain, depending, in the extent of their application, upon the
will and caprice of the judge."[1] The reason why equity "must be arbitrary and uncertain," Powell maintained, was that there could be
no principles of substantive justice. A court of equity, for example,
should not be permitted to refuse to enforce an agreement for simple "exorbitancy of price" because "it is the consent of parties alone,
that fixes the just price of any thing, without reference to the nature
of things themselves, or to their intrinsic value. . . . Therefore," he
concluded, "a man is obliged in conscience to perform a contract

which he has entered into, although it be a hard one."[2] The entire conceptual apparatus of modern contract doctrine — rules dealing with offer and acceptance, the evidentiary function of consideration, and especially canons of interpretation — arose to express this will theory of contract.

Powell's argument against conceptions of intrinsic value and just price reflects major changes in thought associated with the emergence of a market economy. It appears that it was only during the second half of the eighteenth century that national commodities markets began to develop in England. From that time on, "the price of grain was no longer local, but regional; this presupposed [for the first time] the almost general use of money and a wide marketability of goods."[3] In America, widespread markets in government securities arose shortly after the Revolutionary War, and an extensive internal commodities market developed around 1815.[4] The impact of these developments on both English and American contract law was profound. In a market, goods came to be thought of as fungible; the function of contracts correspondingly shifted from that of simply transferring title to a specific item to that of ensuring an expected return. Executory contracts, rare during the eighteenth century, became important as instruments for "futures" agreements; formerly, the economic system had rested on immediate sale and delivery of specific property. And most important, in a society in which value came to be regarded as entirely subjective and in which the only basis for assigning value was the concurrence of arbitrary individual desire, principles of substantive justice were inevitably seen as entailing an "arbitrary and uncertain" standard of value. Substantive justice, according to the earlier view, existed in order to prevent men from using the legal system in order to exploit each other. But where things have no "intrinsic value," there can be no substantive measure of exploitation and the parties are, by definition, equal. Modern contract law was thus born staunchly proclaiming that all men are equal because all measures of inequality are illusory.

THE EQUITABLE CONCEPTION OF CONTRACT IN THE EIGHTEENTH CENTURY

The development of contract, it often has been observed, can be divided into three stages, which correspond to the history of economic and legal institutions of exchange. In the first stage, all exchange is instantaneous and therefore "involves nothing corresponding to 'contract' in the Anglo-American sense of the term. Each party be-

comes the owner of a new thing, and his rights rest, not on a promise, but on property." In a second stage, "exchange first assumes a
contractual aspect when it is left half-completed, so that [only] an
obligation on one side remains." The "third and final stage in the
development occurs when the executory exchange becomes enforceable."[5] According to orthodox legal history, when English judges
declare at the end of the sixteenth century that "every contract executory is an assumpsit in itself," and that "a promise against a promise will maintain an action upon the case," the conception of contract as mutual promises has triumphed and, according to Plucknett, "the process is complete and the result clear."[6] "Damages were
soon assessed," Ames added, "not upon the theory of reimburse-
ment for the loss of the thing given for the promise, but upon the
principle of compensation for the failure to obtain the thing promised."[7]

It is my purpose to demonstrate that, contrary to the orthodox
view, the process was not complete at the end of the sixteenth century. Instead, one finds that as late as the eighteenth century contract law was still dominated by a title theory of exchange and damages were set under equitable doctrines that ultimately were to be
rejected by modern contract law.

To modern eyes, the most distinctive feature of eighteenth century contract law is the subordination of contract to the law of property. In Blackstone's *Commentaries* contract appears for the first
time in Book II, which is devoted entirely to the law of property.
Contract is classified among such subjects as descent, purchase, and
occupancy as one of the many modes of transferring title to a specific
thing.[8] Contract appears for the second and last time in a chapter
entitled, "Of Injuries to Personal Property."[9] In all, Blackstone's extraordinarily confused treatment of contract ideas occupies only
forty pages of his four volume work.

As a result of the subordination of contract to property, eighteenth century jurists endorsed a title theory of contractual exchange
according to which a contract functioned to transfer title to the specific thing contracted for. Thus, Blackstone wrote that where a seller
fails to deliver goods on an executory contract, "the vendee may
seize the goods, or have an action against the vendor for detaining
them."[10] Similarly, in the first English treatise on contract, Powell
wrote of the remedy for failure to deliver stock on an executory contract as being one for specific performance.[11]

The title theory of exchange was suited to an eighteenth century
society in which no extensive markets existed, and goods, therefore,

were usually not thought of as being fungible. Exchange was not conceived of in terms of future monetary return, and as a result one finds that expectation damages were not recognized by eighteenth century courts. Only two reported eighteenth century English cases touch on the question of expectation damages for breach of contract. *Flureau v. Thornhill*[12] (1776) seems to have confronted the question of damages for the loss of a bargain. A purchaser of a lease, who sued for failure to deliver because of a defect in title, sought to recover not only his deposit, but also damages sustained as a result of the lost bargain. The report of the case does not disclose whether the plaintiff attempted to recover the increased value of the lease, or, rather, the loss he had suffered from selling stock to finance the payment. In any event, the court refused to allow him more than restitution of his payment, one judge contemptuously noting that he could not "be entitled to any damages for the fancied goodness of the bargain, which he supposes he has lost."[13]

A second English case involving the damage issue is *Dutch v. Warren*[14] (1720). The case would be irrelevant were it not for the fact that it was regularly cited by later jurists ransacking the English reports for early instances of the recognition of expectation damages.[15] The case represented a buyer's action for restitution of money paid on a stock purchase contract, the price of the stock having fallen by the time delivery was due. Although the court said the case was "well brought; not for the whole money paid, but the damages in not transferring the stock at that time,"[16] the case obviously does not establish the modern rule that one may recover expectation damages in excess of the purchase price for failure to deliver stock in a rising market. Indeed, Lord Mansfield referred to it not as establishing a rule for damages, but as illustrating the equitable nature of the action for money had and received.[17]

One of the handful of executory contracts in America before the Revolution appeared in *Boehm v. Engle*[18] (1767), in which two sellers sued a buyer who, alleging bad title, had refused to accept a deed for land. Since Pennsylvania had no equity court in which a seller could have sued for specific performance,[19] Boehm brought "a special action on the case for the consideration money" or contract price, not, it should be emphasized, for the value of the lost bargain. He was thereby suing, in effect, for specific performance and not for the change in value of the land. The suit was therefore consistent with Blackstone's title theory: the contract had transferred title from seller to buyer and all that remained was an action for the price.

To appreciate the radical difference between eighteenth century and modern contract law, consider a case decided during a period in which the demise of the title theory was becoming plain. *Sands v. Taylor*[20] was an 1810 New York suit against a buyer who had received a part shipment of wheat but had refused to receive the remainder contracted for. Under the old title theory, sellers were apparently required to hold the goods until they received the contract price from the buyer. But in *Sands v. Taylor* the sellers immediately "covered" by selling the wheat in the market and thereafter suing the buyer for the difference between market and contract price. While acknowledging that there were "no adjudications in the books, which either establish or deny the rule adopted in this case," the court ratified the seller's decision to "cover" and allowed him to sue for the difference. "It is a much fitter rule," it declared, "than to require . . . [the seller] to suffer the property to perish, as a condition on which his right to damages is to depend." In reaching this result the court was forced to fundamentally transform the title theory. The sellers, it said, "were, by necessity . . . thus constituted trustees or agents, for the defendants." The trust theory was thus created in order to overcome a result which, though inherent in eighteenth century contract conceptions, was becoming increasingly anomalous in a nineteenth century market economy. Under an economic system in which contract was becoming regularly employed for the purpose of speculating on the price of fungible goods, the old title theory of contract, conceived of as creating a property interest in specific goods, had outlived its usefulness. The demise of the title theory roughly corresponded to the beginnings of organized markets and the transformation of an economic system that had used contract as simply one means of transferring specific property.

The most important aspect of the eighteenth century conception of exchange is an equitable limitation on contractual obligation. Under the modern will theory, the extent of contractual obligation depends upon the convergence of individual desires. The equitable theory, by contrast, limited and sometimes denied contractual obligation by reference to the fairness of the underlying exchange.

The most direct expression of the eighteenth century theory was the well-established doctrine that equity courts would refuse specific enforcement of any contract in which they determined that the consideration was inadequate.[21] The rule was stated by South Carolina's Chancellor Desaussure as late as 1817:

It would be a great mischief to the community, and a reproach to the justice of the country, if contracts of very great inequality, obtained by fraud, or surprise, or the skillful management of intelligent men, from weakness, or inexperience, or necessity could not be examined into, and set aside.[22]

Seven years later, the Chief Justice of New York noted the still widespread opinion of American judges that equity courts would refuse to enforce a contract where the consideration was inadequate.[23]

Supervision of the fairness of contracts was not confined to courts of equity. The same function was performed at law by a substantive doctrine of consideration which allowed the jury to take into account not only whether there was consideration, but also whether it was adequate, before awarding damages. The prevailing legal theory of consideration was expressed by Chancellor Kent as late as 1822, on the very eve of the demise of the doctrine that equity would not enforce unfair bargains.[24] In contract actions at law, he wrote, where a jury determined damages for breach of contract, "relief can be afforded in damages, with a moderation agreeable to equity and good conscience, and . . . the claims and pretensions of each party can be duly attended to, and be admitted to govern the assessment."[25]

Eighteenth century American reports amply support Kent's statement. In Pennsylvania, for example, where no equity court sat, eighteenth century judges instructed juries in actions on bonds that they "ought to presume every thing *to have been paid,* which . . . in equity and good conscience, *ought not to be paid.*"[26] Without an equity court, Chief Justice McKean declared, courts were obliged to turn to juries for "an equitable and conscientious interpretation of the agreement of the parties."[27] As a result, Pennsylvania lawyers often argued that a plaintiff's claim on a contract "should be both legal and equitable before he can call on a jury to execute the agreement,"[28] and that "inadequacy of price, known to the other party, is a ground to set aside a contract."[29]

In Massachusetts, the eighteenth century rule was that a defendant in an ordinary contract case could offer evidence of inadequacy of consideration in order to reduce his damages. At three separate points in his student notes, written around 1759, John Adams indicated that "sufficient Consideration" was necessary to sustain a contract action. "No Consideration, or an insufficient Consideration, a good Cause of Motion in Arrest of Judgment," Adams noted in one of these entries.[30] In *Pynchon v. Brewster*[31] (1766), Chief Justice Hutchinson instructed the jury in an action for a fixed price that they "might . . . if they thought it reasonable, lessen the Charges in the [plaintiff's] Account."[32] One year later Hutchinson observed that "it seems hard that an Inquiring into the Consideration should be denied, and that Evidence should be refused in Diminution of Damages."[33]

In his *Autobiography*, John Adams pondered the implications of the conception of objective value that lay at the foundation of these rules. He wrote:

It is a natural, immutable Law that the Buyer ought not to take Advantage of the sellers Necessity, to purchase at too low a Price. Suppose Money was very scarce, and a Man was under a Necessity of procuring a £100 within 2 Hours to satisfy an Execution, or else go to Gaol. He has a Quantity of Goods worth £500 that he would sell. He finds a Buyer who would give him £100 for them all, and no more. The poor Man is constrained to sell £500s worth for £100. Here the seller is wronged, tho he sell voluntarily in one sense. Yet, the Injustice, that may be done by some Mens availing them selves of their Neighbours Necessities, is not so Great as the Inconvenience to Trade would be if all Contracts were to be void which were made upon insufficient Considerations. But Q. What Damage to Trade, what Inconvenience, if all Contracts made upon insufficient Considerations were void.[34]

Another indication of the equitable nature of damage judgments in the eighteenth century was the almost universal failure of American courts either to instruct juries in strict damage rules or else to reverse damage judgments with which they disagreed. As a result, the community's sense of fairness was often the dominant standard in contracts cases. A commentator, referring to a 1789 Connecticut commercial case, noted that "the jury were the proper judges, not only of the fact but of the law that was necessarily involved in the issue."[35] Whatever they believed about the proper allocation between judge and jury on matters of law, most judges were prepared to leave the damage question to the jury. For example, in a 1786 lawsuit in which the jury's award was lower than the agreed contract price, the South Carolina Supreme Court refused to grant a new trial, since "this is a case sounding in damages, and . . . the jury have thought proper to give a kind of equitable verdict between the parties."[36] Likewise, the Virginia General Court appeared to adopt the position that excessive damages were not sufficient cause for a new trial.[37] Where "positive law, and judicial precedents [are] totally silent on the subject [of damages]," Pennsylvania's Chief Justice McKean remarked, "the principles of morality, equity, and good conscience, would furnish an adequate rule to influence and direct our judgment."[38] And it was entirely clear that it was the jury's sense of equity that would prevail. While trying a case in the United States Circuit Court, Supreme Court Justice Washington found that, by awarding a lesser judgment, the jury had ignored his instruction that the plaintiff was entitled to recover the full amount of the contract. Asked to award a new trial, Washington refused on the ground that "the question of damages . . . belonged so peculiarly to the jury, that he could not allow himself to invade their province."[39]

Further support for the existence of a substantive doctrine of consideration in the eighteenth century is found in American courts' enforcement of the rule that "a sound price warrants a sound commodity."[40] While there is no direct evidence of a substantive doctrine of consideration in eighteenth century England, several unreported trial decisions supported the "sound price" rule,[41] and as late as 1792 Blackstone's successor in the Vinerian Chair at Oxford, Richard Wooddeson, proclaimed the sound price doctrine to be good law.[42] Thus, one may conclude that in both England and America, when the selling price was greater than the supposed objective value of the thing bought, juries were permitted to reduce the damages in an action by the seller, and courts would enforce an implied warranty in actions by the buyer.

What we have seen of eighteenth century doctrines suggests that contract law was essentially antagonistic to the interests of commercial classes. The law did not assure a businessman the express value of his bargain, but at most its specific performance. Courts and juries did not honor business agreements on their face, but scrutinized them for the substantive equality of the exchange.

For our purposes, the most important consequence of this hostility was that contract law was insulated from the purposes of commercial transactions. Businessmen settled disputes informally among themselves when they could, referred them to a more formal process of arbitration when they could not, and relied on merchant juries to ameliorate common law rules.[43] And, finally, they endeavored to find legal forms of agreement with which to conduct business transactions free from the equalizing tendencies of courts and juries. Of these forms, the most important was the penal bond.

The great advantage of the penal bond or sealed instrument was that at common law it precluded all inquiry into the adequacy of consideration for an exchange. In the medieval legal system, the use of "penal bonds with conditional defeasance," as they were called, enabled individuals to impose unlimited penalties on parties who had failed to perform agreed upon conditions. The use of penal bonds declined somewhat in England during the seventeenth and eighteenth centuries at first equity, then common law courts undertook to relieve against the penal feature—the recovery of the entire sum stipulated because of even a minor breach of a specified condition.[44] Although American courts appear to have followed the English and also "chancered" these bonds,[45] virtually all large business transactions in America until the beginning of the nineteenth century took the form of two independent bonds, each of which stipulated damages for failure to perform the agreed act.[46]

Despite the practice of "chancering," the use of bonds may still have avoided an equitable inquiry into the fairness of the exchange in most cases. From the beginning of the eighteenth century English judges had begun to distinguish between penalties — which they would relieve against — and liquidated damages — which the parties were free to stipulate without the interference of courts.[47] By the time Lord Mansfield ascended to the bench, the English courts were predisposed to regard most damage provisions in bonds as liquidated and hence enforceable. "Where the covenant is 'to pay a particular liquidated sum,' " Mansfield declared, "a Court of Equity can not make a new covenant for a man."[48] And summing up developments during the preceding century, Lord Eldon declared in 1801 that he could "not but lament" any supposed principle that even an "enormous and excessive" damage provision, to which the parties had agreed, should be voided as a penalty. "It appears to me extremely difficult to apply, with propriety, the word 'excessive' to the terms in which parties choose to contract with each other. . . . It has been held . . . that mere inequality is not a ground of relief."[49]

It is impossible to determine from court records whether American courts also distinguished penalties from liquidated damages. In Massachusetts, it appears, judges chancered bonds without the aid of juries by simply giving damages for half the stipulated penalty.[50] This practice took account of the widespread custom of writing into bonds a penal sum twice the amount of estimated damages. But above all this allowed the parties to determine their own damages free from judicial intervention. To what extent juries were involved in the chancering process in other states we do not know. If juries were simply instructed to ignore stipulated damages in a bond and to return verdicts for actual damages, bonds could not have represented an important device for avoiding the jury's equitable inquiry into the nature of a transaction. However, it appears that even as late as the last decade of the eighteenth century the number of bonds used to effect business transactions still vastly exceeded the number of ordinary contracts containing mutual promises; this suggests that courts did not have unlimited discretion in cases involving bonds.

The late use of bonds, the absence of widespread markets, and the equitable conception of contract law conspired to retard the development of a law of executory contracts. Indeed, the primitive state of eighteenth century American contract law is underscored by the surprising fact that some American courts did not enforce executory contracts where there had been no part performance. For example, in *Muir v. Key*,[51] a Virginia case decided in 1787, a buyer of tobacco brought an action for nondelivery on a bond containing mutual promises. In the same action, the seller sued for the price. The jury returned a verdict for the buyer, which the court reversed on the ground that unless the plaintiff had paid in advance he could not sue on the contract. Thus, as late as 1787 in Virginia, there could be no buyer's action on a contract without prepayment. Nor, according to one of the judges, could the seller sue without delivery of the tobacco.[52]

The view that part performance was required for contractual obligation seems to have been held elsewhere in eighteenth century America as well. In his study of Massachusetts law, William Nelson states that "as a general rule . . . executory contracts were not enforced . . . unless the plaintiff pleaded performance of his part of the bargain."[53] Thus, in his "Commonplace Book" (1759) John Adams insisted that in "executory Agreements . . . the Performance of the Act is a Condition preecedent to the Payment." For example, if two men agree on a sale of a horse, Adams wrote, "yet there is no reason that [the seller] should have an Action for the Money before the Horse is deliverd."[54] And even as late as 1795 Zephaniah Swift of Connecticut wavered between the view that performance is unnecessary for an action on a contract and the view that without either payment or delivery "the bargain is considered of no force and does not bind either [party]."[55] It is not difficult to understand why some courts did not enforce executory contracts without part performance. The pressure to enforce such contracts would not be great in a premarket economy where contracts for future delivery were rare[56] and where merchants framed most executory transactions that did arise in terms of independent covenants through the use of bonds.

Even where executory contracts were enforced without part performance, the infrequency with which they arose slowed the development of precise legal rules for dealing with them.[57] Eighteenth century courts were regularly confronted instead with commercial cases framed in terms of penal bonds. The legal categories required to enforce independent covenants were radically different from a conception of contracts depending on mutual promises. There was no need to inquire into questions of offer and acceptance to determine whether there had been "a meeting of minds." Nor was there any reason to develop rules for regulating "order of performance" or tender where each covenant was treated as independent.[58] Finally, because of its liquidated damage provision, the bond delayed until the nineteenth century any detailed inquiry into precise rules of damages.

The use of bonds seems to have substantially declined in both England and America during the early decades of the nineteenth century. If, in fact, bonds were still an important vehicle for avoiding inquiry into the fairness of an exchange during the eighteenth century, they became increasingly unnecessary as judges took control of the rules for measuring damages. Furthermore, liquidated damage provisions were not well suited to predicting market fluctuations in an increasingly speculative economy.[59] The result was that the executory contract came gradually to supersede the bond for most nineteenth century business transactions.

Before turning to outright reversals of eighteenth century law, however, it is important to note that there was a period of uneasy compromise between the old learning and the new. The transitional nature of the late eighteenth and early nineteenth centuries is revealed most explicitly in the confused relationship between the common counts, which by the end of the eighteenth century had emancipated the law of contract from the tyranny of the older forms of action, and Blackstone's 1768 division of the field of contract law into express and implied contracts.[60]

By highlighting the express agreement, Blackstone's division was an early indication of a tendency away from an equitable and toward a will theory of contract law. It also represented an effort to create a theoretical framework as a substitute for the older forms of action. However, Blackstone himself placed the common counts in the category of implied contracts,[61] which had the significant effect of identifying them with the still dominant equitable conception of contract. Implied contracts, Blackstone wrote, "are such as reason and justice dictate, and which therefore the law presumes that every man has contracted to perform."[62] For one of the common counts — indebitatus assumpsit for money had and received — Blackstone cited Lord Mansfield's then recent path-breaking decision in *Moses v. Macferlan*[63] (1760), in which the Chief Justice declared: "In one word, the gist of this kind of action is, that the defendant, upon the circumstances of the case, is obliged by the ties of natural justice and equity to refund the money."[64]

As a result of this unrestrained identification of contract with "natural justice and equity," the triumph of the common counts threatened to reinforce the equitable conceptions which Blackstone's distinction between express and implied contracts had appeared to displace as the unifying principle of contract. This persistence of an equitable tradition in English contract law also influenced American courts. *Palfrey v. Palfrey*[65] (1772), for example, involved an action in contract by children against their mother for improper occupation of a house they had inherited on their father's death. Rejecting the defendant's argument that the proper form of action was in trespass, the Massachusetts Superior Court held that the contract action would lie. In a long and elaborate opinion, the normally form-bound and technically oriented Judge Edmund Trowbridge maintained that there was an implied contract by the defendant to pay. Judge Trowbridge noted that "it [was] necessary to know what is at this day intended by an implied contract . . . because . . . 'many of the old cases are strange & absurd, the strictness has been relaxed & is melting down in common sense of late times.' " Since the plaintiffs were "clearly entitled to recover upon the merits & must in another action if not in this," the judges "ought to use [their] utmost sagacity to give them judgment." Judge Trowbridge concluded, using language borrowed from Blackstone and Mansfield:

It seems to be settled that implied contracts are such as reason & justice dictate; Therefore if one is under obligation from the ties of natural justice to pay another money and neglects to do it, the law gives the sufferer an action upon the case, in nature of a bill in equity to recover it; and that mere justice & equity is a sufficient foundation for this kind of equitable action.[66]

Blackstone's interpretation of the common counts as implied contracts did not ultimately secure the dominance of the equitable conception of contract, however, because of unresolved confusions in the pleading system. It appears that the common count of indebitatus assumpsit was variously used both for suing on an express contract price and for suing on an implied contract. When it was used to sue on an express contract, another common count, quantum meruit, was employed to sue on an implied contract. "In an action for work done," a mid-eighteenth century English commentator noted, "it is the best way to lay a Quantum Meruit with an Indebitatus Assumpsit. For if you fail in the proof of an express price agreed, you will recover the value."[67] As late as the turn of the century, it was also the prevailing practice in America to sue in indebitatus assumpsit for an express contract and for counts in both indebitatus and quantum meruit to be "usually joined in the declaration; so that on failure of proof of an express debt or price, the Plf. may resort ad debitum equitatis,"[68] that is, to an equitable action in quantum meruit.

The transitional nature of the late eighteenth century is thus revealed in the failure of eighteenth century lawyers to perceive any latent theoretical contradictions involved in joining counts on express and implied contract. Their failure to do so undoubtedly resulted from the theoretical confusions underlying the common counts themselves. Two very different conceptions of contract were submerged within actions on the common counts. One was based on an express bargain between the parties; the other derived contractual obligation from "natural justice and equity." But in the eighteenth century there was little occasion to see the two doctrinal strands as contradictory. Contract had not yet become a major subject of common law adjudication. The existence of mercantile arbitration, and the predominance of bills of exchange, bonds, and sealed instruments in business dealings meant that few of the legal problems that a modern lawyer would identify as contractual entered the common law courts.

In eighteenth century America, the equitable tradition in the common counts was tied not only to a general theory of natural justice but also to an economic system often based on customary prices.[69] The striking existence of this remnant of the medieval just price theory of value can be seen in two Massachusetts colonial cases. In *Tyler v. Richards*[70] (1765), the plaintiff brought indebitatus assumpsit for boarding and schooling the defendant's son. The defendant argued that indebitatus "will not lye; they ought to have brought a *Quantum Meruit*." For the plaintiffs, John Adams and Samuel Quincy argued that "it had always been the Custom of this Court, to allow" the action "if the Services alledged were proved to have been done. As every Man is supposed to assume to pay the customary Price. Assumpsit is always brought for Work done by Tradesmen, and is always allowed. The Price for Boarding and Schooling is as much settled in the Country, as it is in the Town for a Yard of Cloth, or a Day's Work by a Carpenter." Adams and Quincy were thus attempting to convince the court that if the value of goods or services was "settled" and bore a "customary Price," there was no difference between this action and indebitatus for a "sum certain." The defendant, however, argued that "if this Proof is admitted, there will be an End of any Distinction between *Indebita-*

tus Assumpsit and a *Quantum meruit.*" The court accepted the defendant's argument and dismissed the action.

In *Pynchon v. Brewster*[71] (1766), the plaintiff brought indebitatus "upon a long Doctor's Bill for Medicines, Travel into the Country and Attendance." This time, Adams, for the defendant, argued on the authority of *Tyler v. Richards* that indebitatus would not lie. The Chief Justice, however, distinguished *Tyler* on the ground that "Travel for Physicians, their Drugs and Attendance, had as fixed a Price as Goods sold by a Shopkeeper, and that it would be a great Hardship upon Physicians to oblige them to lay a *Quantum Meruit.*"

What emerges from these cases is that in America suits in indebitatus were sometimes based on a system of fixed and customary prices. Though the *Richards* court denied the analogy between the price of schooling and the "settled" price for a yard of cloth, it never challenged Adams' premise that the prices of most goods and services were conceived of as "settled." Similarly, while acknowledging the "uncertain" price of schooling, Chief Justice Hutchinson had no doubt that the price of a doctor's medicine and services "had as fixed a Price as Goods sold by a Shopkeeper."[72]

Of course, there could not have been a customary rate for every exchange that might be entered into and sued upon; the jury's power to set a reasonable price in quantum meruit was necessary to fill in the gaps. Indeed, it appears that the jury had discretion to mitigate or enlarge the damages even in indebitatus actions.[73] But the concept of customary prices formed the necessary foundation for a legal system which awarded contract damages according to measures of fairness independent of the terms agreed to by the contracting parties. By the end of the eighteenth century, however, the development of extensive markets undermined this system of customary prices and radically transformed the role of contract in an increasingly commercial society.

THE RISE OF A MARKET ECONOMY AND THE DEVELOPMENT OF THE WILL THEORY OF CONTRACT

Early attacks on eighteenth century contract doctrine

For a variety of reasons, it is appropriate to correlate the emergence of the modern law of contract with the first recognition of expectation damages. Executory sales contracts assume a central place in the economic system only when they begin to be used as instruments for "futures" agreements; to accommodate the market function of such agreements the law must grant the contracting parties their expected return. Thus, the recognition of expectation damages marks the rise of the executory contract as an important part of English and American law. Furthermore, the moment at which courts focus on expectation damages rather than restitution or specific performance to give a remedy for nondelivery is precisely the time at which contract law begins to separate itself from property. It is at this point that contract begins to be understood not as transferring the title of particular property, but as creating an expected return. Contract then becomes an instrument for protecting against changes in supply and price in a market economy.

The first recognition of expectation damages appeared after 1790 in both England and America in cases involving speculation in stock. Jurists initially attempted to encompass these cases within traditional legal categories. Thus, Lord Mansfield in 1770 referred to a speculative interest in stock as "a new species of property, arisen within the compass of a few years."[74] In 1789 the Connecticut Supreme Court of Errors held that recovery of expectation damages on a contract of stock speculation would be usurious.[75] And as late as 1790 John Powell concluded that specific performance, and not an action for damages, was the proper remedy for failure to deliver stock on a rising market.[76]

These efforts to encompass contracts of stock speculation within the old title theory were soon to be abandoned, however. Between 1799 and 1810 a number of English cases applied the rule of expectation damages for failure to deliver stock on a rising market.[77] In America the transformation occurred a decade earlier, in response to an active "futures" market for speculation in state securities which rapidly developed after the Revolutionary War in anticipation of the assumption of state debts by the new national government. The earliest cases allowing expectation damages on contracts of stock speculation appeared in South Carolina, Virginia, and Pennsylvania.

In South Carolina, three cases between 1790 and 1794 established the rule of expectation damages in stock cases. The first case, *Davis v. Richardson*[78] (1790), involved a "short sale" of South Carolina indents, or government stock. The defendant had borrowed the stock, promising its return with interest at a future time. "In consequence of the prospect of the adoption of the funding system by Congress," the value of the stock increased and the defendant could only "cover" at a substantially higher price. The South Carolina Supreme Court made no effort to conceal the significance of the damage question before it. "It is of extensive importance to the community, that the principle should *now be settled and ascertained with precision,*" the court declared. "A great number of contracts in every part of the state, depend upon the determination of this question: and it is fortunate, that so respectable a jury are convened for the purpose of fixing a standard for future decisions." And with the aid of advice from a "respectable" merchant jury, the court announced its holding: "Whenever a contract is entered into for the delivery of a specific article, the value of that article, at the time fixed for delivery, is the sum a plaintiff ought to recover."

It is entirely possible, of course, that the defendant in *Davis v. Richardson* was not the stock speculator that I have supposed him to be. In specie-scarce postrevolutionary South Carolina, where bonds and securities were regularly used for money, he may simply have been treating the indents as currency. As a result, he may have been one of the earliest casualties of the almost instant creation of a speculative market for state securities after the establishment of the national government. Prevailing economic and legal conceptions about the true nature of stock transactions were in a state of flux. Twice in the next four years, lawsuits[79] involving expectation damages on stock were carried to the Supreme Court of South Carolina in an attempt to reverse the ruling in *Davis v. Richardson.* The major argument put forth by Charles Pinckney, the leader of the South Carolina bar, was that the allowance of expectation damages was nothing more than the allowance of usury.[80] In *Atkinson v. Scott*[81] (1793), where the disputed securities had appreciated by 850 percent in one year, the Supreme Court admitted that such contracts "must strike every mind at the first blush" as "evidently usurious." If, Pinckney argued, South Carolina stock was to be treated as money, the borrower could only be expected to pay the value at the time of the contract plus interest. But in a world in which a "respectable" jury of merchants had recognized that stocks were traded on speculation, it made no sense for courts to deny the speculative purpose of the transaction. The result was that Pinckney's argument was rejected, and by 1794 the South Carolina legal system applied the rule of expectation damages to what appear to be the first organized markets that had developed in that state.

In Virginia, the transformation of legal conceptions took an identical path. In *Groves v. Graves*[82] (1790), the rule of expectation damages arose in connection with a buyer's action for securities. After a jury had awarded the plaintiff expectation damages, however, Chancellor Wythe, still reflecting eighteenth century moral and legal conceptions, enjoined the enforcement of the judgment on the grounds that the transaction "appeared to have been designed to secure unconscionable profit . . . and to have been obtained from one whom he had cause to believe at that time to be needy." He allowed damages only to the extent of the original value plus interest. But the Virginia Court of Appeals reversed his decree, holding that "the contract was neither usurious, or so unconscionable as to be set aside." And, in marked contrast to the earlier practice of not reviewing jury damage awards, the court held that the jury erred in measuring damages as of the time of trial and not as of the time of delivery.[83] The case thus suggests that judicial supervision of juries' damage awards may have arisen simultaneously with the recognition of expectation damages.

The first published opinion in Pennsylvania allowing expectation damages for failure to deliver stock certificates on a rising market was decided in 1791.[84] The rule was elaborated in a 1795 case, *Gilchreest v. Pollock,*[85] where a seller of stock sued the buyer's surety for failure to accept the transfer of United States securities that had fallen in price after the contract was made. While the merchant jury in South Carolina had had no difficulty in reaching their result, the Pennsylvania court felt compelled to charge its lay jurors that "the sale of stock is neither unlawful nor immoral. It is confessed, that an inordinate spirit of speculation approaches to gaming and tends to corrupt the morals of the people. When the public mind is thus affected, it becomes the legislature to interpose."[86]

The early Pennsylvania case is somewhat anomalous in that it rested on an unpublished opinion, rendered in 1786, which recognized a market price for wheat and announced that "the rule or measure of damages in such cases is to give the difference between the price contracted for and the price at the time of delivery."[87] With this one exception, however, the evidence appears to indicate that the rule of expectation damages first arose in connection with stock speculation both in England and in America.[88] In England the principle of expectation damages was not generalized in cases dealing with sales of commodities until 1825,[89] and Chitty's treatise on contracts, published in 1826, is the first to announce a general rule of expectation damages for failure to deliver goods.[90]

In America the application of expectation damages to commodities contracts correlates with the development of extensive internal commodities markets around 1815. The leading case is *Shepherd v. Hampton*[91] (1818), in which the Supreme Court held that the measure of damages for failure to deliver cotton was the difference be tween the contract price and the market price at the time of delivery. Within the next decade a number of courts worked out the problems of computing expectation damages for commodities contracts,[92] one of them noting that "most of the [prior] cases in which this principle has been adopted, have grown out of contracts for the delivery and replacing of stock."[93]

The absorption of commodities transactions into contract law is a major step in the development of a modern law of contracts. As a result of the growth of extensive markets, "futures" contracts became a normal device either to insure against fluctuations in supply and price or simply to speculate. And as a consequence, judges and jurists began to reject eighteenth century legal rules which reflected an underlying conception of contract as fair exchange.

It has already been noted that in the eighteenth century, commercial classes endeavored to cast their transactions in legal forms which avoided the equalizing tendencies of early contract doctrine. Not surprisingly, the first direct assault upon the equitable conception of contract appeared in adjudications involving one of these forms, the negotiable instrument.

During the second half of the eighteenth century, a movement developed to eliminate the substantive significance of the doctrine of consideration in cases involving negotiable instruments. In 1767 the Massachusetts Superior Court held by a 3-2 vote that even in an action between the original parties to a promissory note, the promisor could not offer evidence of inadequate consideration in mitigation of damages.[94] "People," Chief Justice Hutchinson declared, "think themselves quite safe in taking a Note for the Sum due, and reasonably suppose all Necessity of keeping the Evidence of the Consideration at an End; it would be big with Mischief to oblige People to stand always prepared to contest Evidence that might be offered to the Sufficiency of the Consideration. This would be doubly strong in Favour of an Indorsee."

It was one thing to argue that in order to make notes negotiable a subsequent endorsee would be allowed to recover on a note regardless of the consideration between the original parties. This argument, of course, itself entailed a sacrifice of judicial control over bargains that commercial convenience was beginning to demand. It was, however, quite a different matter to exclude evidence of consideration between the original parties to the note, as the Massachusetts court decided. With this decision, it became possible for merchants to exclude the question of the equality of a bargain by transacting their business through promissory notes.

The Massachusetts decision was handed down two years after
Lord Mansfield's dramatic but unsuccessful attempt to destroy the
doctrine of consideration in the case of *Pillans v. Van Mierop*[95]
(1765), a case between merchants involving a promise to accept a
bill of exchange. "I take it," Mansfield declared in dictum, "that the
ancient notion about the want of consideration was for the sake of
evidence only; for when it is reduced into writing, as in covenants,
specialties, bonds, etc., there was no objection to the want of consideration." While it is impossible to know from this pronouncement
whether Mansfield's *ratio decidendi* was that consideration was
unnecessary for all written instruments or merely for those between
merchants, two conclusions are clear. First, by explaining the requirement of consideration exclusively in terms of its evidentiary
value in proving the existence of a contract, Mansfield had cut the
heart out of the traditional equalizing function of consideration.
Second, whether or not upon reflection Mansfield would have extended these views to cover all written instruments — where the writing was itself sufficient evidence of a contract — he at least meant to
apply the rule to negotiable instruments. Indeed, as Mansfield's
decision was being announced, the second volume of Blackstone's
Commentaries was at the press, also propounding the rule that evidence of lack of consideration would not be admitted in an action
on a negotiable instrument.[96] For thirteen years, English law stood
thus on the verge of rejecting the ancient requirement of consideration. But in *Rann v. Hughes*[97] (1778), the House of Lords reaffirmed the requirement of consideration for written instruments.

The views of Mansfield and Blackstone were to have a greater
effect than the decision by the House of Lords, however. The report
of that decision was unpublished until 1800 and was unknown by
American judges before the early years of the nineteenth century.[98]
Thus, even after Mansfield's opinion was overruled, we find Zephaniah Swift, the first American treatise writer, stating that the
principle that had emerged from negotiable instruments law — he
cited Blackstone — "clearly destroys all distinction between sealed
and unsealed contracts."[99] The result, he concluded, was that a
written contract "precludes an enquiry into the consideration."[100] A
more important factor than the accident of reporting, however, was
the congeniality of Mansfield's and Blackstone's views to American
judges, whose own opinions were gradually inclining toward a conception of contract as a sacred bargain between private parties.

The most persistent American advocate of the Mansfield position
was the able judge of the New York Supreme Court, Brockholst Livingston, whose commercial law practice before he ascended to the

bench was probably second only to that of Alexander Hamilton. In 1804 Livingston reiterated the position that as between even the original parties to a negotiable instrument, the failure of consideration could not be shown. "It is not necessary, as in other simple contracts, to state a consideration in the declaration; the instrument itself imports one, and in this respect partakes of the quality of a speciality [sealed instrument]."[101] Livingston extended the argument to cover simple contracts in a case decided one year later. In *Lansing v. McKillip*[102] (1805), he dissented from the court's opinion requiring that consideration be proved by the plaintiff before he could recover on a contract. At first, he urged only that the traditional burden of proof be altered so that a defendant who wished to negate a contract be required to show lack of consideration. In the process, however, he was moved to attack the very requirement of consideration itself. Ridiculing a rule of consideration that "does not demand an absolute equivalent, but is satisfied, in many cases, with the most trifling ground that can be imagined," he urged the court to "be content in point of evidence, with a declaration . . . that he has received a *valuable one,* without indulging the useless curiosity of prying further into the transaction." Livingston was fully aware that his opinion directly attacked the traditional equalizing function of consideration. "Why," he asked, is a court "so very careful of a defendant's rights as not to suppose him capable of judging for himself, what was an adequate value for his promise? Would it not be more just, and better promote the ends of justice, that one, who had signed an instrument of this kind, should, without further proof, be compelled to perform it, unless he could impeach the validity on other grounds?"[103]

Like Mansfield's earlier effort, this attack on consideration initially failed, but in its most important respect it ultimately succeeded. It was part of a movement, which had begun in England during Mansfield's tenure and continued throughout the nineteenth century, toward overthrowing the traditional role of courts in regulating the equity of agreements. The underlying logic of the attack on a substantive doctrine of consideration came to fruition in America with the great New York case of *Seymour v. Delancey*[104] (1824), in which a sharply divided High Court of Errors reversed a decision of Chancellor Kent, who had refused to specifically enforce a land contract on the ground of gross inadequacy of consideration between the parties. "Every member of this Court," the majority opinion noted, "must be well aware how much property is held by contract; that purchases are constantly made upon speculation; that the value of real estate is fluctuating." The result was that there "exists an honest difference of opinion in regard to any bargain, as to its being a beneficial one, or not." The court held that only where the inadequacy of price was itself evidence of fraud would it interfere with the execution of private contracts.[105]

The nineteenth century departure from the equitable conception of contract is particularly obvious in the rapid adoption of the doctrine of caveat emptor. It has already been noted that, despite the supposed ancient lineage of caveat emptor, eighteenth century English and American courts embraced the doctrine that "a sound price warrants a sound commodity." It was only after Lord Mansfield declared in 1778, in one of those casual asides that seem to have been so influential in forging the history of the common law, that the only basis for an action for breach of warranty was an express contract,[106] that the foundation was laid for reconsidering whether an action for breach of an implied warranty would lie. In 1802 the English courts finally considered the policies behind such an action, deciding that no suit on an implied warranty would be allowed.[107] Two years later, in the leading American case of *Seixas v. Woods,*[108] the New York Supreme Court, relying on a doubtfully reported seventeenth century English case,[109] also held that there could be no recovery against a merchant who could not be proved knowingly to have sold defective goods. Other American jurisdictions quickly fell into line.[110]

While the rule of caveat emptor established in *Seixas v. Woods* seems to be the result of one of those frequent accidents of historical misunderstanding, this is hardly sufficient to account for the widespread acceptance of the doctrine of caveat emptor elsewhere in America. Nor are the demands of a market economy a sufficient cause. Although the sound price doctrine was attacked on the ground that there "is no standard to determine whether the vendee has paid a *sound* price,"[111] the most consistent legal theorist of the market economy, Gulian Verplanck, devoted his impressive analytical talents to an elaborate critique of the doctrine of caveat emptor.[112] The sudden and complete substitution of caveat emptor in place of the sound price doctrine must therefore be understood as a dramatic overthrow of an important element of the eighteenth century's equitable conception of contract.[113]

The synthesis of the will theory of contract

The development of extensive markets at the turn of the century contributed to a substantial erosion of belief in theories of objective value and just price. Markets for future delivery of goods were difficult to explain within a theory of exchange based on giving and receiving equivalents in value. Futures contracts for fungible commodities could only be understood in terms of a fluctuating conception of expected value radically different from the static notion that lay behind contracts for specific goods; a regime of markets and speculation was simply incompatible with a socially imposed standard of value. The rise of a modern law of contract, then, was an outgrowth of an essentially procommercial attack on the theory of objective value which lay at the foundation of the eighteenth century's equitable idea of contract.

We have seen, however, that there was a period during which vestiges of the eighteenth century conception of contract coexisted with the emerging will theory. It was not until after 1820 that attacks on the equitable conception began to be generalized to include all aspects of contract law. If value is subjective, nineteenth century contracts theorists reasoned, the function of exchange is to maximize the conflicting and otherwise incommensurable desires of individuals. The role of contract law was not to assure the equity of agreements but simply to enforce only those willed transactions that parties to a contract believed to be to their mutual advantage. The result was a major tendency toward submerging the dominant equitable theory of contract in a conception of contractual obligation based exclusively on express bargains. In his *Essay on the Law of Contracts* (1822), for example, Daniel Chipman criticized the Vermont system of assigning customary values to goods that were used to pay contract debts. Only the market could establish a fair basis for exchange, Chipman urged. "Let money be the sole standard in making all contracts," for "if, therefore, it were possible for courts in the administration of justice, to take this ideal high price as a standard of valuation, every consideration of policy, and a regard for the good of the people would forbid it."[114]

Nathan Dane's *Abridgment* (1823) and Joseph Story's *Equity Jurisprudence* (1836) also contributed to the demise of the old equitable conceptions. But nowhere were the underlying bases of contract law more brilliantly and systematically rethought than in Gulian C. Verplanck's *An Essay on the Doctrine of Contracts* (1825).

Verplanck was the first English or American writer to see in the "different parts of the system" of contract law "clashing and wholly incongruous" doctrines.[115] He emphasized "the singular incongruity" of a legal system that "obstinately refuses redress in so many, and such marked instances of unfairly obtained advantages" and yet "occasionally permit[s] contracts to be set aside upon the ground of inadequacy of price."[116] There were, he asserted, many "difficulties and contradictions" to be found in existing legal doctrine over "the question of the nature and degree of equality required in contracts of mutual interest,"[117] as well as over the standards of "inadequacy of price" and "inequality of knowledge." "Where," he asked, "shall we draw the line of fair and unfair, of equal and unequal contracts?"[118]

Verplanck's *Essay* was written as an attack on the doctrine of caveat emptor, which had then only recently been adopted by the United States Supreme Court in *Laidlaw v. Organ*[119] (1817), one of the first cases to come before the Court involving a contract for future delivery of a commodity. The case, Verplanck wrote, raised "the important and difficult question of the nature and degree of equality in compensation, in skill or in knowledge, required between the parties to any contract . . . in order to make it valid in law, or just and right in private conscience."[120] He attacked caveat emptor on the ground that it should be fraudulent to withhold "any fact . . . necessarily and materially affecting the common estimate which fixes the present market value of the thing sold."[121]

In refusing to separate law and morals,[122] Verplanck was boldly independent of other theorists of the market economy. But at its deepest level, Verplanck's *Essay* marks the triumph of a subjective theory of value in a market economy. Wishing to base legal doctrine on "the plainer truths of political economy,"[123] he insisted that although just price doctrines bore "the impression of a high and pure morality,"[124] they were "mixed with error" and arose "from the introduction of a false metaphysic in relation to equality."[125] Thus, he disputed the view of "lawyers and divines . . . that all bargains are made under the idea of giving and receiving equivalents in value."[126] There could be no "such thing in the literal sense of the words, as adequacy of price [or] equality or inequality of compensation," since "from the very nature of the thing, price depends solely upon the agreement of the parties, being created by it alone. Mere inequality of price, or rather what appears so in the judgment of a third person, cannot, without reference to something else, be any objection to the validity of a sale, or of an agreement to sell."[127]

Verplanck's *Essay* represents an important stage in the process of adapting contract law to the realities of a market economy. Verplanck saw that if value is solely determined by the clash of subjective desire, there can be no objective measure of the fairness of a bargain. Since only "facts" are objective, fairness can never be measured in terms of substantive equality. The law can only assure that each party to a bargain is given "full knowledge of all material facts."[128] Significantly, Verplanck defined "material facts" so as not to include "peculiar advantages of skill, shrewdness, and experience, regarding which . . . no one has a right to call upon us to abandon. Here, justice permits us to use our superiority freely."[129]

All know what a wide difference exists among men in these points, and whatever advantage may result from that inequality, is silently conceded in the very fact of making a bargain. It is a superiority on one side—an inferiority on the other, perhaps very great, but they are allowed. This must be so; the business of life could not go on were it otherwise.[130]

Thus, while he refused in theory to separate law and morality, Verplanck confined fraud to a range sufficiently narrow to permit the contract system to reinforce existing social and economic inequalities.

Though Verplanck's reconsideration of the philosophical foundations of contract law was by far the most penetrating among the American treatise writers, Nathan Dane and Joseph Story were more influential in contributing to the overthrow of an equitable conception of contract. In the very first chapter of his nine volume work, Dane elaborated some of the principles of contract law. One of his most important themes involved the "difference between morality and law." He explained that while "in some special cases the *law of the land and morality* are the same," they differ in most cases, "when policy, or arbitrary rules must, also, be regarded." " '*Virtue is alone the subject of morality*,' " he continued, but "law has, . . . often, for its object, the peace of society, and what is practicable: Hence, though every . . . undue advantage in a bargain, to the hurt of another party, practised by one, is an act of injustice in the eyes of *morality*; yet it is not the mean [*sic*] of restitution in the eyes of the *law*; because [it is] often, impracticable *in every minute degree*."[131]

Dane also attacked all conceptions of a substantive theory of exchange. Equity decisions, Dane exclaimed, had become "trash" since they were "the productions of inferior lawyers" and "ignorant and indolent judges" who offered "no rule of property or conduct."[132] "Inadequate price in a bargain," he wrote, "does not defeat it, merely because inadequate." But Dane remained willing to regard an unequal bargain as evidence that a "person did not understand the bargain he made, or was so oppressed, that he thought it best to make it." Indeed, in his characteristic style, he continued to repeat the substance of the old learning while contributing to its overthrow. "When an agreement appears very unequal, and affords any ground to suspect any imposition, unfairness, or undue power or command, the courts will seize any very slight circumstances to avoid enforcing it."[133]

Dane was still reflecting an eighteenth century world view in which unequal bargaining power was conceived of as an illegitimate form of duress and in which lack of understanding was not yet identified only with mental disability. And yet in a world of speculation and futures markets, in which all value must simply turn on "an honest difference of opinion,"[134] legal doctrine eventually renounced all claims to make judgments about oppression. With the publication of Joseph Story's *Equity Jurisprudence* (1836), American law finally yielded up the ancient notion that the substantive value of an exchange could provide an appropriate measure of the justice of a transaction. "Inadequacy of consideration," Story wrote, "is not then, of itself, a distinct principal of relief in Equity. The Common Law knows no such principle. . . . The value of a thing . . . must be in its nature fluctuating, and will depend upon ten thousand different circumstances. . . . If Courts of Equity were to unravel all these transactions, they would throw every thing into confusion, and set afloat the contracts of mankind."[135]

The replacement of the equitable conception of contract with the will theory can be seen in Dane's assault on the eighteenth century practice of suing on a theory of implied contract where there had been an express agreement. In a long and unusually polemical technical discussion, Dane argued that once there is an express contract there could be no quantum meruit recovery off the contract on a theory of natural justice and equity.[136] Dane's attack on quantum meruit becomes comprehensible only as an effort to destroy an equitable conception of exchange in light of a newly emerging theory of value based on the subjective desires of contracting parties. Without a socially imposed standard of value, implied contracts make no sense. Where "there is no fixed or unchangeable comparative value between one piece of property and another" and all value "depends on the wants and opinions of men,"[137] it becomes impossible to measure damages by reference to customary value. The only basis for measuring contractual obligation, then, derives from the "will" of the parties, and the crucial legal issue shifts to whether there has been a "meeting of minds."

The victory of the emerging will theory of contractual obligation was not at first complete. When Theron Metcalf delivered his lectures on contracts in 1828 he still reflected the tension between the old learning and the new.[138] Implied contracts, he wrote, were "inferred from the conduct, situation, or mutual relations of the parties, and enforced by the law on the ground of justice; to compel the performance of a legal and moral duty."[139] In support of this, he cited Chief Justice Marshall's statement that implied contracts "grow out of the acts of the parties. In such cases, the parties are supposed to have made those stipulations which, as honest, fair, and just men, they ought to have made."[140] Though both Metcalf and Marshall were beginning to pretend that contractual obligation derives only from the will of the parties, their predominant form of expression continued to recognize standards of justice external to the parties. Indeed, Metcalf still maintained that "in sound sense, divested of fiction and technicality, the only true ground, on which an action upon what is called an implied contract can be maintained, is that of justice, duty, and legal obligation."[141]

By the time William W. Story's *Treatise on the Law of Contracts* appeared in 1844, however, the tension between the two theories had dissolved. "Every contract," he wrote, "is founded upon the mutual agreement of the parties." Both express and implied contracts were "equally founded upon the actual agreement of the parties, and the only distinction between them is in regard to the mode of proof, and belongs to the law of evidence." For implied contracts, he concluded, "the law only supplies that which, although not stated, must be presumed to have been the agreement intended by the parties."[142] From the perspective of modern legal categories, Story had thus completely obliterated "implied in law" contracts by insisting that all implied contracts were those "implied in fact." Since the only basis for the contractual obligation was the will of the parties, Story now maintained, implied promises "only supply omissions, and do not alter express stipulations"; he was thus prepared to announce the "general rule" that there could be no implied contract where an express agreement already existed.[143]

With Story's announcement of the "general rule," the victory of the will theory of contractual obligation was complete. The entire conceptual apparatus of modern contract doctrine — rules dealing with offer and acceptance, the evidentiary function of consideration, and canons of construction and interpretation — arose to articulate the will theory with which American doctrinal writers expressed the ideology of a market economy in the early nineteenth century.

The application of the will theory of contract to labor contracts

Thus far, we have seen the changes in contract law which were necessary to meet the needs of the newly emerging market economies in
England and America. There is evidence, however, that the change
from the eighteenth to the nineteenth century also involved a pervasive shift in the sympathies of the courts. In the eighteenth century
the subjection of individual bargains to the extensive supervisory
powers of courts and juries expressed the legal and ethical culture of
the small town, of the farmer, and of the small trader. In the nineteenth century, the will theory of contract was part of a more general process whereby courts came to reflect commercial interests.
The changing alliances are painfully obvious in nineteenth century
courts' discriminatory application of the recently discovered chasm
between express and implied contracts.

The most important class of cases to which this distinction applied was labor contracts in which the employee had agreed to work
for a period of time — often a year — for wages that he would receive
at the end of his term. If he left his employment before the end of
the term, jurists reasoned, the employee could receive nothing for
the labor he had already expended. The contract, they maintained,
was an "entire" one, and therefore it could not be conceived of as a
series of smaller agreements. Since the breach of any part was therefore a breach of the whole, there was no basis for allowing the employee to recover "on the contract." Finally, citing the new orthodoxy proclaimed by the treatise writers, judges were led to pronounce the inevitable result: where there was an express agreement
between the parties, it would be an act of usurpation to "rewrite"
the contract and allow the employee to recover in quantum meruit
for the "reasonable" value of his labor.[144]

Courts in fact seemed driven to resolve all ambiguity in contracts
in favor of the employer's contention that they were "entire." It
made no "difference . . . whether the wages are estimated at a gross
sum, or are to be calculated according to a certain rate per week or
month, or are payable at certain stipulated times, provided the
servant agree for a definite and whole term." Under these circumstances, it should be emphasized, the assumption that the agreement was "for a definite and whole term" was simply a judicial construction not required by the terms of the agreement. Moreover, it
did not "make any difference, that the plaintiff ceased laboring for
his employer, under the belief that, according to the legal method
of computing time, under similar contracts, he had continued la-

boring as long as could be required of him." Nor did it matter that the "employer, during the term, has from time to time made payments to the plaintiff for his labor."[145] The result of the cases was that any employee not shrewd or independent enough to demand immediate payment for his work risked losing everything if he should leave before the end of the contract period. The employer, in turn, had every inducement to create conditions near the end of the term that would encourage the laborer to quit.

The disposition of courts ruthlessly to follow conceptualism in the labor cases was not, however, quite matched in cases involving building contracts. Building contracts are similar to labor agreements in that there is no way of restoring the status quo after partial performance. Nevertheless, nineteenth century courts allowed builders to recover "off the contract" when they had committed some breach of their express obligation. The leading case is *Hayward v. Leonard*[146] (1828), in which the Supreme Judicial Court of Massachusetts held that a builder could recover in quantum meruit "where the contract is performed, but, without intention, some of the particulars of the contract are deviated from." If there was "an honest intention to go by the contract, and a substantive execution of it," the court held, it would not decree a forfeiture. It should be noted that the Massachusetts court in *Hayward v. Leonard* expressly rejected Dane's view that the existence of an express contract barred recovery in quantum meruit. There was, Chief Justice Parker declared, "a great array of authorities on both sides, from which it appears very clearly that different judges and different courts have held different doctrines and sometimes the same court at different times."[147] The result was that in Massachusetts and in most other states two separate lines of cases were developed, one dealing with service contracts, for which recovery in quantum meruit was barred, and another applying to building contracts, for which recovery "off the contract" of the reasonable value of the performance was permitted.

Few courts attempted to rationalize what Theophilus Parsons was later to call these "very conflicting" decisions.[148] The leading explanation came from *Hayward v. Leonard* itself. In the labor cases the employee usually broke his contract "voluntarily" and "without fault" of his employer. Breach of building contracts was often "without intention" and compatible with an "honest intention" to fulfill the contract.[149] Thus, it was not that courts had abandoned an underlying moral conception of contracts, but that the morality had fundamentally changed. The focus had shifted from an emphasis on the role of quantum meruit in preventing "unjust enrichment." The express contract had become paramount; denial of quantum meruit recovery was now employed to enforce the contract system. It was now regarded as just for the employer to retain the unpaid benefits of his employee's labor as a deterrent to voluntary breach of contract. But it was still unjust for the beneficiary of a building contract to enrich himself because of an honest mistake in performing the contract.[150]

While the judges who adhered to the distinction between labor and building contracts never acknowledged an economic or social policy behind the distinction, it seems to be an important example of class bias. A penal conception of contractual obligation could have deterred economic growth by limiting investment in high risk enterprise. Just as the building trade was beginning to require major capital investment during the second quarter of the nineteenth century, courts were prepared to bestow upon it that special solicitude which American courts have reserved for infant industry. Penal provisions in labor contracts, by contrast, have only redistributional consequences, since they can hardly be expected to deter the laboring classes from selling their services in a subsistence economy.

Although nineteenth century courts and doctrinal writers did not succeed in entirely destroying the ancient connection between contracts and natural justice, they were able to elaborate a system that allowed judges to pick and choose among those groups in the population that would be its beneficiaries. And, above all, they succeeded in creating a great intellectual divide between a system of formal rules—which they managed to identify exclusively with the "rule of law"—and those ancient precepts of morality and equity, which they were able to render suspect as subversive of "the rule of law" itself.

CUSTOM AND CONTRACT

The growth of commerce and industry confronted American courts in the nineteenth century with the fundamentally new question of how to account for the legal significance of recently established commercial customs and trade practices too varied and diverse to be incorporated into general rules of law. The medieval legal learning that still dominated English and American law books early in the century was hardly adequate to explain why commerical practices should bear the obligatory character of legal rules. Nathan Dane's *Abridgement of American Law* (1824), for example, simply repeated the old learning and thereby underlined the inapplicability of traditional categories for the newly emerging questions involving commercial custom. In his chapter on "Assumpsit, Customs and Prescriptions," Dane reiterated the orthodox view that "a custom must have been time out of mind; for if any one can shew when it began, it is not a good custom." And although this chapter purported to deal with the writ of assumpsit or contract, its entire conceptual framework emphasized the medieval association of custom with prescription, and these in turn were related primarily to categories of property law. "The foundation of a custom," Dane wrote "is *consent.*" The best evidence of consent, he wrote, were "actions repeated or continued by the same rule" or, in other words, ancient usage. Thus, he concluded "Custom and prescription are all one."[151]

Yet, by this time it had already become a familiar part of commercial litigation in America to decide marine insurance and other commercial cases in terms of "the custom of merchants" or "the usage of a trade." And Dane was hardly unaware of the fact that the binding authority of commercial custom could not be justified in terms of the traditional category of "immemorial usage." Indeed, though he wrote that "the true test of a commercial usage is its having existed a sufficient time to have become generally known, and to warrant a presumption that the contracts are made in reference to it,"[152] one looks in vain for any general observations on the nature and limits of the binding authority of commercial custom.

The problem for Dane, as it was for his successors, was how to relate law and custom. So long as common law and immemorial usage were treated as synonymous, the problem would rarely arise. Indeed, medieval law conceived of custom as law precisely because it was inherently binding and obligatory. "It ought to be compulsory," Dane wrote, "and not left to each one's option to use or obey it or not."[153] Despite his acknowledgment of recent commercial customs, therefore, Dane was still operating largely out of an earlier conception of custom as settled law which the parties were not free to alter by their own practices or agreement.

During the second half of the eighteenth century, however, under the influence of Lord Mansfield, English courts had sharply reversed their traditional antipathy to commercial interests and turned to mercantile custom as a source of law.[154] Since the authority of commercial custom could not easily be absorbed into the traditional category of ancient usage, Mansfield instead emphasized its universal character and its correspondence with the dictates of natural reason. It was, Mansfield frequently reiterated, not the law of any one country but of the entire civilized world. Proof of particular commercial customs, therefore, served only as evidence of universal mercantile custom that already had ripened into a rule of law. "The point of law is here settled," Mansfield declared in a commercial case involving tender of proof of usage "and when once solemnly settled, no particular usage shall be admitted to weigh against it: this would send everything to sea again."[155]

Under Mansfield's influence American judges at the turn of the century still also conceived of law and commercial usage as in a state of conflict and tension. While early nineteenth century judges were eager to absorb established English commercial rules into the general law of nations, they were often reluctant to delegate the law-making function directly to commercial interests by declaring that particular mercantile customs should be elevated to the dignity of general law. "Though usage is often resorted to for explanation of commercial interests," Kent wrote, echoing Mansfield, "it never is, nor ought to be, received to contradict a settled rule of commercial law."[156] Since "the law upon this subject is settled," Supreme Court Justice Washington declared in an 1807 case involving a bill of exchange, "it would therefore, be improper to let a contrary usage be proved, which is only proper in doubtful cases."[157] In another case, he also drew a sharp distinction between law and mercantile use. "You may examine witnesses to prove a particular course of trade, or other matters in the nature of facts; but not to show what the law is. Nothing could be more dangerous, than to fix the law upon the opinions of particular men."[158]

Unlike Dane, who viewed custom as immemorial usage, Justice Washington and his contemporaries were attempting to apply Mansfield's notion that mercantile usage was evidence, in doubtful cases, of what the universal law of merchants was. Once this law was known and thereby settled, courts would no longer entertain further evidence of commercial custom in succeeding cases. Mansfield's conception, however, depended for its practical implementation on a relatively small commercial community with fundamentally homogeneous interests. In the field of marine insurance, for example, where eighteenth century English merchants served on different days as both insurer and insured, there was a common interest in settling legal rules and little resulting fear that the law would discriminate in favor of particular groups. It was thus fashionable for Mansfield to declare that it was more important that a rule be settled than that it be settled correctly. In nineteenth century America, by contrast, the mercantile interest gradually becomes less homogeneous and deals more and more often with noncommercial classes. The establishment of incorporated insurance companies during the last decade of the eighteenth century, for example, for the first time created permanently antagonistic interests between American insurers and merchants.

The reaction of the dominant commercial interests was to insist with ever greater enthusiasm that mercantile custom should either be completely absorbed into the general law or else be recognized as a separate and independent body of law. In the first published American book on commercial law in 1802, George Caines insisted that "in what appertains to trade, let it be constantly remembered, that custom alone is a law."[159] In 1810, in a treatise on bills of exchange, Zephaniah Swift carried this argument still further to maintain that mercantile transactions were "governed by the customs and usages of nations, and not by municipal law."[160] The year before, as Chief Justice of Connecticut, Swift had held that the custom of merchants could be admitted to change the general common law.[161]

Nevertheless, most courts at the turn of the nineteenth century were still unwilling to yield to a conception of mercantile law that allowed commercial interests to define entirely the scope and substance of the rules by which they would be governed. For example, at a 1784 trial in Massachusetts "evidence of eminent merchants in Boston" was offered to the jury of "a *mercantile usage there*" which involved the rules governing nonpayment of bills of exchange. "But the Court directed the jury" that "a clear rule of law" already existed and that it should therefore ignore evidence of mercantile usage.[162] Yet in an 1813 insurance case, the Massachusetts Supreme Court, still citing Mansfield's view that evidence of custom would not be admitted where "the law . . . is plain, well settled, and generally understood," had begun nevertheless to reinterpret Mansfield in light of the new economic relationships that were beginning to emerge. "The usage of no class of citizens," the court continued to declare, "can be sustained in opposition to principles of law." Yet, anticipating the newly emerging "will" theory of contract, the judges acknowledged that "evidence of custom and usage is useful in many cases to explain the intent of parties to a contract." While the court was thus unwilling simply to convert commercial custom into law, it was prepared to bring about a similar result through an interpretation of the parties' intent.[163]

The shift to an internal trading market after 1815 brought commercial interests more frequently into legal relationship with farmers, planters, and consumers. As the reality of a small and insular trading community thus began to be replaced by a large and powerful, though occasionally internally divided, mercantile class, Mansfield's conception of a universal custom of merchants began also to wane. It was becoming clearer that the effort to turn to mercantile interests for evidence of commercial custom often simply meant that "the usage of [one] class of citizens" would be sustained "in opposition to principles of law."

One result was that by the end of the first quarter of the nineteenth century a different theory of the importance of commercial custom had begun to emerge which regarded mercantile usage not as an evidentiary source of established law that would be determined and applied by courts but as a question of fact whose function would simply be to illuminate the intention of the contracting parties. If usage could not be sustained "in opposition to principles of law" it could nevertheless serve "in many cases to explain the intent of parties to a contract."

By the time Theron Metcalf delivered his lectures on the law of contracts in 1828, judges and jurists had already begun to see proof of commercial custom not as evidence of a universal custom of merchants but as part of the established will theory of contracts. The resort to custom, Metcalf explained, was entirely for the purpose of construing the intention of the parties. "The purpose of construction," he wrote, "is to *find* the meaning of the parties; not to *impose* it." Thus, "mercantile contracts are construed according to the sense attached by mercantile usage to the terms employed by the parties. And so of other contracts, not strictly mercantile, if there be a usage which the parties must be supposed to have had in view, when their contracts were made."[164]

By disguising the problem of class legislation this effort to conceive of custom simply as evidence of contractual intent relieved courts of the embarrassing task of bestowing on commercial usage the dignity of law. That such a problem was at the forefront of legal controversy over the proper scope of mercantile custom can be seen in the great case of *Gordon v. Little*,[165] decided by the Pennsylvania Supreme Court in 1822. The case involved the economically crucial question of whether the common law rule of strict liability for common carriers would be applied to boats carrying freight on the Ohio and Mississippi rivers. The plaintiff sued for the value of goods destroyed when the defendant's boat sank. Maintaining that he should be liable only for negligence, the defendant offered evidence "of a general usage or custom in relation to the liability of carriers on the

western waters," which amounted to "an implied contract." The
trial court rejected the evidence, except so far as such usage served
to explain "the common and commercial meaning of the words 'the
unavoidable dangers of the river,' " in the bill of lading.[166] The
question for the Supreme Court was whether the trial court had correctly rejected the defendant's offer of evidence of custom and
usage.

Writing for the majority, Chief Justice Tilghman managed to
straddle between the old conception of a general commercial custom and a still latent view of custom as merely establishing the intention of contracting parties. He first affirmed the trial court's ruling that the proper construction of the words of the contract did not
support the defendant's position. "If the case had rested solely on
the written contract, there would have been much to say in favour of
the decision of the [trial] Court, because, be the common law what
it may, the parties have a right to alter or modify it by special contract, and when they have done so, the question is, what is the construction of the contract?"[167]

Having thus acknowledged that the parties were free to alter legal
rules by contract—an important and new conception of the relationship between legal rules and contractual powers—Tilghman
nevertheless felt bound to abide by the trial court ruling that the
parties had not actually agreed to contract out of common law liability. He was thus thrown back on the central argument that commercial custom itself was the source of general legal rules. And he
concluded that it should have been open to the defendant to prove
that a local custom existed on the Ohio and Mississippi Rivers that
suspended the common law rule and limited carrier liability only to
damages arising from negligence.[168]

Tilghman thus sought to apply Mansfield's conception of the relationship between commercial custom and law to a fragmented economy in which diverse and varied commercial practices had eroded
the social and economic basis for a general custom of merchants. As
Judge John Gibson observed in a powerful dissent, Tilghman was
prepared to bestow on local customary usage "the dignity of a law of
local obligation . . . superseding the common law within the district
where it is supposed to prevail." "What is the custom relied on
here?" Gibson asked.

Not a general one, pervading the State; for such would be part of the common law, and determinable by the Judges. . . . With us, particular customs
have no force. I know not a greater or a more embarrassing evil than a law
of merely local obligation. The rule of the carrying business of the Ohio

ought to be that of the Juniata, the Susquehanna, the Delaware, and their tributary streams. Suppose a different usage to exist in respect to each—is there to be different law in respect to each? . . . It is impossible to get away from the conclusion, that by giving the usage any further effect than that of a convenient subject of reference, to explain a latent ambiguity in the expressions of the parties, where their meaning would be otherwise doubtful, we repeal an established principle of the common law, a matter which I apprehend is not open to us.[169]

Gibson saw another problem in permitting customs of particular trades to vary the general rules of law. "If we go by the common law," he wrote,

we shall have a definite, known rule; which, applied to the facts by the Court, will produce as much certainty of result, as legal proceedings are susceptible of; if we go by usage, the whole matter will have to be determined by the jury, on evidence of the common practice and understanding on the subject; which would be to go by no rule at all. So that the right to compensation will, in every instance, depend on what the jury may think the proper degree of diligence. We should be perpetually inquiring by a jury as to what is the law of the land; and the degree of diligence required of the carrier, would be as fluctuating as the opinions of the witnesses called to establish it.[170]

Without an insulated and homogeneous commercial class from which Mansfield, in his day, could draw merchant jurors, a law allowing proof of local usage, Gibson saw, would give lay jurors power to control commercial rules. But even beyond his fear that anticommercial juries might ultimately come to determine the substantive rules of commercial law, Gibson believed that such a system could not provide "a definite, known rule, which . . . will produce as much certainty of result as legal proceedings are susceptible of." The "embarrassing evil" of "a law of merely local obligation" seemed inevitable if the majority's view of custom were allowed to prevail. As each special interest sought to endow its own particular custom with "the dignity of . . . law," the claims of the legal system to impartiality were drawn into question. Commercial custom thus threatened not only to establish various "law[s] of local obligation" which fragmented the economy; it also enabled powerful interests to impose their usage on those outsiders with whom they dealt.

Gordon v. Little underlines the pitfalls that American judges faced in attempting to follow Mansfield's conception of mercantile custom. In a society in which commercial interests were becoming both increasingly powerful and diverse, any overt attempt to give commercial custom the force of law was both impolitic and impractical. It exposed the judiciary to the charge of class legislation while it endowed lay jurors with the discretionary authority to decide which customs were binding. Indeed, by 1836 David Hoffman was prepared to acknowledge in his *Course of Legal Study* "the numerous legal doubts" that beset every attempt to absorb the law of merchants into the general law.[171] Yet, in *Gordon v. Little* we see that a newer conception has already come to the fore. If there is disagreement about whether custom can change the general law, all agree that contract can. If the established common law rule "be not the most convenient," even Justice Gibson agreed, "the parties have, in every case, power to establish a particular measure of responsibility for themselves."[172] The result was that by the time Theron Metcalf delivered his lectures on contract in 1828 judges had begun to move away from the late eighteenth century notion that courts could convert mercantile custom into law. Their only function, Metcalf maintained, was simply to determine "if there be a usage which the parties must be supposed to have had in view, when their contracts were made."[173] While both sides in *Gordon v. Little* still conceived of usage as performing the very limited function of "explain[ing] words of doubtful import," by 1844 W. W. Story could declare that in the interpretation of a contract, usage or custom should be resorted to, not only to explain the meaning of technical or ambiguous terms "but also to supply evidence of the intentions of the party in respect to matters, with regard to which the contract itself affords a doubtful indication, *or perhaps no indication at all.*"[174]

Judges, indeed, were not unaware that a contractarian justification for the recognition of mercantile custom appeared to relieve them of considerable "embarrassment." The "difficulty," the Massachusetts Supreme Court observed in 1847,[175] was that commercial "usages differ essentially from those more general customs, which are known and exist as part of the general law of the land." They "may be of comparatively recent origin, and may be limited to a single city or village." As a result, "Learned jurists have often expressed their regret at the extension of this species of evidence . . . as impairing, in some degree, the symmetry of the law, and tending to uncertainty and embarrassment in the administration of justice." The court, in addition, recognized that deference to mercantile custom was "also liable to the serious objection, that the knowledge, by the party to be affected by it, of the existence of such usage, is a mere legal presumption, which may often be unfounded in reality." These objections, however, were thought to be overcome where the authority of mercantile custom was said to derive from the will of the contracting parties. "Usages of this character," the Massachusetts court concluded, "are only admissible upon the hypothesis that the parties have contracted in reference to them."

Whatever their misgivings about mercantile usage as an appropriate source of law, most judges in the first half of the nineteenth century were eager to yield their objections before the new contractarian rationale. If the contracting parties were free to change or modify legal rules, as the court in *Gordon v. Little* acknowledged, they could surely be permitted to incorporate commercial custom into their agreement. Thus, under the guise of enforcing the party's intentions, courts by the middle of the century had begun, in fact, to follow W. W. Story's view that they were free to turn to custom "to supply evidence of . . . intentions . . . with respect to which the contract itself affords . . . no indication at all."

This unacknowledged shift back to a theory of preexisting custom under the guise of construing the parties' intentions would shortly shade into an explicit assertion of an "objective" theory of contract itself. If a "subjective" theory of contract served its historic function of destroying all remnants of an objective theory of value, it had the drastic limitation of making legal certainty and predictability impossible. Once contractual obligation was founded entirely on an arbitrary "meeting of minds," it endowed the parties with the power totally to remake law. Though it once had the important advantage of disguising judicial lawmaking, the contractarian justification for custom created an even more individualized and random "law of local obligation" deriving only from the parties' will. To the extent that it was seriously followed, it made every contract a unique event depending only on the momentary intention of the parties. Once the subjective theory of contract had performed the function of enabling judges and jurists to destroy the connection between contract law and a conception of objective value, they felt free once again to revive an objective theory of contract and to reintroduce its intellectual companion, a conception of general mercantile custom.

It was Joseph Story who first saw the disintegrating effects of a subjective theory of contract on commercial law. "I own myself no friend to the almost indiscriminate habit, of late years," he wrote in a case decided in 1837, only five years before his decision in *Swift v. Tyson*, "of setting up particular usages or customs in almost all kinds of business and trade, to control, vary, or annul the general liabilities of parties under the common law as well as under the commercial law."

It has long appeared to me, that there is no small danger in admitting such loose and inconclusive usages and customs, often unknown to particular parties, and always liable to great misunderstandings and misinterpretations and abuses, to outweigh the well-known and well-settled principles of law. And I rejoice to find, that, of late years, the courts of law, both in England and America, have been disposed to narrow the limits of the operation of such usages and customs, and to discountenance any further extension of them.[176]

Others, like Pennsylvania's Chief Justice Gibson, complained that proof of custom to explicate the parties' intentions "transfer[s] the functions of the judge from the bench to the witness-stand." If custom could be shown "only through evidence of witnesses," he concluded, it would undermine the law-declaring function of judges, who were the only "constitutional expositors" of "the general law."[177]

The connection between a subjective theory of contract and its corrosive effects on predictability in commercial law was first elaborated in Theophilus Parsons' *Law of Contracts* (1855). For the first time, Parsons drew away from the infatuation of earlier treatise writers with the will of the parties, and what is most important, he emphasized that the construction and interpretation of contracts was not a question of fact for the jury but a question of law. "The importance of a just and rational construction of every contract . . . is obvious," he began. "But the importance of having this construction regulated by law, guided always by distinct principles, and in this way made uniform in practice, may not be so obvious, although we think it as certain and as great." A just construction of an individual contract is desirable, he acknowledged. But it was more important that there be rules regulating "the rectitude, consistency and uniformity of all construction," so that parties will be able to take "precautions" in their own behalf. "And hence arises the very first rule; which is, that what a contract means is a question of law. It is the court, therefore, that determines the construction of a contract."[178]

In order for the court and not the jury to have the power of construction, it was necessary to dispense with the subjective theory of intention. Thus, Parsons moved on to articulate the objective theory that would thereafter dominate American law. "The rule of law," he declared, "is not that the court will alway construe a contract to mean that which the parties to it meant; but rather that the court will give to the contract the construction which will bring it as near to the actual meaning of the parties as the words they saw fit to employ, when properly construed, and the rules of law, will permit." "So," he concluded, "the rules of law, as well as the rules of language, may interfere to prevent a construction in accordance with the intent of the parties."[179] Thus, in the interest of "rectitude, consistency, and uniformity," Parsons was prepared to overthrow the orthodox view, propounded earlier by Chancellor Kent in his *Commentaries* (1832), that "the plain intent" of the parties to a contract should prevail even "over the strict letter of the contract."[180]

Parson's "objective" theory of contracts develops at roughly the same time as the social significance of the market was once more undergoing a major transformation. Between 1790 and 1850 the overwhelming emphasis in legal and economic thought was on the random and fluctuating nature of value that had been introduced by a market economy. "Nothing can be more variable than price . . . ," Verplanck had declared in 1825. "The market rate of all goods is constantly changing. . . . There is, then, it is evident, nothing like a permanent value of one commodity, or kind of property, as compared with another."[181] In short, the social meaning that legal writers derived from the early stages of the growth of markets was that value was arbitrary and subjective and that only individual preferences could correctly determine the price of a thing.

There was, of course, another meaning to be derived from the development of markets: that stable market price can provide an "objective" measure of social value. Even Verplanck occasionally saw the possibility of using the market price as a measure of the fairness of a transaction. "If a merchant were to pay me eight dollars a barrel for a cargo of flour which he could have bought off any one else for four . . . no one would hesitate to say that here is an unequal bargain. . . . Thus it is that inadequacy in price may afford strong presumptive evidence . . . of some error or fraud in the substance of the contract."[182] Yet, despite such occasional lapses, the overwhelming message of Verplanck and all other legal writers and judges before mid-century was that price could not provide a social measure of value because it radically fluctuated due to "the speculation, the tastes, wants, or caprices of purchasers."[183]

As markets became more "mature" by becoming more extensive and uniform, due largely to improved transportation, the social experience of price stability tended to reintroduce, though in substantially diluted form, the power of a conception of objective value.[184] During the 1850s, for example, courts no longer defended the rule of caveat emptor in terms of the subjectivity of value but rather in the entirely different terms of whether the contracting parties had an equal opportunity to learn the objective market price of a commodity.[185] As a result of this shift, Theophilus Parsons became the first treatise writer to complain that the rule of caveat emptor was "severe, and . . . sometimes works wrong and hardship; and it is not surprising that it has been commented upon in terms of strong reproach, not only by the community, but by members of the legal profession." In fact, the notion of objective value began to resurface in Parsons' work. While he at first seemed inclined to overthrow caveat emptor in all cases "where an article is sold of which the value is materially affected by some defect which the buyer cannot know or discover . . . ," his orthodox disposition ultimately led him to conclude that the existing rule "may perhaps be regarded as upon the whole well adapted to protect right [and] to prevent wrong."[186]

Nevertheless, the impact of developing national markets seems strongly to have affected the character of Parsons' work. As the Civil War approached, "the organization of both foreign and domestic trade had reached a degree of specialization and a country-wide integration typical of modern national economies. . . . Cotton was sold by sample in New York and New Orleans, and tremendous quantities of grain changed hands in Chicago and Buffalo merely on the basis of recognized grades."[187] The need for a uniform and consistent set of essentially impersonal commercial rules had begun to overwhelm legal doctrines such as caveat emptor, which still presupposed an economy based on face-to-face dealings.

One of the most interesting expressions of this tendency toward depersonalization is also typical of the way in which, under changed circumstances, a minor exception to a common law rule often eventually undermines the rule itself. As early as 1816 the Massachusetts Supreme Court held that a sale by sample warranted that the article sold was of the same kind as the sample, thus constituting a small, but clear, exception to the rule of caveat emptor.[188] In 1831 the Pennsylvania Supreme Court, remarking on "the constant change, which is daily taking place in the course of trade, and commercial dealings," relaxed the rule still further by holding that all articles sold through bills of sale included a warranty.[189]

This limited revival of judicially enforced warranties also reflected the increasing standardization of goods made possible by extensive markets. Courts were more often willing to enforce warranties of quality in a bill of sale where there was a clear prevailing commercial standard of quality, although many cases still distinguished between enforceable warranties of kind and warranties of quality, which were regarded as too uncertain and indefinite for judicial enforcement.[190] Courts, in essence, were developing a series of new doctrines designed to apply to large impersonal business dealings between commercially sophisticated insiders which, we have seen, were rapidly replacing the face-to-face transaction as the dominant mode of trade.[191]

By the time Samuel Williston published his treatise on *Contracts* in 1920, American jurists, still under the influence of Parson's treatise, were bent on establishing the historical validity of the objective theory of contractual obligation. "In the formation of contracts," Williston wrote, "it was long ago settled that secret intent was immaterial, only overt acts being considered in the determination of such mutual assent as that branch of the law requires." Though driven to establish the historical *bona fides* of the objective theory, Williston was nevertheless compelled to acknowledge that "during the first half of the nineteenth century there were many expressions which seem to indicate the contrary, chief of which was the familiar rubric, still reechoing in judicial dicta that a contract requires the meeting of the minds of the parties."[192]

Williston's followers spared no effort in their attempt to discredit the earlier subjective theory. "The historical fallacy which has misled some American courts," one of them wrote, "was borrowing from . . . civil law doctrine . . . which was alien to Anglo-American theories of contract. . . . Of the channels through which this infusion of alien doctrine took place, the writings of Story and Kent had the widest influence."[193]

But what in fact accounted for the initial triumph of the "will" or "subjective" theory? Far from representing any "alien" infusion of legal ideas, it arose from the basic structure of legal change in the early nineteenth century. The earlier contract doctrine, which posited objective standards of value, was hostile to a market economy. The leading intellectual weapon employed to destroy this theory depended on the view that all value was subjective and that the only basis of legal obligation was an arbitrary convergence of individual wills or "meeting of the minds." Rules for construction of contracts, Chancellor Kent had thus maintained, existed solely to ascertain "the mutual intention of the parties. . . . To reach and carry that intention into effect," Kent argued, "the law, when it becomes necessary, will control even the literal terms of the contract, if they manifestly contravene the purpose."[194]

Radical subjectivism thus became the banner under which modern contract law triumphed during the first half of the nineteenth century. Yet, it soon became apparent that the new ideology had succeeded only too well. If a subjective theory of contract served its historical function of destroying all remnants of an objective theory of value, it had the drastic limitation of making legal certainty and predictability impossible. Once contractual obligation was founded entirely on an arbitrary "meeting of minds," it endowed the parties with a complete power to remake law. To the extent that it was seriously followed it made every contract a unique event depending only on the momentary intention of the parties. National markets, however, required uniformity and standardization, which inevitably entailed a sacrifice, at least in theory, of the individual's power to contract.

The emergence of the objective theory, then, is another measure of the influence of commercial interests in the shaping of American law. No longer finding it necessary to enter into battle against eighteenth century just price doctrines, they could devote their energies to establishing in the second half of the nineteenth century a system of objective rules necessary to assure legal certainty and predictability. And having destroyed most substantive grounds for evaluating the justice of exchange, they could elaborate a legal ideology of formalism, of which Williston was a leading exemplar, that could not only disguise gross disparities of bargaining power under a facade of neutral and formal rules of contract law but could also enforce commercial customs under the comforting technical rubric of "contract interpretation."

TORT AND CONTRACT

As contract became permanently detached from its eighteenth century roots in natural justice and customary ideas of fair exchange, the legal system was required for the first time to come to terms with a widening gulf between contractual and noncontractual legal duties. During the eighteenth century duties imposed by the common law were still generally regarded as both logically and normatively prior to obligations undertaken through private agreement. Indeed, until the ideological triumph of the will theory of contracts after 1825, jurists did not yet perceive any fundamental conflict between contractual and customary duties. Even cases involving commercial custom and usage were usually not yet conceived of as involving any fundamental collision between preexisting legal duties and commercial usage but rather as turning on the quite different question of the extent to which new legal rules should be derived from established mercantile custom. But as it became clear during the first quarter of the nineteenth century that in certain instances courts were prepared to allow private agreement to suspend obligations imposed by law, lawyers faced the question of determining the extent to which legal duties should be enforced despite a private determination to modify them by contract.

The problem first arose in connection with insurance contracts. To what extent could individuals employ the contract of insurance in order to protect themselves from losses resulting from their own negligence? Until around 1830, the view generally prevailed that there could be no insurance for injury caused by negligence. "An agreement by one party to indemnify another against losses voluntarily incurred," Phillips wrote in his *Treatise on the Law of Insurance* (1823),

seems to be so obviously opposed to the general interest of a community, that it could hardly be enforced by any legal tribunal. And there is the same objection, in a smaller degree, against sustaining a contract to indemnify a man against the consequences of his own negligence. By such an agreement one man would consent to put himself wholly in the power of another, and it could operate only to the injury of the parties, and of the community of which they were members.[195]

Likewise, in the first edition of his *Commentaries* (1828), Chancellor Kent stated as "the better opinion" the principle that "the insurer is not liable for . . . damage [which] may be prevented by due care, and . . . is within the control of human prudence and sagacity."[196] He cited the leading New York case of *Grim v. Phoenix Insurance Co.*[197] (1816), which held that marine loss by fire arising from negligence could not be covered by insurance.[198]

After 1830, however, judicial opinion abruptly shifted to allow individuals to insure against losses arising from their own negligence. Reflecting this change, Chancellor Kent added a new chapter on fire insurance to the second edition of his *Commentaries* (1832). Though it had been questioned, Kent now wrote, "whether the negligence and frauds which the insurance of property from fire has led to, did not counterbalance all the advantages . . . the public judgment in England, and in this country, has long since decided that question with perfect satisfaction."[199] In a series of cases during the 1830s, the United States Supreme Court for the first time upheld recovery of fire losses brought about through the negligence of the insured.[200] Yet, it was quite some time before legal writers could fully rationalize this new doctrine. In his treatise on the law of insurance in 1854 Joseph Angell reiterated the "general rule" that "the law will not enable a party to recover compensation for an injury, of which his own negligence and want of due caution, or the misconduct of his agents, have been the primary cause." The contract of insurance, he simply concluded, "forms an exception."[201]

Nevertheless, under the growing influence of the will theory of contracts, judges and legal writers after 1830 gradually came to accept the proposition that many legal duties imposed by the state could be modified or abrogated by contract. An 1830 article entitled "Customs and Origin of Customary Law" appearing in the influential legal journal, *The American Jurist*, clearly articulated the newly emerging intellectual system. "Under a free and natural system of laws," the writer declared, "men have a right to make any agreement they please, however contrary to the usual custom," and where not contrary to the original social contract, "this agreement will be recognised as the law to them."[202] Only "an unnatural and artificial extension" of public institutions could create a "power to overrule the express agreements of individuals . . .," since "whatever men have consented to, that shall bind them, and nothing else." All custom originated in consent, and therefore legal rules established by custom "have no authority except as rules of circumstantial evidence of the intention of the parties, with respect to those things concerning which they made no express agreement."[203]

For instance, a common carrier or an innkeeper is liable for all accidents which happen to property in his custody, except those occasioned by the acts of God or the king's enemies. This is a rule drawn from the policy of the transaction; and therefore, as it only has this presumptive force of law, as being generally the most expedient agreement between the parties, it is not the law in respect to those cases in which it is proved that the contrary was expressly agreed to.[204]

Thus, he continued, "suppose . . . that . . . the policy . . . with respect to innkeepers and common carriers, should become wholly useless; and suppose that constant repetitions of a stipulation to the contrary, among the parties interested, should present to the magistrate unequivocal evidence of public sentiment, that such was the fact; would not the magistrate be bound, in a case where such stipulation was not made, to suppose it was intended, and to presume that the custom was so notoriously the other way, that it sufficiently proved that the parties must have known or supposed so?"[205]

Thus, contract, reflecting the "self-legislative instinct of society," could legitimately abrogate all preexisting common law duties. "Considered as emanating from . . . courts, it is a system of paternal, regulating, construing, preservative legislation, which does not aim to restrain or control, but only systematizes, harmonizes, and combines the enterprises and inventions which men voluntarily undertake." For one of the "chief excellencies" of this "system of political economy" is that it "leaves men to make their own arrangements, and contents itself with regulating them."[206]

Under the influence of the contractarian theory and its laissez-faire underpinnings, common law judges after 1830 began to allow private agreements to suspend long-standing customary duties. The foremost beneficiaries of the new legal doctrine were common carriers — mainly railroads and steamboat shippers — who sought to escape from the centuries old rule that carriers were strictly liable for the safe transit of goods, unless they could show that injury was caused by "the act of God or the public enemies."

There were, before 1830, intimations by courts that common carriers were free to contract out of legal liability.[207] There were also occasional efforts, finally unsuccessful, to distinguish water from land carriage and to exempt the former from the strict obligations imposed upon common carriers.[208] After 1830, however, American courts fell into line in allowing carriers to limit their strict liability by contract and, in many cases, even to immunize themselves from liability for negligence.

Under the influence of contractarian ideas, English courts had begun at the turn of the century to allow common carriers to limit liability by simple notice.[209] These English decisions, however, assumed the status of "settled" law in America only with the publication of Kent's *Commentaries* (1827) and Story's treatise on *The Law of Bailments* (1832). It was "strenuously urged" in some of the English decisions, Kent noted, that there was "no sound distinction" by which courts could allow carriers to contract out of strict liability and nevertheless refuse to allow them to limit liability arising from "negligence and misfeasance." The only "well settled" limitation on contracting out, he concluded, was that there could be no agreement to limit liability due to "gross negligence."[210]

Five years later, Joseph Story proclaimed the end of all debate over the question whether carriers were free to alter their liability by contract. While acknowledging that "it was formerly a question of much doubt," he nevertheless was prepared, solely on the authority of English decisions, to declare the rule "now fully recognised, and settled beyond any reasonable doubt." The only bar to altering common law liabilities, Story concluded, was that carriers could not "exempt themselves from responsibility in cases of gross negligence and fraud."[211]

American courts immediately began to fall into line. "It was formerly a question of much doubt," the Pennsylvania Supreme Court declared in 1835, echoing Story, "how far common carriers, on land, could by contract limit their responsibility. But that they have the power seems now to be settled, although, many learned judges have expressed some regret that the validity of *notices* restricting their liability, was ever recognised." While even the binding effect of notices "must now be admitted," carriers still could not by contract "free themselves from *all* responsibility, particularly where there is gross negligence or fraud."[212] Despite the tone of resignation, resistance to the contractarian ideology continued in Pennsylvania. "It is not too late," Chief Justice Gibson argued in 1839, "to say that the policy which dictated the rule of the common law, requires that exceptions to it be strictly interpreted."[213] And as late as 1848 the Pennsylvania court reiterated the "frequently expressed regret by many of our judges, that a common carrier was ever permitted to limit . . . responsibility." In fact, "were the question an open one in Pennsylvania," Judge Thomas Bell declared, "I should, for one, unhesitatingly . . . repudiat[e] a principle which places the bailor absolutely at the mercy of the carrier, whom, in a vast majority of instances, he cannot but choose to employ."[214] Yet, despite continuing expressions of "great reluctance" by Pennsylvania judges to allow common carriers to contract out of common law liability, after 1851 it was settled that carriers could limit liability for all but negligently caused injuries.[215]

In South Carolina and New York, the judges went still further. By 1838 the South Carolina Supreme Court "remind[ed] the owners of steamboats that they have but to give public notice that they will not be liable. . . . They may and will relieve themselves, whenever essential to their interests, by special acceptances."[216] And in 1846 the court disputed the view "that a carrier cannot, by notice, and of course by agreement, divest himself of his liability for negligence or want of care." "I am wholly unable to see any reason," Judge Evans concluded, "why, on this, as on most other subjects, men may not be left to take care of their own interests."[217]

Despite occasional intimations to the contrary,[218] the New York courts during the 1830s and 1840s held firm in refusing to allow carriers to contract out of liability.[219] "There are no principles in the law better settled," Justice Cowen declared in 1838,[220] "than that whatever has an obvious tendency to encourage guilty negligence, fraud or crime, is contrary to public policy. Such, in the very nature of things, is the consequence of allowing the common carrier to throw off or in any way restrict his legal liability."

The traveller and bailor is under a sort of moral duress. . . . My conclusion is, that [the carrier] shall not be allowed in any form to higgle with his customer and extort one exception and another, not even by express promise or special acceptance any more than by notice.

The result was that in the leading case of *Gould v. Hill*[221] (1842) the New York Supreme Court held that common carriers could not modify their common law duties by contract.

But under the influence of a ruling by the United States Supreme Court in 1848 that allowed carriers to limit liability by contract,[222] judicial resistance in New York began to crumble. Beginning in 1850 New York judges overruled *Gould v. Hill* and allowed carriers to restrict their liability.[223] One court felt "compel[led]" to overrule *Gould v. Hill* because of "the great importance of the question to a commercial people, especially the importance of uniformity between the courts of the State and Union, in the rules of law regulating commercial transactions."[224] To refuse to allow carriers to contract out, another declared, "would . . . be an unwarrantable restriction upon trade and commerce, and a most palpable invasion of personal right."[225]

The United States Supreme Court had, in fact, implied that it might allow a carrier to immunize itself from liability even for gross negligence.[226] Following this cue, the New York court in 1853 held that by a special contractual provision a carrier could bar liability even for gross negligence.[227] Another court was prepared to allow a railroad to contract out of liability for personal injuries to passengers "resulting from carelessness or mere *non-feasance*."[228] Finally, by 1859 a New York judge, summing up the experience of a decade, was ready to allow a carrier to limit liability arising "from any degree of negligence," if the parties "with a full consciousness of [their] rights" had consented.[229] It was only in 1873, under the influence of the Granger Movement, that the United States Supreme Court began to stem the contractarian tide by finally holding that a common carrier could not contract out of liability for negligence.[230]

The treatise writers of the 1840s and 1850s were generally hostile to the efforts of carriers to restrict their liability. Yet it should be emphasized that their inability to break out of the dominant contractarian ideology made their efforts largely futile or ineffective. In 1846 Simon Greenleaf sought to bestow his approbation only on those contracts which would "qualify" the carrier's liability by placing a ceiling on the extent of responsibility, while he disapproved of those which "limit, restrict or avoid the liability."[231] Joseph Angell and Theophilus Parsons devoted their energies to attempting to limit the binding effect of carrier's notices while conceding, as Angell put it, "the power of making an *express* contract."[232] And following Parsons and Angell, the Chief Justice of Vermont, Isaac Redfield, focused entirely in his treatise on railroads on the question of whether notices, as opposed to express contracts, could limit liability. He concluded that while carriers might limit "extraordinary" or strict liability by notice they could not exonerate themselves from "ordinary" liability for negligence.[233] The English decisions, he acknowledged, had gone further and allowed carriers to use notices to claim "exemption from liability, even for gross neglect, and wilful misconduct." As a consequence, "the entire business population of the realm almost was at the mercy of these . . . carriers." This result, he concluded, illustrated a "tendency of judicial administration . . . to be seduced into the belief, that it is safe to follow any theory or abstraction, however specious" even after the results no longer "commend themselves to our sense of justice."[234]

But in focusing entirely on the always more congenial technical question of notice, the treatise writers themselves had also been "seduced" by contractarian "theory" and "abstraction." The unilateral notice issued by a carrier, it is true, could be subjected to the orthodox criticism that there was no real "meeting of the minds" sufficient to make a binding contract. Yet, it is no small indication of the apologetic and rationalizing character of the treatise tradition that none of these writers was willing to reach back only one generation to a time when orthodox doctrine still recoiled at the idea that any common law duties could be suspended by contract. As a result, they no longer found any basis for arguing against limitations on liability brought about, as Angell put it, "by *positive* agreement in each individual."[235] Nor did they ever seek to explain why notices that suspended "extraordinary" liabilities could be absorbed into contract theory, while those that annulled "ordinary" liabilities could not be. Their theoretical commitments as well as their "sense of justice," in short, seemed to have been entirely submerged by the practical necessity of finding a limited formula that would not place "the entire business population . . . at the mercy of these . . . carriers."

The process of redefining all economic relationships in terms of contract received its classic formulation in the first half of the nineteenth century not in the common carrier cases, which were theoretically confused, but in those involving the liability of employers for the negligence of their employees.

During the colonial era, the apprenticeship system was the primary model for relations between laborers and their masters. In the early period "the master's role, and responsibilities, was indistinguishable from the father's, and the servant's obligations were as total, as moral, and as personal as the son's."[236] In the course of the eighteenth century,

officially, legally, the assumption continued that the master stood *in loco parentis*, that his duties included all those of an upright father, and that the obligations of apprentices remained, as sanctified in law and tradition, filial in scope and character. But both sets of obligations were increasingly neglected. . . . Moral indoctrination, Christian training, and instruction in literacy seemed encumbrances upon a contractual arrangement of limited purpose.[237]

In his treatise on family law, published in 1816, Tapping Reeve nevertheless still continued to propound the classical view that the master's relation to the apprentice stood "*in loco parentis.*"[238] And while courts throughout the first half of the nineteenth century continued to invoke the filial theory, in practice most of the moral content of the earlier system was replaced by a purely monetary relationship that grew out of the factory system. Reflecting this change, Chief Justice Gibson in 1845 held that a Pennsylvania statute of 1770 did not require the apprentice to live in the master's house. Though he noted that "in the country" the apprentice was "still a part of the family," this was no longer so in the city. Invoking a "duty to interpret statutes so as to fit them . . . to the business and habits of the times," he felt no longer able to ignore an economic situation in which apprentices were usually paid wages in lieu of board and lodging.[239]

This gradual decay of a paternalistic and hierarchical relationship among employers and workers explains in part the absence of American cases before the 1840s in which employees sue their employers for injuries resulting from negligence. There were surely many instances of worker injuries before 1840, but many were probably compensated out of benevolence or charity, depending on the extent of personal relationship between master and servant. The introduction of railroads after 1830, however, not only magnified the risk of serious employment injuries; it also seems to have established the first really impersonal system of employment in America. Only after this breakdown of paternalistic relationships between workers and employers in the developing corporate economic structure do we find laborers turning to courts to seek compensation for injuries arising from their employment.

It is worthy of note that in both England and America the question of the legal liability of employers for the negligent injuries of workers is first posed at virtually the same moment. In 1837 an English employee for the first time sues his employer for on-the-job personal injuries.[240] In America the issue is first presented in a South Carolina railroad case in 1841.[241] And in the landmark case of *Farwell v. Boston and Worcester R. R.*[242] (1842), Chief Justice Shaw of Massachusetts settled the issue for the rest of the century by elevating the paradigm of contract to its supreme place in nineteenth century legal thought.

In all these cases, the judges were torn between competing conceptions of legal relations. First, there was the ancient customary principle of *respondeat superior* that held a master liable for the tortious acts of his servants. Thus the South Carolina court acknowledged that there could be "no question, that, in general, the principal is liable for acts of the agent, performed in the execution of his agency."[243] Before the nineteenth century, jurists would not have conceived of this principle as deriving from private agreement, but as imposed by a normatively superior customary law. Now, however, the South Carolina court was prepared to hold that the principle of *respondeat superior* was suspended for those in a bargaining relationship, that the liability of the employer "can, in general, be readily ascertained from the object of the contract, and the relative position of the parties."[244] In short, both the liability of a railroad and the legal rights of an injured "fellow servant" were to be determined exclusively by contract.

Is it incident to the contract that the company should guarantee him against the negligence of his coservants? It is admitted he takes upon himself the ordinary risks of his vocation; why not the extraordinary ones? Neither are within his contract.[245]

But why did the contractarian framework necessarily require the court to assume that employer liability for "extraordinary" risks was excluded from the contract? "No prudent man," one judge answered, "would engage in any perilous employment, unless seduced by greater wages than he could earn in a pursuit unattended by any unusual danger."[246]

Thus, the contractarian ideology above all expressed a market conception of legal relations. Wages were the carefully calibrated instrument by which supposedly equal parties would bargain to arrive at the proper "mix" of risk and wages. In such a world the old ideal of legal relations shaped by a normative standard of substantive justice could scarcely coexist. Since the only measure of justice was the parties' own agreement, all preexisting legal duties were inevitably subordinated to the contract relation.

Chief Justice Shaw's opinion in *Farwell,* more than any other nineteenth century case, reveals the triumph of these intellectual premises. And the pervasiveness of those premises could not be more clearly underlined than by the astonishing fact that even the injured employee's counsel conceded at the outset that the case should be decided only on principles of contract. "The plaintiff does not put his case . . . upon the general principle which renders principals liable for the act of their agents," he acknowledged, "but on the ground, that a master, by the nature of his contract with a servant, stipulates for the safety of the servant's employment."[247]

Rarely in the history of American law has so significant a case of "new impression," as Chief Justice Shaw called it, been so thoroughly determined by the intellectual impoverishment of counsel. "It was conceded," Shaw thus began his opinion, that the doctrine of *respondeat superior* could be applied only to the liability of a master to compensate *strangers* for his servant's negligence. In the present case, however, between parties in a bargaining relationship, "the claim, therefore, is placed, and must be maintained at all, on the ground of contract." Having arrived at this point, the result was inevitable. "The general rule, resulting from considerations as well of justice as of policy," Shaw concluded, "is, that he who engages in the employment of another . . . for compensation, takes upon himself the natural and ordinary risks and perils incident to the performance of such services, and in legal presumption, the compensation is adjusted accordingly."[248]

The doctrine of "assumption of risk" in workmen's injury cases expressed the triumph of the contractarian ideology more completely than any other nineteenth century legal creation. It arose in an economy which already had all but eradicated traces of an earlier model of normative relations between master and servants. And without the practice of enforcing preexisting moral duties, judges and jurists could no longer ascribe any purpose to legal obligations that were superior to the expressed "will" of the parties. As contract ideology thus emasculated all prior conceptions of substantive justice, equal bargaining power inevitably became established as the inarticulate major premise of all legal and economic analysis. The circle was completed; the law had come simply to ratify those forms of inequality that the market system produced.

VII The Development of Commercial Law

T HE triumph of a contractarian ideology by the middle of the nineteenth century enabled mercantile and entrepreneurial groups to broadly advance their own interests through a transformed system of private law. Having destroyed or neutralized earlier protective or regulatory doctrines at the same time as they limited the power of juries to mete out rough and discretionary communal standards of justice, a newly established procommercial legal elite was able to align itself with aggressive business interests.

But there remained throughout the first half of the nineteenth century strong elements in American society opposed to the expanding values of a market economy. Reflecting a still dominant precommercial consciousness of rural and religious America, these groups, though consistently on the defensive throughout the century, continued to constitute a powerful political force to be reckoned with. As nineteenth century courts increasingly absorbed procommercial doctrines into American private law, these groups offered intermittent resistance in state legislatures. Indeed, one important strand of the Codification Movement of the 1820s and 1830s reflected the view that American courts were becoming the servants of the wealthy and the powerful.

The history of commercial law in both England and America has often been written as if it had nothing to do with such political and economic struggle. Though "Whig history" is now widely condemned in constitutional historiography, there remains a strong residue of the "Whig" tradition in areas outside constitutional law. In the historiography of commercial law, for example, there is a still generally shared assumption that "modernization" is an unqualified good and that the development of legal doctrines to reflect the triumph of a market system was both inevitable and desirable. Indeed, conceptions like "commercial uniformity" and "legal certainty and predictability" are often uncritically put forth as explanations of the rise of commercial law, without the slightest understanding that each of these disembodied constructs conceals a whole set of political and economic values which were, in fact, resisted from the beginning of the century.

This chapter attempts to show some of the ways in which the American judiciary—especially aggressively procommercial federal judges—managed to overthrow earlier anticommercial legal rules by the time of the Civil War. While they certainly strove for legal "uniformity" as well as legal "certainty and predictability," the substantive rules and doctrines they developed were primarily meant to protect those groups that stood to benefit from an expanding market economy.

The Rise of Negotiability

No development had a more shattering effect on American conceptions of the nature of contract than the necessity of forging a body of commercial law during the last decade of the eighteenth century. At the heart of all commercial problems lay the question of negotiable instruments and of whether the American legal system could assimilate the principle of negotiability into a conception of contract that had been forged in a precommercial society.

Negotiability challenged a whole range of accepted legal notions. First, it ran completely contrary to the ancient common law hostility to assignment. This policy of refusing to allow one man to assign his right to sue another arose, wrote Wooddeson, "to avoid a multiplicity of suits, by preventing those who would not prosecute their right themselves, from transferring it to others of a more litigious disposition; and particularly to prohibit the granting of pretended titles to *great men*" who had the resources to tie up small landowners in endless lawsuits.[1] By the beginning of the nineteenth century, however, every American state had modified the common law rule to allow for some form of assignment.

Another characteristic of negotiable instruments that challenged traditional conceptions of contract was its break with the doctrine of privity of contract. How could *A*, who had given a promissory note to *B*, be sued by *C*, to whom *B* had transferred the note, when nothing passed between *A* and *C*. Since they were not in privity, the common lawyers reasoned, neither promise nor consideration moved from one to the other. While statutes allowing assignments tended to solve this problem, they did not solve another very similar one. Where *C* endorsed a note over to *D*, could *D* sue *B*, a prior endorsor, if the original promisor defaulted? Here, the common law requirement of privity presented an insurmountable conceptual barrier to negotiability.

A third problem went to the very heart of the principle of negotiability. Could *C*, a subsequent endorsee, receive a better title to an instrument than *B* had? In a normal contractual relationship between *A* and *B*, if *B* sued on a promise, *A* could offer various defenses to the action. For example, he could maintain that there was no consideration for his promise to pay, which, if established, would void the obligation. He might also offer a variety of other defenses — infancy, payment, set-off of a prior debt, fraud, discharge under an insolvency law, illegality of the contract, usury — any of which would be sufficient to discharge his obligation to pay the note. The ideal of negotiability, however, required that notes should circulate as freely as money, so that subsequent innocent purchasers of an instrument might depend on payment regardless of any unknown defects in the obligation arising out of the original transaction between distant parties.

The central question that joined all these legal issues was whether contract categories would control the known intent of the parties to a transaction. Would the original policy of preventing assignments be continued regardless of the parties' intentions? Would the doctrine of privity, the conceptual foundation of all contract law, be permitted to yield to the apparent intention of men to bind themselves to pay notes that were negotiable from hand to hand? What was the status of the rule that one could not receive a better title to property than his grantor had to convey?

Finally, lurking behind all these considerations was a major issue of public policy. Would the legal system sanction private arrangements whose effect was to increase the supply of money by allowing individuals to agree to substitute their own notes for currency designated by the state?

All these issues, of course, were not new by the last decade of the eighteenth century. Many had been fought out in England earlier in the century and some had been settled even before that. By the beginning of the eighteenth century, the negotiability of bills of exchange between English merchants, which had been recognized for at least 150 years in merchants courts, was already regularly enforced in the central courts as well. Promissory notes, however, confronted greater hostility in the English courts because of their more recent vintage. They had only come into use late in the seventeenth century, and in *Clerke v. Martin* (1702) Chief Justice Holt refused to recognize their negotiability. "Such notes," he declared, "were innovations upon the rules of the common law . . . unknown to the common law, and invented in Lombard Street, which attempted in these matters of bills of exchange to give laws to Westminster Hall."[2] Two years later, in response to that decision, the English Parliament passed the Statute of 3 & 4 Anne making promissory notes negotiable.[3]

The English statue of 1704 did not, however, establish the principle of negotiability in America. Before the Revolution, in fact, only three colonial legislatures expressly adopted the English statute.[4] Even in England the full elaboration of the principal of negotiability only began during Lord Mansfield's tenure as Chief Justice (1765-88) and many points were still to be decided when he left the bench.

Though bills of exchange were in wide use in eighteenth century America,[5] there appears to be a relatively small number of cases of negotiability of promissory notes before the Revolution. In New York, it appears, promissory notes were frequently assigned in the eighteenth century, though we have no evidence of whether courts were willing to enforce the true test of negotiability—limitation of the defenses of the promissor.[6] In Massachusetts, however, it is clear that colonial judges refused to recognize the full negotiability of promissory notes. In *Russell v. Oakes* (1763), a divided Massachusetts high court held that the maker of a promissory note was not liable to an innocent purchaser of the note if he paid the obligation before the note was endorsed. To refuse an innocent purchaser recovery under such circumstances, Chief Justice Hutchinson protested in dissent, denied the very principle of negotiability of notes. The result was "that one half of the Trade must be extremely precarious, for it rests upon such Bills, whose Credit must be destroyed. It destroys the Distinction between Notes negotiable and not."[7] Again, in *Tuttle v. Willington*[8] (1772), the Massachusetts court reaffirmed the rule that a purchaser of a note assumes the risk that it may already have been paid.

Whether because of these decisions or not, there appears to have been little use of negotiable promissory notes in Massachusetts before the Revolution. "Even as late as 1770," Nathan Dane wrote, "all the English law books then published upon the subject of commerce and negotiable contracts might have been read, and even studied in a few weeks by an English or an American lawyer. These in forty years since have probably increased tenfold, and are still rapidly increasing in bulk and value."[9]

The major factor bringing the question of negotiability to the fore was the universal scarcity of specie after the American Revolution. But the problem of scarcity of money, though intensified after the Revolution, was also a persistent complaint of colonial merchants. They, however, had conducted most of their business by purchasing bills of exchange drawn on large English mercantile houses. The effect of the Revolution was to destroy for a time this network of international trade and thereby to dry up even this important source of liquidity. Promissory notes thus rapidly began to replace bills of exchange after the Revolution. In addition, until there were banks ready to discount promissory notes, there was no ready market for these notes and hence little occasion to transfer them from hand to hand.[10] In the years between 1785 and 1800 judges in virtually every state were forced to come to terms with the problems of negotiable instruments and, in the process, with the challenge to traditional contract conceptions they entailed.

In 1790 only three American states recognized full negotiability of promissory notes; by 1800 only two more states had adopted the negotiability principle.[11] And when in 1810 Zephaniah Swift published *A Treatise on Bills of Exchange and Promissory Notes,* he acknowledged that there was still "great embarrassment in the negotiation of notes" because "the assignee takes it subject to every legal objection, and ground of defence attached to it in the hands of the original parties."[12] In Virginia and Pennsylvania, for example, judges regarded the absence of a statute as posing an insurmountable obstacle to negotiability. Colonial legislation in both states had authorized the assignee of a note or bond to sue the maker in his own name, thus dispensing with the necessity at common law of securing the cooperation of the original promisee. After the Revolution, however, lawyers pressed the judges to go further and to construe these statutes as having conferred full negotiability, thereby cutting off all the promisor's original defenses. However, the high courts in both states refused, holding that the statutes did no more than make assignment of notes easier.[13] The result was that in Virginia and Pennsylvania, as well as in Kentucky,[14] which followed Virginia law, promissory notes at the turn of the century were still a long way from serving as a substitute for currency as they had for almost a century in England.[15]

In Massachusetts, where the judges eventually established negotiability without the authority of a statute, endorsed promissory notes first began to appear with some frequency in the 1780s. Yet, in 1785, the high court still allowed a promisor to defeat recovery by an innocent endorsee on proving want of consideration in the original transaction.[16] And even as late as 1809 the court reiterated the colonial rule that the maker could prevail against an innocent endorsee by proving payment before the note was endorsed.[17] Only gradually, then, did full negotiability come to be established in Massachusetts.

Even in South Carolina, which had established the negotiability principle by colonial statute, the extreme scarcity of money thrust the problem of negotiability suddenly to the fore in the decade after the Peace of 1783. During that period "bonds [sealed promissory notes] to the amount of several hundred thousand pounds sterling had been passed away from man to man, in a great variety of transactions. . . . In sales and purchases, they had served as a kind of circulating medium, which relieved the distresses of the country exceedingly, at a period when the cash had been drained out of it, or exported to foreign countries." The question of whether these bonds were negotiable came before the Supreme Court of South Carolina in *Parker v. Kennedy*[18] (1795), "a case of great expectation and consequence to the community," since "more money depended upon the determination of this question, than any other which had ever been discussed in South Carolina."

Most of these bonds were negotiated in the period after 1783, "while parties were prohibited from recovering their debts by installment laws, and other acts throwing impediments in the way of creditors, [and] many of the obligors had gone off and left the state without leaving funds for discharging their obligations; many others died in insolvent circumstances; while others again, had become bankrupts." As a result, "very little more than one half (if so much) had proved good." On the other hand, "Many of these bonds had been negotiated at one-half of their nominal value . . . and in many cases, had become objects of speculation to adventurers." Under the circumstances, the court was forced to decide "a question of great magnitude": whether the intermediate endorsers were liable where the original promisor was unable to pay.[19]

By a 3 to 2 vote the court held that the bonds were not negotiable under the South Carolina statute and that, as a result, the holders could not recover. *Parker v. Kennedy* thus raised a number of extraordinary issues concerning the extent to which courts would be bound by a society's commercial practices. On the one hand, it was acknowledged by all the judges that at common law a bond was itself not a negotiable instrument and that an endorsement by the promisee merely represented an assignment of his right to another.

In short, in suing the promisor, the endorsee continued to be subject to any defenses that could be raised between the original parties. But what of the legal relationship between endorsor and endorsee? Did the transfer of the bond merely constitute a traditional assignment of interest in which the assignor concurrently renounced all obligations, or did it constitute a continuing guarantee (as in the case of an endorsor of a negotiable instrument) to pay if the promisor did not?

It was over this last question that the court divided. For the dissenting judges it was urged that "if the sense of the whole community was taken, it would be found that most men have understood that, when they put their names in this way on bonds, they made themselves ultimately liable if the obligors were unsolvent at the time."[20] In a period when "current coin had been gradually drained out of the country, so there was scarcely any circulating medium left for the payment of debts . . . . necessity . . . gave rise to an offer of these bonds in payment." The only criterion for judging the legal effect of an assignment was "commercial use" or "commercial convenience and utility," since "the intention of the parties, when it can be discovered, ought to govern in all contracts."[21]

In writing the leading opinion for the majority, Judge Grimke declared that he had not "paid the least attention to the arguments, *ad inconvenienti,* nor to those pointing out the facility with which negotiations could be carried on by bonds." Even though "the court have been told what the intention of the legislation was with respect to bonds and notes," it "considered themselves as bound down to the technical terms of the law, and to abide by their legal import, though directly contrary to the precise intention of the legislature." Bonds, he concluded, had never been regarded as negotiable instruments, and thus an assignment could not make an endorsor liable.[22]

Parker v. Kennedy thus represents a special South Carolina variation on the theme of negotiability that was repeated in many other states. For in that state, which had copied the English statute, the first postrevolutionary cases dealing with the problems of negotiability arose not over ordinary promissory notes but over bonds or sealed instruments. That the issue of negotiability first arose in connection with bonds seems to confirm the suspicion that until the nineteenth century bonds were widely used to prevent judicial interference with commercial transactions. Whether or not they were used in order to prevent juries from setting damages, bonds almost surely could successfully immunize usurious contracts from legal attack.

But most American courts refused to alter the common law distinction between bonds and notes.²³ As a result, whatever other advantages a creditor might gain from taking a bond, he still could not sell the bond as a negotiable instrument. Even the scarcity of money would not induce most American judges to recognize the negotiability of bonds.

Besides the technical problems of reconciling negotiability with common law conceptions, there were several major issues of policy that contributed to the resistance to negotiability in America. First, there was a continuing fear that through negotiability of notes legislatures would lose control over the money supply. In 1777, for example, the Virginia Assembly barred all notes payable to bearer on the ground that they might, in the language of the statute, "tend to the deception and loss of individuals, as well as to the great injury of the publick, by increasing the quantity of money in circulation." The reason for the law, George Mason conjectured, was that "private bills of credit have been issued by merchants who were exasperated with the depreciating state and continental currency issues and thus resorted to issuing promissory notes."²⁴

The most important source of resistance to negotiability was derived, however, not from considerations of governmental monetary policy but from a precommercial conception of the requirements of "common justice and honesty" among individuals.²⁵ Even without the benefit of a Vermont negotiability statute Judge Nathaniel Chipman in 1793 sought to derive negotiability from "principles of right, arising from the nature of the transaction itself."²⁶ Yet, he finally was unwilling to concede the essence of negotiability: the right of an endorsee to be free from a promisor's original defenses. This idea, he wrote, was established under statutes, "which had more regard to public policy, than private honesty."

That no one should benefit himself, or legally charge an innocent man, by a fraudulent act, is a law of nature, of reason, and common honesty, which ought not to be less regarded in the administration of justice, than acts of legislation.²⁷

Why, Chipman asked, in a case before the Vermont Supreme Court, should judges "introduce an arbitrary custom, to bind a man contrary to his express agreement, and the real equity of the case?" Negotiability, "it is said, is established by the course of trade, and is for the benefit of commerce. This is at least problematical. But as this State is not, and from its local situation cannot be greatly commercial, this may be laid out of the question."²⁸

In addition to public policy considerations, the Virginia statute of

1777 regarded negotiability as tending "to the deception and loss of individuals." Virginia judges also refused on similar grounds to adopt the principle of negotiability. "It is urged, as a reason for the rejection of former opinions," one judge wrote, "that they tended to impose deceptions upon the public, and to cramp commerce, by destroying . . . negotiability. . . . As it strikes me, they rather tend to prevent, than to countenance those frauds, and . . . it is preferable to sacrificing a majority of the public, to the avarice and injustice of a few."[29]

Indeed, many instances were cited in the early nineteenth century to confirm this fear that negotiable instruments might be used to oppress commercially unsophisticated groups. "Thousands have been lost in this state," a Kentucky lawyer reported in 1822, "to merchants and others on the sea board and in other parts of the union," due to "ignorance" of the Kentucky law preserving all of the promisor's defenses.[30] Equally important, negotiable instruments seem to have been widely used to evade the restrictions of state usury laws, since many American courts had already erected a major barrier to proof of usury in negotiable instruments cases. In *Walton v. Shelley*[31] (1786), the English judges, under the influence of Lord Mansfield, had established the rule that in an action by an endorsee against the maker of a note, the intermediate endorsor would not be permitted to testify that the original transaction was usurious. While the English judges did not pretend that this rule was simply an application of traditional common law doctrine barring witnesses who had an interest in the outcome of a suit, many believed it to be an extension of orthodox law. In fact, the rule had quite the opposite effect, since if the original transaction was usurious the endorsor would himself be exposed to a separate lawsuit by the endorsee. Realizing that the rule of *Walton v. Shelley* was an entirely novel one in English law, the judges overruled it only twelve years later.[32]

Nevertheless, as with so many other doctrines propounded by Mansfield, the rule of *Walton v. Shelley,* which had the practical effect of freeing negotiable instruments from the restraints of the usury laws, found a congenial resting place in nineteenth century America even after it was disavowed in England. Massachusetts adopted the rule in 1795[33] and New York in 1802. In *Winston v. Saidler,* a sharply divided New York Supreme Court acknowledged that the rule had nothing to do with technical questions of evidence but rather was grounded in a broad commitment to commercial interests. Allowing a party to a negotiable note to testify, Justice

Thompson wrote, "would greatly embarrass trade and commerce, and almost entirely prevent the circulation of this species of paper."[34] In Massachusetts, Chief Justice Parsons in 1808 defended the rule on identical grounds. "The circulation of negotiable paper is extremely useful to trade, as it multiplies commercial credit, and the notes pass from man to man as cash. Any rule of law, tending unnecessarily to repress this circulation, is therefore against public policy."[35]

The only reason why procommercial judges so committed to negotiability should have continued to recognize the defense of usury at all was that it was a legislative creation. As a result, while the courts of Massachusetts and New York persisted in entertaining the defense of usury, they were able, under the protective facade of technical rules of evidence, to make usury virtually impossible to prove. There was a sound basis, then, for the widespread suspicion among noncommercial groups that negotiable instruments were becoming the vehicle through which oppressive agreements could be enforced in American courts of law.

By 1842, when the United States Supreme Court decision in *Swift v. Tyson* allowed the federal courts to enforce a "general commercial law," a majority of state courts, many of them in Western states, had in fact rejected the rule in *Walton v. Shelley* and allowed an endorsor to impeach a note as usurious.[36] But by this time the United States Supreme Court had itself adopted the rule,[37] thus enabling commercial interests to employ the federal courts to enforce usurious notes under the general commercial law even in the majority of states that had rejected the doctrine of *Walton v. Shelley*.[38]

That the rule in *Walton v. Shelley* could be enforced in the federal courts, though contrary to the common law rule of the state in which the transaction arose, provides an important measure of the influence of the federal judiciary in advancing commercial interests during the nineteenth century. The history of negotiability in the federal judiciary is an especially interesting tale, since it allows us not only to trace the progress of general conceptions of negotiability but also to determine the extent to which federal courts promoted commercial interests through a "general commercial law" that overrode state decisional law.

The first really important cases involving negotiability to come before the federal courts were in fact efforts at "forum shopping" precisely for the purpose of evading state laws that refused to adopt the principle of negotiability. For example, at the turn of the century, the Virginia courts did not recognize negotiability of notes, holding that a statute of 1748, which allowed assignments, nevertheless preserved the right of the promisor to set up all defenses against a remote assignee.[39] And while the Virginia courts departed from the limited common law conception of assignment, by interpreting ordinary contract rules to permit an assignee to sue his immediately prior assignor,[40] they stood firm against allowing assignees to sue remote assignors.[41]

Having failed in their efforts to get the Virginia courts themselves to establish the negotiability principle, holders of notes turned to the federal courts. In two cases decided in 1801 and 1802, a divided federal circuit court in the District of Columbia held that after the maker of a note failed to pay, an endorsee could recover against a prior endorsor.[42] Not only was this result contrary to an express decision of the Virginia Supreme Court, but it could be justified only on the already rejected principle that notes were negotiable.

These decisions were immediately brought before the United States Supreme Court in *Mandeville v. Riddle*.[43] Reversing the lower court, Chief Justice Marshall stated that without a Virginia statute making notes negotiable, the federal courts could not entertain the action. Since the Virginia statute did not authorize the action, he asked, were there common law principles on which it could be maintained? And drawing on Virginia cases that were decided during his own practice in that state, Marshall declared: "In the language of the books, there is a privity between the assignor and his immediate assignee; but no privity is perceived between the assignor and his remote assignee." In short, a statute was necessary to authorize negotiability because it formed an exception to ordinary common law contract principles. Most important of all in terms of the much debated issue of whether the early federal judiciary enforced a general commercial law, Marshall treated the question as simply one of applying Virginia law, offering not the slightest suggestion that the federal courts had any independent power to establish rules of commercial law.

At the same time, however, Circuit Judge William Cranch, who had formed part of the overruled majority in the case below, wrote a long and famous essay on the history of negotiable instruments, which, as Supreme Court reporter, he incorporated as an appendix to the volume in which Marshall's opinion appeared.[44] The gist of Cranch's argument was that negotiability of notes was a common law principle that required no statutory authorization. Thus, he disputed Lord Holt's famous decision that negotiability was unrecognized at common law and maintained that the Statute of 3 & 4 Anne, which reversed Holt's decision, was merely an affirmance of a pre-

existing common law rule. "The custom of merchants," he concluded,

. . . ought not to be considered as a system contrary to the common law,
but as an essential constituent part of it, and . . . it always was of coequal
authority. . . . The truth seems to be, that the principles of the common
law have not been changed, nor innovated upon, by the introduction of
. . . commercial principles, but that these principles have existed from the
earliest times, even from the rudest state of commerce, and the only reason
why we do not find them in the ancient books, is, that the circumstances
had never occurred which rendered it necessary to draw them forth into
judicial decision.[45]

Cranch's strategy, it should be noted, was to dispute orthodox
notions of the foundations of this particular common law doctrine,
not to insist more broadly on an independent commercial law,
which, only a decade before, James Wilson had declared was "not
the law of a particular country, but the general law of nations."[46] In
the interim, the idea of a general commercial law had begun to be
resisted, not only in connection with the debate over federal common law crimes, but also because it was increasingly suggestive of
special class legislation. Cranch's most powerful argument, therefore, was to maintain that the principle of negotiability "existed
from the earliest times."

At the very end of his essay, however, Cranch introduced a new
argument without even a semblance of historical support. Even if
negotiability did not exist at common law, he casually asserted,
"there never has been a doubt," that it was always recognized in
equity.[47] And despite the weight of his lengthy historical claim of
common law support, it was this last argument that eventually
found favor in the United States Supreme Court. Two years later, in
the course of a Supreme Court argument having little to do with the
question of negotiability, one of the lawyers mentioned in passing
that an endorsee might sue his remote endorsor. Chief Justice Marshall stopped him and denied that he could. And with William
Cranch present and ready to include a footnote in his later report of
the case, the Chief Justice declared that he had "always been of
opinion, that in such cases a suit in chancery can be supported;
though I do not recollect any case in which the point has been decided."[48] Sure enough, this was a sufficient cue to bring Mandeville
and Riddle back for another round.

In his second decision of the case in 1809, Chief Justice Marshall,
as expected, reversed himself and held that an endorsee could re-

cover against a remote endorsor in equity. Even though "the action . . . is not given by statute," Marshall declared, "such a contract must, of necessity, conform to the general understanding of the transaction." "General opinion," he maintained, determined the credit risk of a note on the basis of the credit of the endorsors, "and two or more good indorsors are deemed superior to one." Thus, there needed to be some way for a subsequent endorsee to reach back to protect his interest, and "equity will of course afford a remedy."[49]

Despite the contrary view of Virginia judges, then, the United States Supreme Court, out of the blue, had invoked an independent equity power to establish the principle of negotiability in the federal courts. The "general understanding of the transaction," to which Marshall ultimately appealed, was, of course, contrary to the expressed understanding of the Virginia courts. Negotiability, they had held, was not a substantive common law right, and, undoubtedly, they meant to exclude as well equitable remedies to vindicate any such supposed right. Within just six years, then, Marshall's own conception of the underlying source of negotiability had radically shifted from the position that it arose only through legislative command to the view that it was founded on "the general understanding" of the nature of the contract itself. Henceforth, suitors on negotiable instruments who could not prevail at home would gain relief in the federal courts.

Riddle v. Mandeville was the first clear assertion in the federal courts of the idea that a general commercial law existed independently of the decisional law of the states. It was applied again in *Bank of United States v. Weisiger*[50] (1829), where the high court ignored Kentucky's constantly reiterated rule against negotiability and once more sustained an equitable suit by an endorsee against a remote endorsor.[51] It is of more than passing interest that this first assertion of the power of the federal courts to enforce a general commercial law despite a contrary state rule arose in order to establish a major procommercial legal doctrine that had consistently been opposed at the state level.

This reversal of the common law view that in the absence of privity there could be no suit between endorsor and endorsee to a note was, however, only one step in the movement toward recognizing a special negotiable status for promissory notes. There still remained the ultimately crucial question of whether a promisor's defenses would continue to be available against a bona fide purchaser of the note.

When Joseph Story published his treatise on promissory notes in

1845, the law of negotiable instruments still remained in a confused and chaotic state. It was, of course, in the very nature of the treatise tradition to exaggerate the uniformity of American law by stressing doctrines—such as full negotiability of notes—to which commercial interests and their legal allies were devoted. Yet, though in somewhat disguised and underplayed form, even Story's treatise revealed the still contradictory state of American commercial law as mid-century approached. "In some States," Story acknowledged, "the circulation of Promissory Notes still remains clogged with positive restrictions, or practical difficulties, which greatly impede their use and value, and circulation."[52] Indeed, at the time Story wrote, ten states continued to bar full negotiability through a combination of limits on assignments and refusal to cut off defenses.[53]

In 1850 the United States Supreme Court was first called upon to decide the status of one of these state laws that allowed the maker of a note to assert all his original defenses against a remote endorsee. An Alabama statute had allowed a promisor "the benefits of all payments, discounts, and set-offs" against an endorsee which he could have claimed against his promisee. As in *Riddle v. Mandeville,* the question once more was the extent to which the federal courts would recognize a general commercial law of negotiability free from the interfering restrictions imposed by anticommercial states. This time, however, the legal issue could no longer be framed simply as one of finding an appropriate remedy for a supposed legal right, since the very essence of full negotiability lay in the power of endorsees to treat notes as unfettered currency. Though under the immediate glow of Justice Story's recent recognition of a general commercial law in *Swift v. Tyson,* the Supreme Court in *Withers v. Greene*[54] refused to override the Alabama statute.

By this time, it was to be expected that endorsees would argue that the power of negotiability was simply an inherent part of the right of contract itself. Indeed, William Cranch had argued, as early as 1804, that the negotiability could be deduced from simple contract principles. "Every man," he wrote, "has a natural right to make such contracts as he pleases," provided they were contrary neither to positive nor moral law. "And all contracts entered into without fraud or force, are legally and morally obligatory, according to their spirit and intent."[55] Yet in *Withers v. Green,* Justice Daniel rejected the argument that preserving the promisor's equities was an unconstitutional interference with contracts. "We cannot regard" the Alabama statute, he declared, "as changing the rights of the parties arising out of the contract itself, nor as conferring new rights on others not inherent in such original obligations."

There was simply too much legal history to be overcome for negotiability to be elevated to the status of an inherent common law right. Statutes providing for assignment of notes could not be confounded with full negotiability, Daniel acknowledged. "Contracts at common law, to which this simple power of assignment is extended by statute, differ essentially from those which arise out of and are governed by the law merchant, or from such as are placed on the footing of the law merchant by express legislative enactment."[56] Thus, despite all the attempts of his predecessors to create commercial uniformity by rewriting the history of negotiability, Daniel in part returned to the orthodox view that it was strictly a legislative creation. Even so, the status of negotiability had been subtly transformed during the preceding half century. Without the positive intervention of a state statute limiting defenses, the United States Supreme Court would now invoke a "general commercial law" to enforce complete negotiability despite a contrary state judicial policy. Through the efforts of the federal judiciary as well as the treatise writers, negotiability had been elevated to the rule, deviations from it had become the exception, even as there remained strong currents of the old anticommercial mentality in southern and western states.

The story, however, does not end here. Only five years later, Justice Daniel, citing *Swift v. Tyson,* held that a Mississippi statute regulating the negotiability of bills of exchange was "a violation of the general commercial law, which a state would have no power to impose and which the courts of the United States would be bound to disregard." "Any state law," Daniel argued, "must be nugatory and unavailing" if it denies "the rights . . . secured . . . under the commercial law."[57] The Supreme Court thus appeared to overrule the one limitation imposed by *Swift v. Tyson:* that federal courts were bound at least to follow state statutes on commercial subjects. Later, building still further on this decision, the Supreme Court in 1879 drew into question the very same provision of the Alabama statute that had been reluctantly enforced in *Withers v. Greene,* now claiming only that it was "for the Federal courts to determine upon their own judgment" what the "rule and doctrines" of commercial law would be.[58] After this case, it appears, it was no longer possible for states to resist the rule of full negotiability that by now had come to be enforced by the federal courts as part of the general commercial law. "The question is one of general law," the Supreme Court concluded, "and depends nowise for its solution upon local laws and usages."[59]

THE LAW OF INSURANCE: THE DEVELOPMENT
OF ACTUARIAL CONCEPTIONS OF RISK

No area of commercial law consumed more attention of American judges and elite lawyers during the years between 1790 and 1820 than the law of marine insurance. It was marine insurance litigation involving enormous sums that enabled America's first generation of commercial lawyers to become wealthy exclusively through the practice of law. After leaving the federal government in 1795, Alexander Hamilton devoted much of his New York commercial practice to insurance cases as did Brockholst Livingston before he ascended the New York Supreme Court bench.

Horace Binney, who also had an extensive commercial practice, rejoiced at the "unparalleled harvest to the bar of Philadelphia" resulting from the Atlantic sea war then in progress. "Insurance cases were probably never so numerous or important as in Philadelphia from 1807 to 1817," Binney's biographer points out. "That city was the first commercial port in the United States, and her insurers were as active as her merchants."[60]

We can get some rough measure of the importance of marine insurance cases simply by counting their proportion in the published reports. The New York reports, which appear somewhat earlier than others, illustrate the general trend. Between 1799 and 1807 the proportion of reported marine insurance cases decided by the New York Supreme Court varied between 8 percent and 18 percent and averaged around 12 percent during those nine years. The average then fell to 2.5 percent during 1808-10, reflecting the decline in shipping due to the embargo. Between 1811 and 1814 the yearly average rose to between 5 percent and 8 percent. Between 1815 and 1820, however, it constituted only between 1 percent and 3 percent of all cases. Finally, there are no reported cases in 1821 and 1822, reflecting the end of marine insurance as a major subject of litigation.[61]

In terms of its influence on American law after 1790, one of the most important characteristics of the development of English insurance law is that it had itself only reached maturity during the previous thirty or forty years. In his influential treatise on insurance, published in 1787, Sir James Park noted that fewer than sixty cases on insurance had ever been decided by the English courts before Mansfield became Chief Justice.[62] The 150 years before the appointment of Mansfield to the King's Bench in 1756, another writer has concluded, were "almost barren waste as far as the history of the development of insurance law is concerned."[63]

The substantive principles of English insurance law first elaborated by Mansfield reflected the relatively closed and homogeneous English mercantile community of the eighteenth century. "The assurers were still, for the most part, merchants ordinarily engaged in other business. . . . It is clear that a great deal of underwriting was still done by men mainly engaged in other branches of commerce."[64] At mid-century the insurance business centered on Lloyd's Coffee-House, which "was nothing but a coffee-house to which all the world had access." Even after 1769, when "a rather more elaborate organization began to take shape . . . anyone in the place (without regard to his means) could write any risk on any voyage for any amount."[65]

Out of this system, as one contemporary remarked, there developed a "practice so prevalent amongst merchants who are also underwriters, of interchanging their risks with each other for their reciprocal benefit."[66] Marine insurance, in short, depended on circulating and reciprocal relationships among "a casual group of unorganized, uncontrolled men," who would alternately serve as both insurer and insured on different voyages.[67]

Most of the marine insurance doctrines developed during Mansfield's tenure expressed one fundamental premise: any failure to disclose a material fact that affected risk was sufficient to void an insurance policy. In the small homogeneous eighteenth century English trading community, legal doctrines requiring full disclosure fully accorded with private expectations of appropriate business behavior. The transatlantic migration of English insurance doctrine, however, introduced several complications.

In America, a decisive change in the nature of marine insurance underwriting began in the 1790s with the establishment of incorporated insurance companies. Before that time, underwriting seems to have followed, though on a much smaller scale, the informal and unorganized patterns set by Lloyd's in England. "In the American trading centers the merchants were the underwriters. It was they who supplied the capital; multiple subscription of a policy was usual."[68] Indeed, a student of eighteenth century marine insurance found that marine underwriting in colonial Boston "had never been intended for profit" but rather for mutual protection of commercial interests by merchants who served alternately as insurers and insured.[69]

One of the most important consequences of the chartering of American insurance corporations was the destruction of that eighteenth century system of circulating and reciprocal relations between underwriters and merchants and the creation in its place of a permanent conflict of interests between the two groups. This in turn brought about a series of challenges to orthodox and previously agreed upon doctrines in insurance law which, with striking regularity, courts at first resolved in the companies' favor. In order to protect insurance companies from merchant-oriented juries, it became necessary not simply to promulgate legal doctrines favoring their interests but also to reduce the power of juries. One of the great American transformations in the relations between judge and jury arose out of marine insurance cases at the beginning of the nineteenth century, when courts forged new procedural weapons that enabled them to reverse damage awards by juries.[70]

The development of American insurance law during the nineteenth century reflected a gradual acceptance of what we might call an actuarial conception of social risk. I do not mean to suggest by this that formal actuarial theories first appeared only at this late date — in fact they can be traced at least as far back as the mid-seventeenth century development of probability theory and its subsequent application to mortality statistics.[71] Rather, I wish to emphasize instead the development of a social consciousness that comes to conceive of a greater and greater portion of activity as appropriately within the realm of chance.

A close analogy is to the change in attitudes toward bankruptcy legislation among American businessmen during the nineteenth century. Especially before the Panic of 1819, the bulk of respectable business opinion stood firmly against bankruptcy laws, regarding them as providing an illegitimate windfall to profligate and irresponsible merchants and entrepreneurs.[72] But as a result of increasing "depersonalization and bigness" in business relations, Peter Coleman shows in his study of the history of bankruptcy legislation, there was a gradual shift away from this eighteenth century ideal of mercantile "respectability." "Customers became names on pieces of paper rather than faces and personalities." This, in turn, contributed to "a more formal, legalized perception of relationships."

The logic of this formalization and legalization of relationships between debtors and creditors led directly to the acceptance of bankruptcy relief as an equitable, rational and systematic way of writing down bad debts. It represented the full flowering of the bookkeeper mentality. . . . And so the pendulum of opinion swung from hostility to bankruptcy relief to an attitude that mixed indifference with tolerance and outright approval.[73]

Accompanying this change in attitude was a strikingly new perception among respectable businessmen and lawyers, who campaigned to enact bankruptcy legislation. Business failure, they now frequently argued, was essentially a random consequence of uncontrollable economic forces.[74] In 1832 the author of a treatise on insolvency could offer a long list of the "great and extensive . . . hazards, which, attend the enterprise and undertakings of merchants, and so frequently plunge them into misfortune. . . . Instead of converting every thing they touch, like Midas, into gold, they find all their hopeful expectations disappointed, and themselves surrounded by one unbroken scene of widespread ruin and desolation."[75] What was once regarded as fundamentally subject to individual control and responsibility came to be viewed — no doubt for self-interested reasons — as beyond moral and legal control.

The rise of insurance law in the nineteenth century came gradually to reflect similar changes in social conceptions of risk. Marine insurance — by far the most important type of pre-nineteenth century insurance — bore for the longest time many traces of its preactuarial origins. In the first place, the informal and unorganized structure of eighteenth century private underwriting precluded any extensive pooling of risks. This in turn encouraged men to think of each individual insurance policy as a unique and personalized transaction having no more general statistical significance. Furthermore, during its rise to prominence in eighteenth century England, marine insurance law was preoccupied with limiting the scope of its coverage in order radically to set insurance policies apart from legally proscribed gambling contracts. As a result, the standard marine insurance policy, reflecting the moralistic atmosphere from which it first emerged, covered only "extraordinary perils of the sea," and not those "ordinary" risks whose inclusion would have converted it into a general policy of indemnity that looked too much like gambling. Two defenses most successfully used by underwriters to defeat recovery — unseaworthiness and barratry — underlined the limited scope of the risks covered by marine insurance policies.

Barratry provides the easiest point of contrast with nineteenth century developments. Under most marine policies, it was open to the insurer to defend against recovery on the ground that the loss was caused by the neglect of the captain or some other agent of the insured. This defense of barratry was just another application of a more general principle that one could not insure against his own misfeasance. It was also one basis for distinguishing between "extraordinary" (and hence insurable) and "ordinary" (or uninsurable) risks. In the early nineteenth century, however, we begin to see courts gradually confine the defense of barratry to what amounted to willful acts that brought about a marine loss. In fire insurance cases after 1830 courts extended this development and began to reject the defense of negligence put forth by insurers. Thus, in this doctrinal area, as in several others, judges began to include larger and larger categories of risk within the terms of coverage of a policy. The courts, in short, were beginning to express what would become a more general change in nineteenth century legal consciousness: conceiving of greater numbers of business risks simply as "costs of doing business," largely outside the control of individual responsibility.

The most significant changes in the American law of insurance were brought about as fire insurance began to outdistance marine insurance as the leading form of underwriting after 1840.[76] Before the 1840s, in fact, there were only a handful of judicial decisions involving fire insurance. "The whole law of [fire] insurance," a Massachusetts commission noted in 1837, " . . . may now be stated in a text not exceeding thirty pages of the ordinary size."[77] And even as late as 1852 the Massachusetts Supreme Judicial Court observed that "Fire insurance is . . . in its rudiments. The cases on marine insurance throw little, if any, light on the present question."[78]

The late appearance of fire insurance is something of a mystery. While policies covering fire risks had been issued in the eighteenth century by marine insurance or mutual stock companies, "up to the great New York conflagration of 1835, fire insurance had been written on a small scale in the United States, largely, though not exclusively, on a local basis."[79] Early nineteenth century law may have played a role in discouraging the use of fire insurance. Until the 1830s judges and jurists denied that one could recover for fires brought about by his own negligence or that of his agents. "It has been made a question by some persons," Chancellor Kent wrote in 1832, "whether the negligence and frauds which the insurance of property from fire has led to, did not counterbalance all advantages and relief which such insurances have afforded in cases of extreme distress." Though he claimed that "the public judgment . . . has long since decided that question" in favor of fire insurance, even as late as 1854 Joseph Angell continued to wonder in his treatise on *Insurance* "whether in a general, or national point of view, the benefits resulting from [fire insurance] are not more then [*sic*] counter-balanced by the mischief it occasions" such as "carelessness and inattention which security naturally creates, and temptation to arson." Yet, he concluded — as by 1854 virtually all American judges had concluded — that insurance against fire was "so essential to the present state of the country for the protection of the vast interests embarked in manufactures, and on consignments of goods in warehouses."[80]

As fire insurance came to be thought of as an important instrument for promoting economic security and growth, American judges became active in distinguishing rules governing fire insurance from those that had prevailed in the law of marine insurance. While virtually all marine insurance doctrines were stacked in favor of the insurer, most of the major changes in doctrine resulting from the rise of fire insurance were favorable to the insured. When the Massachusetts Court declared in 1852 that "the cases on marine insurance throw little, if any, light," on litigation over fire insurance, it was reflecting a judicial policy of gradually disentangling fire from marine insurance doctrines.

One major change in insurance doctrine will serve to illustrate the underlying shift to an actuarial consciousness brought about as a result of increased litigation over fire insurance. Marine insurance law had drawn a radical distinction between warranties involving the condition of a ship or nature of the voyage and mere representations. Breach of a warranty resulted in a complete voiding of the policy even without proof that the breach increased the risk under the policy. By contrast, simple misrepresentations would only defeat recovery under a policy if they were made knowingly and only then if they materially increased the risk. In accordance with the overwhelming emphasis on full disclosure, marine insurance law had stringently enforced a variety of implied warranties, which had the effect of defeating recovery for even an insignificant breach. For example, any deviation from the stipulated route of a marine voyage would void a policy even without a showing that it had increased the risk of loss.

On the other hand, American courts from the beginning tended to be substantially more liberal in allowing the insured to depart from the strict terms of fire insurance policies. In perhaps the first fire insurance case decided by an American court, the Supreme Judicial Court of Massachusetts in 1808 rejected the marine insurance doctrine that any deviation was sufficient to void a policy of insurance. Though the plaintiff had made certain alterations in his building, adding an extension without informing his insurer, the court held that the only question affecting recovery was whether the alteration increased the risk of fire.[81] It thereby moved the rules governing deviation out of the traditional warranty category and into the more liberal class of rules governing representations. Perhaps of greater importance, the crucial inquiry had shifted from one of law to one of fact, so that plaintiff-oriented juries were left to decide as a question of fact whether a particular alteration actually increased the risk.

The usual judicial explanations for relaxing the strict warranty doctrine in cases of fire insurance emphasized that, unlike marine risks, insurance companies themselves were as able to determine the extent of fire risks as the insured.[82] Yet, it is also clear that by the second quarter of the nineteenth century many courts had come to attach a special importance to insurance in providing stable expectations necessary for economic growth. It was necessary to relax "the strictness and nicety" that had been adopted for marine insurance, the Maryland high court declared in 1827, for to do otherwise "would . . . render worse than useless those most useful and indispensable institutions in populous cities—fire insurance companies, —and give a fatal stab to our enterprising manufacturers."[83]

Some courts in fact went still further and overthrew all traditional insurance doctrine in deciding fire insurance cases. We have seen that virtually all courts held that a subsequent alteration of buildings would not defeat recovery under a fire insurance policy unless the change increased the risk. This shift, at least, was consistent with existing doctrine concerning representations. But in *Stebbins v. Globe Ins. Co.*[84] (1829), the New York superior court broke entirely with traditional doctrine and held that unless a change in the structure of a building actually caused fire, the insured could recover under his policy. It was "immaterial" the court said, that the alteration had abstractly increased the risk.

This change in the rules governing "deviations" was part of a more general movement toward conceiving of insurance as a general policy of indemnity. The major institutional development that encouraged this perception was the chartering of incorporated insurance companies with large pools of capital during the 1790s. As a result, judges began gradually to treat individual litigation as no longer involving the personal destinies of the individual merchants and underwriters before them, each of whose livelihood might depend on the outcome, but as just one small component of a more general risk pool. The movement toward conceiving of insurance policies as a form of general protection of all risks becomes pronounced as fire replaces marine insurance as the major form of underwriting after 1840, but one can see this tendency toward an actuarial consciousness much earlier.

One of the earliest examples of an emerging actuarial consciousness appears in Alexander Hamilton's argument in the New York case of *Barnewall v. Church*[85] (1803). The case involved an underwriter's effort to defeat recovery by establishing that the insured vessel was unseaworthy, which, according to established doctrine, included all latent defects, even those unknown to the ship owner. Arguing for the insured, Hamilton put forth an entirely novel argument, based on actuarial conceptions, for treating the marine policy as covering all risks. "The underwriter," Hamilton argued, "in forming his calculation, considers the quantity of losses in proportion to the safe arrivals. On this datum he forms his estimates; seaworthiness must, therefore be included. Of the number [of ships] foundered at sea, many must have perished from latent defects. . . . Therefore, these must have constituted part of the risk calculated. If, then, the calculation be founded upon this, latent defects are paid for, in premiums actually received for them by the underwriter."[86]

Hamilton was not only arguing against the traditional inclusion of latent defects in the class of unseaworthiness; he was also attempting, through statistical argument, to break down the orthodox distinction between ordinary and extraordinary risks and to convert the marine insurance policy into one of general indemnity. Though this argument failed, it seems to have accorded more generally with mercantile thinking of the time, for the insurers' counsel conceded that "it has now become almost a maxim for [merchant] juries never to find a verdict for a defendant when unseaworthiness . . . [is] relied on in defense."[87]

Hamilton's argument was, of course, incapable of proof by pure logic. Though the Chief Justice of New York acknowledged that his "first impression" of Hamilton's "novel position" was "favorable to its correctness, notwithstanding the force of authority against it," he eventually recognized that judicial enforcement of the defense of unseaworthiness "diminishes the number of losses, and thus reduces the chances against [the insurer]."[88] As a result, the aggregate number of losses on which the premium is calculated, he saw, was reduced.

Barnewall v. Church illustrates both the relative unfamiliarity of early nineteenth century judges with statistical thinking and the gradual transformation in modes of thought brought about by the aggregation of risks made possible by corporate insurers. While English marine insurance doctrine had already crystallized before an actuarial consciousness had completely dawned, and for that reason it stood relatively unchanged against the desires of merchants to insure against all marine risks, the subsequent development of fire insurance doctrine more fully reflected the growth of statistical thinking in the nineteenth century.

One of the consequences of the growth of an actuarial consciousness was that courts tended to treat insurance provisions as more or less general policies of indemnity. The clearest shift in doctrine reflecting this change is judicial treatment of the distinction between a warranty and a representation. In the nineteenth century jurists began gradually to react against the highly penal consequences that flowed from breach of warranty by construing many provisions in insurance contract as simple representations, whose breach would void a policy only if it resulted in an increase in the risk. For example, in his influential treatise on *Marine Insurance* (1846), John Duer announced his "conviction that on the subject of representation and warranty, the [continental] law of insurance is, for many reasons, preferable to our own." He attacked the Anglo-American tradition of construing warranties "as a condition, on the literal truth or fulfillment of which, the validity of the entire contract depends." He proposed instead that they be considered simply as representations whose "substantial truth or performance" was sufficient for recovery under a policy and, most important, that these provisions should be "construed, with great liberality, in favor of the assured."[89]

Duer's technical discussion of warranties and representations reflected a more general tendency to erode the distinction between coverage of extraordinary and ordinary risks in insurance policies. For example, in 1824 the great Philadelphia commercial lawyer, Peter S. DuPonceau, protested as "contrary to every principle of sound jurisprudence" various English insurance doctrines that narrowed the scope of coverage.[90] And in 1842 Chief Justice Gibson of Pennsylvania maintained that although "from the language of the books, it would seem that an opinion has sometimes been entertained that there is a distinction between those perils which are extraordinary, and those which are only ordinary, . . . this distinction, if it ever existed, has been nearly, if not altogether obliterated by . . . later cases." He cited, among others, those cases which allowed recovery on an insurance policy for damage brought about "even by the negligence of those who had the injured vessel in charge."[91] Similarly, at the same time as the tort law was beginning to invoke the technical distinction between remote and proximate cause in order to reduce entrepreneurial liability for accidents, judges were extending coverage by rejecting this same distinction in insurance law. "If there be any commercial contract which, more than any other, requires application of sound common sense and practical reasoning in the exposition of it . . .," Justice Story wrote, in one of these cases which upheld recovery for a collision at sea, "it is certainly a policy of insurance; for it deals with the business and interests of common men, who are unused to deal with abstractions and refined distinctions."[92] Those "abstractions and refined distinctions" which Story sought to overthrow in the law of insurance generally represented earlier doctrines that stood in the way of treating the insurance policy as extending to all risks.

Although by the time Duer and Story wrote marine insurance law had become too settled for any extensive overhaul, judges, who had begun to perceive insurance doctrine through actuarial lenses, already had initiated a thorough transformation of fire insurance law. Thus, the Chancellor of New York in 1836 decided against extending to fire insurance "the principle of construing every matter of mere description contained in the body of the policy, although not material to the risk, into an express warranty which is to be literally complied with."[93]

By the middle of the nineteenth century, judges had, in fact, relaxed the penal framework of warranty doctrine. Even in the law of marine insurance, they gradually eroded the remarkably strict rules governing seaworthiness. Despite the opposition of merchants, during the first decade of the nineteenth century, the New York Supreme Court had established the remarkable rule that if a ship foundered at sea for no apparent cause, it would be conclusively presumed to have been due to the vessel's unseaworthiness.[94] This rule seems to have reflected a more general pattern of early nineteenth century insurance decisions which assumed that "the underwriting interests would be preferred over the shipping interests."[95] By the middle of the century, however, many courts had shifted ground and required the insurer to prove unseaworthiness and, in addition, had converted the earlier conclusive legal presumption into a mere evidentiary presumption, thereby returning the task of finding unseaworthiness to juries, which had always been hostile to this restriction on recovery.[96] The warranty of seaworthiness, which had been used prominently by insurance companies as a major technical obstacle to recovery, was thus gradually diluted in its impact.

As a result of these changes, we can see an almost complete shift to an actuarial consciousness in Theophilus Parsons' *Treatise on the Law of Marine Insurance* (1868). The entire law of insurance, he wrote, rests upon "the great law of average . . . a law, or a fact, of vast and wide power. Only of late years has it become the subject of scientific investigation," although it was even now "not yet well understood."

No longer was marine underwriting conceived in terms of insuring individual risks. "It would be impossible to predict with any certainty" whether a single ship would return from a voyage, Parsons acknowledged. But "take ten ships, and there arises some probability as to all of them. Take a hundred, and the probability becomes far greater. Take a thousand, and this probability becomes a rule." As a result, "premiums are not specifically adjusted to each case, except in some slight degree. They must, on the large scale, be founded upon general rules and estimates."

The "ideal perfection" of insurance, Parsons maintained, "would become actual, if all maritime property were covered by insurance, and the risks of this insurance were widely distributed." Short-run "departures" from the ideal were brought about by the law of supply and demand, as insurers and underwriters brought premiums and risk into equilibrium. Nevertheless, the law of insurance also played a crucial role in this process of movement toward a market equilibrium. Parsons offered the rule governing seaworthiness as an example. "Let us first suppose that it is too liberal, or lax, in favor of the insured." This would encourage insurance by "the careless, if not the fraudulent ship-owner," thus increasing pay-outs, driving premiums up, discouraging the careful from insuring, which in turn would drive premiums up still higher. On the other hand, if the rules governing seaworthiness "are too severe and strict," it "would exclude from insurance many vessels which might still be usefully employed" and also impose a burden on "the lower and poorer classes."

Even though it was "impossible" to prescribe a "positive rule or a precise definition" which would bring about equilibrium, Parsons maintained, the major object of insurance law was "to find a just medium." Courts had fluctuated between rules which "for a certain period, have gone in a direction which favors one of the parties to the contract of insurance at the expense of the other," and, then, out of "their judgement . . . of expediency," have manifested "a tendency to go to the opposite extreme. . . . But nothing can be more certain than the law of insurance, as a science and a system, will advance and improve in proportion as it reaches the just medium in all questions of this kind."[97]

Parsons' was one of the earliest systematic statements of an actuarial conception of law that was to receive much more extensive elaboration at the end of the century when Holmes declared that "the man of the future is the man of statistics and the master of economics."[98] For Parsons, as for Holmes, the legal rules of insurance were entirely conventional. They were developed and changed primarily for the purpose of carefully establishing a balance between the money going into and out of an insurance fund. In this conception, there was hardly any room for the idea that legal rules sought to do justice in the individual case. Indeed, it was appropriate to change the scope of a legal rule where it created a serious imbalance between supply and demand.

It is important to see, however, that Parsons did not conceive of the law simply as providing a passive but regular framework that merely reflected the results reached through private bargains. He did not, in short, adopt the view that it was more important that a rule be certain than that it be correct, a view which corresponded to the belief that private bargains over premium and risk would bring supply and demand into harmony if only the law would provide a clear and predictable background of rules. Rather, in his view, the legal system stood as a traffic policeman actively adjusting the rules in order to regulate the flow of supply and demand.

There is a striking parallel here between the underlying assumptions of Parsons' "objective" theory of contract and his actuarial conception of the law of insurance. Both theories sought to subordinate individual bargains to a socially efficient system of legal rules. Both emphasized the aggregate effects of legal rules rather than individualized justice in the particular case. Both, in somewhat different ways, reflected the increase in economic concentration and the resulting desire for standardization, characteristic of the American economy after the middle of the nineteenth century.

USURY

The movement to abolish usury laws provides another excellent barometer of legal conceptions of the nature of economic exchange. Before 1825 there was no great attack in America on statutes outlawing usury. Although Jeremy Bentham's *Defense of Usury* (1796) was to provide the intellectual foundation for later efforts urging repeal of usury laws, it had little impact on American thought until much later. Indeed, before 1825 there were many more published tracts written in defense of usury laws than in defense of usury.

One of the earliest attacks on usury legislation in America was William Sheldon's *Cursory Remarks on the Laws Concerning Usury* (1798). Written in opposition to the actions of a society established in Norwich, Connecticut, for bringing indictments against usurious moneylenders, Sheldon's tract noted that the European war had expanded American trade and that "this increase of trade has occasioned a great demand for money." It was necessary therefore to have "recourse . . . to rich individuals, who, thinking their money, like merchandise, should bear a price in proportion to the demand for it," discounted bills and notes at usurious rates "in the large towns, such as Philadelphia, New York and Boston." Usury laws, Sheldon argued, were "both immoral and unjust," especially since they encouraged "a species of persecution against men of property." He quoted New York Governor John Jay, who complained that "men of otherwise irreproachable character are guilty of the practice of Usury."

Sheldon was the first American, it appears, to follow Bentham and argue that money, like any other commodity, should follow the law of supply and demand. "Where then," he asked,

is the necessity or propriety of preventing the man who possesses money from taking advantage of this favourable time, any more than there is for preventing the Farmer or Merchant from availing themselves of a good time for the sale of their produce or Merchandise—or the Jobber in Susquehanna or other lands, from selling his speculations at 50, 100, or perhaps 500 percent profit if he can?[99]

These arguments, however, were largely ignored for another three decades. Until around 1825 eighteenth century just price conceptions continued to dominate American legal doctrine. In 1820, for example, Virginia lawyer Francis Gilmer defended usury laws on the still orthodox ground that they protected individuals of unequal bargaining power from being oppressed. "If there be a case imaginable . . . in which the inequality of the contracting parties amount to oppression," he wrote, "the loan of money at enormous interest is one. Every reason which sanctions the interference of law in other cases, applies more strongly to this." Even without statutes barring usury, Gilmer added, there was "no doubt" that usurious bargains "would share the same fate" as "the numerous cases I have mentioned, in which courts of equity have rescinded contracts, because unequal and unfair."[100]

Another reason for the late appearance of pro-usury arguments is that before 1825 speculation in land and commodities had not yet become completely rationalized in American legal or economic thought. For example, the still powerful "just price" underpinnings of anti-usury laws were reflected in a 1789 decision of the Connecticut Supreme Court of Errors holding unenforceable a speculative "futures" contract for the delivery of stock. The fact that stock certificates were highly fluctuating in value, the court asserted, was irrelevant, "for there is no article whatever, that can be loaned but what may and frequently does change its relative value." Where "a higher price is given for goods than their actual value, for the purpose of concealing the usurious intent of the contract," the court held, it would refuse enforcement.[101]

There were also occasional efforts before 1825 to justify usury legislation in terms of utilitarian theory. In 1811 Jeremiah Mason, soon to be appointed United States Senator from New Hampshire, corresponded with the Reverend Jesse Appleton, the president of Bowdoin College, about usury laws. He was skeptical of Appleton's assertion that such laws were designed for "the protection of the poor from oppression," since "such laws have been in use in many countries where the rights of the poor were little respected." The "principal object" of usury legislation, Mason wrote, was "to induce the rich capitalist to use his own stock and be industrious."

It is more advantageous to society that the rich capitalist should use his own industry in the employment of his stock, than that he should sit idle and take the benefit of the industry of others. The loan of money therefore at a high rate of interest, which would encourage the capitalist to be idle, has always been discouraged.

But the object of usury laws, Mason continued, was not to prevent all loans, since the capitalist "ought to lend that part which he cannot employ to advantage, and *that part only*. If he lends the whole he must become idle himself." In order to allow this to occur, Mason concluded, "the legal rate [of interest] ought to be nearly the same with the market rate, or the rate it would fix at if not regulated by law."[102] Thus, while he set out to use market theory to justify usury laws, Mason concluded by establishing the market rate as the only appropriate measure for the legal rate of interest. Yet, his identification of moneylending with the encouragement of "idleness" reveals a still dominant precommercial mentality that was attempting to use market theory to make sense of a world that was rapidly fading away.

After 1825 a major movement to repeal regulations on interest was launched. Between 1825 and 1838 a flood of books, pamphlets, and speeches denouncing usury laws was produced. In 1826 John Ramsey M'Culloch argued in *Interest Made Equity* that "every principle of natural justice, and of sound political expediency" required the repeal of usury legislation. "If a landlord is to be allowed to take the highest rent he can get offered for his land—a farmer the highest price for his raw produce—a manufacturer for his goods—why should a capitalist be restricted and fettered in the employment of his stock?"[103] It was typical of writing in both political economy and law after 1825 to convert the issue of unequal bargaining power into one of simple intellectual comprehension and then to "presume" that, short of mental disability, all persons were equally endowed with sufficient understanding. For example, in his *Lectures on the Elements of Political Economy* (1826), Thomas Cooper of South Carolina signaled a revival of interest in Bentham's "profound investigation" of usury. "Usury laws," he wrote, "are the result of that mischievous intermeddling so constantly attendant on ignorance in the seat of power. When two persons meet to contract for their common benefit, if they are of reasonable age and understanding, they can make their own bargain better than any one else can make it for them. . . . Men should be presumed possest of a common and reasonable share of understanding."[104]

In 1828 Willard Phillips argued in *A Manual of Political Economy* not only that usury laws were "manifestly unjust" but that "in commercial places these laws have very little force," since "no one can avail himself of them without becoming infamous."[105] In addition, a series of four articles in the *Journal of Law* for 1830 signaled the beginnings of a major legal campaign against usury legislation.[106]

First, it was necessary to dispose of the religious sanction. "Whatever may have been the mandates of the Jewish lawgiver on the subject of usury," the writer began, "the Christian dispensation contains no such doctrines."[107] Next, it was important to identify money as simply another commodity and to sing the praises of unregulated commodity markets. "A free market for money would produce a fair and regulated competition for it, as it does in relation to every other article of trade." Finally, it was argued that the law could not effectively intervene to stem the inexorable spirit of the age. Usury laws "afford no effectual protection to the prodigal and weak, while they impose upon the industrious and the enterprising inconvenient restraints."[108]

It would be tedious to recount in detail the arguments of each opponent of the usury laws. Suffice it to say that in general they reflect each stage in the ultimate triumph of a market ideology. "There is nothing in the nature of money or money transactions," wrote Thomas Dew, professor of political economy at the College of William and Mary in 1834, "calling for the special interference of the legislator, and justifying the restrictions on the profits made by one class of capitalists, while all others are left free to make what gains the state of the market will allow."[109] "No reasoning," another commentator declared, "can be used with regard to this species of contract which will not equally justify interference with every other form of contracting."[110] Many opponents of usury laws emphasized that free competition would lower the interest rate. "OTHER THINGS BEING EQUAL," one writer proclaimed, "MONEY WILL BE PROCURED MORE EASILY, AND AT A LOWER ACTUAL COMPENSATION WITHOUT THE RESTRICTION OF USURY LAWS, THAN IT CAN BE WHERE THE RATE OF INTEREST IS PRESCRIBED BY STATUTE."[111]

In 1834 a group of Massachusetts citizens petitioned the state legislature for repeal of the laws against usury. Usury laws, they argued, "are at variance with the commercial spirit of the age."

We are firm in the opinion that all money transactions should be regulated, like those in other articles, of trade, only by this spirit of competition; and that no greater evils would or could, in the present age, arise from the traffic in money being thus unrestricted, than are now felt from the perfect freedom allowed to traffic in other commodities.

While the committee of the Massachusetts legislature that considered the petition had "no hesitation in saying that they entirely concur[red]" in the petitioners' views, they nevertheless considered "any sudden and extensive changes in the laws" to be "generally inexpedient." However, they did recommend a repeal of usury laws as they applied to negotiable instruments, since "the evils of the existing system are felt more strongly in reference to this class of contracts, than to any other."[112]

John Bowles, who had joined with A. H. Everett, in making the report of the committee of the Massachusetts legislature public, himself published *A Treatise on Usury Laws* in 1837. He further elaborated upon the now familiar "historical proofs, and . . . argument[s], to show the inutility and positive mischiefs" of laws regulating the amount of interest charged. Usury laws, he wrote, were "a violation of the best principles of both politics and political economy . . . the fundamental political doctrine . . . that government is best, which by the smallest machinery, and the simplest process, and the least infringement of individual liberty, effects the purpose for which government is intended, —the general welfare." Not only was "the prevention of prodigality and imprudent speculation" an insufficient justification for legal interference; laws against usury penalized "the class of men . . . composed of the fairest, the most scrupulous, and the most humane."[113]

The laissez-faire assumptions behind the attack on usury laws were further underlined in Thomas Cooper's notes to his edition of the *Statutes of South Carolina* published in 1837. "The public notions on the subject of usury," Cooper wrote, "have been totally changed by Jeremy Bentham's brief treatise on that subject; and in a few years it is to be hoped that legislatures will cease somewhat from their mischievous propensity of governing too much; and leaving men who have arrived at the age of discretion, to make their own contracts as circumstances dictate the convenient terms of the mutual bargain." Money, Cooper observed, "is really worth more at one time than another," and "every one acknowledges that compensation ought to be proportioned to risk. . . . The whole business of the stock exchange is founded on this principle, because it is a principle of common sense." Finally, he concluded: "The less legislatures interfere with the common business concerns of those who ought to know what they are about, the better."[114]

In the period between 1825 and 1838 only two major defenses of usury laws were undertaken. The first, by John Whipple of Rhode Island, represented a general attack on Bentham's free market conception of money. Money, he argued, "is totally *unlike* any thing else," since "being in the hands of the few, the facilities for creating an artificial scarcity are much greater, than for creating an artificial scarcity of merchandise." Finally, Whipple emphasized "that the parties do not stand upon equal grounds" when contracting for money "any more than a prisoner contracting with his creditor." Yet, he acknowledged that "in dealing for other articles, men do stand on equal grounds." "In most other countries," he concluded, "there exists a check —a *moral* restraint upon these excesses, which under a free trade system, would be inefficacious."[115]

The second important defense of usury legislation was put forth by Chancellor Kent, who in 1837 remarked on the movement to repeal usury laws. Citing the bar on usury as "a principal of moral restraint of . . . uniform and universal adoption . . . founded on the accumulated experience of every age," Kent argued that they grew "out of the natural infirmity of man, and the temptation to abuse inherent in pecuniary loans."[116]

By attempting to distinguish money from other merchandise, both Whipple and Kent yielded to the spirit of the age and detached the argument against usury from more general historical roots in just price theories of value. They felt that it was no longer possible to maintain either that commodities bore an inherent or customary value or that fear of the consequence of inequality of bargaining power could justify special rules to regulate only the price of money. By conceding the supremacy of the market in general, they tried to do what Chancellor Kent ardently denied he was doing—to take advantage of "monkist prejudices" against money, while reenforcing the more general belief that the market could be the only arbiter of value.

Despite these widening intellectual attacks on usury legislation, by the time of the Civil War every state except California continued to regulate the legal rate of interest. Yet, the forms of regulation had changed dramatically, thus virtually erasing in most states all effective legal deterrence against usury.

Before 1820 the overwhelming majority of American states had followed the pattern of eighteenth century English usury statutes, which declared that contracts extracting unlawful interest were void. The effect of such statutes was to cause lenders to forfeit not only the agreed interest but also the entire loan. In 1819, for example, sixteen states had one or another form of this stringent forfeiture provision on their books, including a Virginia statute which went still further and made the lender liable for a penalty of twice the debt. In only eight states was the penalty for usurious contracts something less than forfeiture. This included the stringent statutes of New Hampshire, which made the lender liable for three times the usurious excess above legal interest, and of Vermont, which subjected the creditor to a penalty of 25 percent of the entire debt. Indiana's statute had the effect of causing the creditor to forfeit twice the interest. In only five states, then, was usury not substantially deterred by penalties. Illinois, Mississippi, and Louisiana had no penal provisions. In Missouri the lender was subjected to the trifling penalty of being unable to recover the usurious excess.[117] And in Pennsylvania the Supreme Court in 1785 had emasculated that state's stringent sixty year old anti-usury statute by holding that the creditor simply forfeited the usurious excess.[118]

By the time of the Civil War, however, the majority of states had eliminated effective deterrence of usury from their statute books. Only seven states along the Atlantic Coast now continued to provide either that usurious contracts were void or that the entire debt should be forfeited.[119] And even in two of these states, negotiable instruments had been excepted from the operation of the usury law. Nineteen states, by contrast, had established mild penalties for usury, limiting debtor remedies to recovery of either the usurious excess above legal interest or the entire interest. California had no penalty at all. Somewhere in between, Massachusetts, New Hampshire, and Wisconsin provided for a penalty of three times the usury, while Indiana continued to allow the debtor to recover a penal sum of twice the usurious excess.[120]

In many states where usury was no longer a ground for voiding a contract, courts also held by this time that usury was not a good defense against a bona fide purchaser of a negotiable instrument.[121] And in those states that allowed bonds to be negotiated, there was also the additional common law barrier against inquiry into the consideration of a sealed instrument.[122] Thus, it was widely possible by the time of the Civil War to arrange usurious transactions in such a way as to entirely avoid running afoul of the usury laws.

Despite the concentration of anti-usury law writing during the 1830s, legislative change was spread out fairly evenly after 1819. Kentucky was the first to eliminate the provisions making usurious contracts void in 1819. Rhode Island and Massachusetts followed in 1822 and 1825, respectively. Statutes in South Carolina (1830), Tennessee (1835), Maine (1840), Georgia (1845), and Ohio (1848) all eliminated the effective deterrence of earlier usury provisions.

In addition to changing the all-important remedial provisions, there was a substantial trend by the time of the Civil War toward increasing the legal rate of interest. California set no legal limit on interest. Minnesota and Texas allowed 12 percent while Arkansas, Illinois, Indiana, Iowa, Michigan, and Wisconsin placed the legal ceiling at 10 percent.

The clearest pattern that emerges from this analysis is that southern and western states, where credit was difficult to obtain, were at the forefront of breaking down restrictions on usury between 1820 and 1860. No newly admitted state after 1820 felt free to imitate the eastern ban on all usurious transactions. On the other hand, the only major commercial state that continued to void usurious contracts as late as 1860 was New York, where, paradoxically, even at the beginning of the nineteenth century it was common to observe that it had "become almost a maxim" for merchant juries "never to find a verdict" for the debtor "when . . . usury [is] relied on in defense."[123] Nevertheless, even the New York legislature had, by 1833, eliminated the usury defense for actions on negotiated notes.[124]

Beginning in the 1820s, then, the assumptions of a market system led to widening attacks on traditional "just price" conceptions of value. The ideal of unregulated exchange transformed not only judicial but also legislative attitudes as well, so that by the time of the Civil War it was substantially easier to engage in usurious transactions. Nevertheless, no legislature actually repealed all restrictions on usury. Rather, penalties for usury were weakened and exceptions for commercial transactions were greatly expanded. By this time, those restrictions that remained were invariably perceived as entirely arbitrary, for it was no longer possible to recapture an earlier and more coherent system of premarket morality.

Swift v. Tyson: The Rise of a General Commercial Law

One of the most interesting and puzzling developments in all of American legal history is the appearance of the Supreme Court's decision in *Swift v. Tyson*[125] in 1842. Standing for almost 100 years until it was overruled in 1938, it proclaimed the rule that in the exercise of its diversity jurisdiction, the federal judiciary was not bound by state judicial decision but rather was free independently to decide a case on the basis of "the general principles of commercial law."

Though it was clear under the 34th section of the Judiciary Act of 1789 that the federal courts were required to follow state "law," Justice Story in *Swift v. Tyson* nevertheless held that "in the ordinary use of language, it will hardly be contended that the decision of courts constituted laws. They are, at most, only evidence of what the laws are, and are not, of themselves law." According to Story's interpretation, therefore, federal courts in diversity cases were required to follow state statutes, but were not generally bound to follow state judicial decisions.

Swift v. Tyson has been regularly identified as expressing the so-called "declaratory" theory of law or the view that courts simply discovered and declared preexisting rules. It is supposed to have reflected, in the famous words of Mr. Justice Holmes, a conception of the common law as a "brooding omniprescence in the sky."[126]

It was critical to my argument in Chapter I to demonstrate that such a view of law was actually in the process of eroding in the decades after 1780. Could Story really still have believed it when *Swift v. Tyson* was decided in 1842? Fortunately, we have a rough indicator of changing jurisprudential assumptions as measured by the development of the closely related field of conflicts of laws, to which Mr. Justice Story contributed a major treatise in 1834. At the outset, therefore, it is crucial to see the relationship between the decline of the declaratory theory of law and the rise of a "conflicts" approach to incompatible state legal rules. As Professor Crosskey saw, "the now prevailing conflicts-of-laws technique ha[d] little application" in the eighteenth century and "was the slow development of a later time," roughly after 1820.[127] The shift to a "conflicts" approach reflected the erosion of the orthodox view that, since judicial decisions were mere "evidence" of a "true" legal rule, a conflict of decisions inevitably meant that one of these rules was simply mistaken. The field of conflicts of laws, then, arose to express the novel view that incompatible legal rules could be traced to differing social policies and that the problem of resolving legal conflicts could not be solved by assuming the existence of only one correct rule from which all deviation represented simple error.

Before 1820 it had already become clear to legal thinkers that certain kinds of conflicting legal rules could be explained only by differences in social policy. This view, we have seen,[128] was precisely the basis of John Milton Goodenow's 1819 attack on common law crimes and his insistence that only statutory law could provide individuals with sufficient notice of essentially arbitrary political decisions. But even Goodenow stopped short of extending his analysis to all common law decisions. "The principles and objects of civil jurisprudence," Goodenow wrote, "are dissimilar and distinct from those of criminal law. . . . Public wrongs, crimes and punishments, depend on the legislative will for their existence as such: private rights and private wrongs are founded on and measured by the immutable principles of natural law and abstract justice." Thus, even as self-conscious a critic of the common law as Goodenow could still conclude in 1819 that "natural justice and reason are the same in all countries and in all ages."[129]

Out of this perspective, it is not surprising that conflicting common law rules should have continued to have been regarded simply as evidence of some departure from "immutable principles of natural law and abstract justice," requiring only a more intensive application of reason to bring about a "solution." Reflecting this view, the first American treatise on conflicts of laws, Samuel Livermore's *Dissertation on Questions Which Arise from Contrariety of the Positive Laws of Different States and Nations* (1828), dealt, as its title suggests, only with those problems that arose from applying conflicting statutes of different states. But such an inquiry represented no departure from an orthodox legal theory that had long recognized that statutes of different states could be in conflict because they were mere expressions of the sovereign will. The subject matter of Livermore's treatise, in fact, reflected a sharp increase in the number of insolvency statutes after the Revolution, a development which raised with much greater intensity the question of whether a statute of one state would constitute a good defense to action in another jurisdiction.

The change from Livermore's modest treatise in 1828 to Story's monumental work on *Conflict of Laws* in 1834, however, reflects recognition of a fundamental erosion of the declaratory theory of law. "It is plain," Story begins, "that the laws of one country can have no intrinsic force, *proprio vigore,* except within the territorial limits and jurisdiction of that country. . . . Whatever extra-territorial force they are to have, is the result not of any original power to extend them abroad, but of that respect, which from motives of public policy other nations are disposed to give them."[130]

The decision to enforce a conflicting law of another jurisdiction, then, arose not because the law had any "intrinsic force" of obligation but from "motives of public policy." "It is an essential attribute of every sovereignty," Story continued, "that it has no admitted superior, and that it gives the supreme law within its own domain on all subjects appertaining to its sovereignty. What it yields, it is its own choice to yield; and it cannot be commanded by another to yield it as a matter of right." Therefore, a court "may with impunity disregard the law pronounced by a magistrate beyond his territory."[131]

And how did Story explain why even the judge-made law of different jurisdictions could conflict? He echoed the opinion of a contemporary court. "Where so many men of great talents and learning are . . . found to fail in fixing certain principles we are forced to conclude that they have failed, not from want of ability, but because the matter was not susceptible of being settled on certain principles." Differences in laws depended upon "the particular nature of [a state's] legislation, her policy and the character of her institutions."[132] A state, therefore, was not bound to recognize laws that "would be prejudicial to its own interests"; recognition of the laws of another jurisdiction was a matter of prudence, "utility," and "comity" "and not of absolute paramount obligation, superceding all discretion on the subject." In short, the obligation to enforce laws turned on a state's "exclusive right to regulate persons and things within its own territory according to its own sovereign will and policy."[133]

Besides this unyielding insistence on a "will" theory of law, the most important transformation in Story's treatment of conflicts of laws was that he did not hesitate to include within his theory the problem of reconciling contradictory common law doctrines. His entire treatment of differing common law rules of contract and commercial law, for example, emphasized that, like statutory conflicts, any solution turned on questions of "convenience" and "the necessities" of giving due respect to the laws of others, "for otherwise, it would be impracticable for [states] to carry on an extensive intercourse and commerce with each other."[134] No longer, in short, was the legal world divided between statutes, which were regarded simply as arbitrary commands of the sovereign, and common law rules, which were thought to be discovered from "immutable principles of natural law and abstract justice." Nor was there any longer a fundamental dividing line between criminal laws, conceived of as arbitrarily derived from social policy, and civil law rules, still believed to be deducible from "natural justice and right reason." Rather, the full flowering of the field of conflicts of laws in Story's treatise represents a complete recognition of the view that legal rules could no longer be conceived of as deriving from a "brooding omniprescence in the sky."

In light of these jurisprudential changes, what are we to make of Mr. Justice Story's opinion handed down only eight years later in *Swift v. Tyson?* Story, it will be recalled, decided that federal courts were not required to follow New York decisions on a point of commercial law because they constituted only mere "evidence" of law, a classic restatement of the declaratory theory. Suppose, instead, that the case had come before Justice Story sitting in a different role, this time as a judge in a state judicial system. What guidance would Justice Story's treatise on *Conflicts* have provided?

There is nothing, of course, in the jurisprudence of Story's *Conflicts* that would have required a judge of another state either specifically to follow or not follow New York decisional law, even though the relevant transactions took place in New York. That determination, the treatise tells us, turns on questions of prudence and policy. Yet it is perfectly clear that a state judge, following Story's treatise, could never refuse to follow New York decisions simply because they were not "law " but only "evidence" of a "general commercial law." The jurisprudence of the treatise on *Conflicts* shows no trace of the view that differing rules of commercial law can be reconciled by reference to one overriding general law. Indeed, the treatise is written precisely because such a view of law can no longer satisfactorily explain the existence of a growing number of conflicting legal rules. In short, the conception of law put forth by Mr. Jus-

tice Story in *Swift v. Tyson* stands sharply opposed to the jurisprudence of his treatise on *Conflict of Laws* written just eight years before.

Before we turn to other, nonjurisprudential explanations of the
decision in *Swift v. Tyson*, it is worth observing that the view of law
put forth in that case may have had a more decisive influence on
subsequent judicial treatment of conflicts of laws than did Story's
treatise. It is one of the grim ironies of American legal history that
state judges in the second half of the nineteenth century tended to
undermine Story's quest for uniformity of legal doctrine by following Justice Story's decision and asking whether conflicting decisions
of other states should be followed or, instead, be treated as "wrong"
from the standpoint of the "general law."[135] Needless to say, it was
virtually impossible to expect a state judge to concede that a common law rule of another jurisdiction, though at variance with that
of his own state, should be enforced because it nevertheless represented a correct application of the "general law." It would have
been infinitely easier to persuade judges simply that prudence, prac-
ticality, and necessity required the enforcement of extraterritorial
common law rules, which is the decided purpose of Story's treatise.
In short, this unexpected extension of the jurisprudence of *Swift v.
Tyson* to the field of state conflicts of laws in the second half of the
nineteenth century disastrously undermined whatever contributions
to legal uniformity that decision itself is supposed to have contemplated.

Swift v. Tyson thus represents the beginnings of a more general
resurgence of legal formalism as the middle of the nineteenth century approached, a trend that will be discussed more thoroughly in
the last chapter. Suffice it to say for now that its unexpected extension to the field of conflicts of laws is one of the most dramatic instances of the indeterminacy of particular results under formalism.
To the extent that state judges were induced to believe once again in
a "general jurisprudence" and in a "declaratory" theory of law, they
were left completely unprepared to deal with the realities of a legal
system that was growing ever more complex and contradictory. The
only hope left, then, for creating uniformity was for the triumph of
the dogmatism of the treatise tradition, which might suppress all
centrifugal legal tendencies before they could even be conceived.

If *Swift v. Tyson* cannot be taken seriously as a simple application
of orthodox legal theory, how should it be understood? First, it
should be pointed out that both the United States Supreme Court
and Mr. Justice Story had been performing similar functions for

quite some time, although they had never before crystallized into so grandiose a statement of legal theory.

We have already seen how as early as the first decade of the nineteenth century the Supreme Court came to allow endorsees of promissory notes to sue their remote endorsers "in equity," even where state decisional law had refused to recognize the principle of negotiability.[136] As a result, procommercial interests were able to use the federal courts to finesse the precommercial decisional law of their states. Since these early Supreme Court decisions were framed in the limited terms of mere recognition of "remedies," however, the judges were able for a time to avoid having to inquire too closely into the specific source of the legal right. By distinguishing between legal rules and remedies, in short, the federal courts were initially able to provide relief to commercial interests while still purporting to follow a state's decisional law.

We have seen that by the time *Swift v. Tyson* was decided, there had developed a clear split between pro- and anticommercial forces over such questions as negotiability and usury. Thus, the much heralded quest for legal uniformity that *Swift v. Tyson* is supposed to have represented can also be seen more concretely as an attempt to impose a procommercial national legal order on unwilling state courts. Indeed, it was not too long before the one major restriction on the pursuit of uniformity originally built into *Swift v. Tyson*—the requirement that state statutes were binding on federal courts because they alone constituted "law"—was also eroded or ignored. In some cases the Supreme Court bluntly declared that state statutes had to yield before the "general commercial law"; in many other cases the federal courts simply refused to be bound by an anticommercial state court "interpretation" of its own statute.[137]

Not only had the Supreme Court furthered commercial interests and overriden state policies long before *Swift v. Tyson* through manipulation of legal remedies, but Justice Story had also tried from the beginning to advance these interests through an expansion of federal jurisdiction. Story's 1815 circuit court decision in *DeLovio v. Boit*[138] provides an early example of the underlying procommercial spirit that eventually came to fruition in *Swift v. Tyson*. Relying on a surprisingly expansive interpretation of the admiralty jurisdiction of the federal courts, Story held that all disputes concerning maritime contracts—and especially the much litigated field of marine insurance contracts—were within the jurisdiction of the federal courts. He derived his claim for an expansive federal admiralty power from the extensive jurisdiction that the English Admiralty had attained during the early seventeenth century, when English commercial interests had used Admiralty as an alternative forum to avoid the anticommercial attitudes of the common law courts. Even though the jurisdiction of the English Admiralty had been substantially cut back by the eighteenth century, Story nevertheless regarded this early precedent as providing the opportunity to establish a federal forum for commercial disputes. Before 1815, it should be emphasized, commercial law revolved almost completely around maritime transactions, so that the effect of *DeLovio* was to create a federal commercial forum.

The most obvious advantage that commercial interests saw in federalizing marine insurance litigation under the admiralty jurisdiction was that it could proceed without juries. Story himself remarked that "to my surprise, I have understood that the opinion [in *DeLovio*] is rather popular among merchants. They declare that in mercantile causes, they are not fond of juries; and in particular, the underwriters in Boston have expressed great satisfaction at the decision."[139]

In fact, as we saw in an earlier discussion of insurance law, hostility to juries was becoming a standard position of marine insurance companies, which regarded jurors as inevitably biased in favor of the insured. Thus, the effort by Story to banish jurors from federal marine insurance litigation probably favored insurance companies at the expense of more general mercantile interests. At a time when this once consensually based field of commercial law was breaking down under the stress of conflicting interests within the commercial community, there did exist a strong independent pressure within the legal profession to reimpose legal uniformity once again. But the establishment of uniformity through the admiralty power, to the extent that it succeeded, was a clear triumph for the insurers at the expense of merchants.

In practice, however, the decision in *DeLovio v. Boit* had only insignificant consequences.[140] Some of Story's Supreme Court brethren, on circuit, refused to construe the federal admiralty power so broadly,[141] and the court itself managed to avoid resolving the question until after the Civil War, when it finally accepted Story's position.[142] The question, in fact, turned out not to be one of great practical consequence primarily because Story had failed to go the full length of proclaiming exclusive federal jurisdiction over maritime contracts, which in this early period would surely have brought on a most intense antinationalist reaction. Finally, after 1815 maritime commerce began to constitute an ever smaller portion of American commercial transactions, so that until the Supreme Court in

1851 extended its admiralty jurisdiction still further to cover internal navigable waters,[143] the reach of the admiralty power was no longer coextensive with commercial life. As a result, as Story himself saw, the case had little practical importance because most insurance litigation continued to brought in the state courts.

It has not been fully appreciated that as a practical matter Story's opinion in *DeLovio v. Boit* attempted to do almost precisely the same thing as he later tried to accomplish in *Swift v. Tyson*. Before 1815 or 1820, most questions of commercial law arose out of maritime transactions and hence every effort to apply a uniform federal rule to these transactions was tantamount to establishing a general commercial law. Only after commercial law expanded to accommodate internal economic growth during the second quarter of the nineteenth century was it no longer possible to control the development of virtually all commercial law through the federal admiralty power. It is in the context of these changed economic circumstances that *Swift v. Tyson* attempts to use federal diversity jurisdiction to accomplish what *DeLovio* had unsuccessfully sought to bring about much earlier.

Neither *DeLovio* nor *Swift* could ever fulfill the grandest aspirations of its author — the desire to establish an exclusive federal forum for commercial disputes, which would not only provide uniformity and certainty but would also take these disputes out of what might otherwise be an uncongenial anticommercial environment often found in state courts. Neither decision was able to erect a complete monopoly over commercial law, since other well-established institutional claims and legal doctrines stood in the way.

VIII The Rise of Legal Formalism

F OR seventy or eighty years after the American Revolution the
major direction of common law policy reflected the overthrow
of eighteenth century precommercial and antidevelopmental common law values. As political and economic power shifted to merchant and entrepreneurial groups in the postrevolutionary period,
they began to forge an alliance with the legal profession to advance
their own interests through a transformation of the legal system.

By around 1850 that transformation was largely complete. Legal
rules providing for the subsidization of enterprise and permitting
the legal destruction of old forms of property for the benefit of more
recent entrants had triumphed. Anticommercial legal doctrines had
been destroyed or undermined and the legal system had almost completely shed its eighteenth century commitment to regulating the
substantive fairness of economic exchange. Legal relations that had
once been conceived of as deriving from natural law or custom were
increasingly subordinated to the disproportionate economic power
of individuals or corporations that were allowed the right to "contract out" of many existing legal obligations. Law, once conceived
of as protective, regulative, paternalistic and, above all, a paramount expression of the moral sense of the community, had come to
be thought of as facilitative of individual desires and as simply re-
flective of the existing organization of economic and political power.

This transformation in American law both aided and ratified a
major shift in power in an increasingly market-oriented society. By
the middle of the nineteenth century the legal system had been reshaped to the advantage of men of commerce and industry at the
expense of farmers, workers, consumers, and other less powerful

groups within the society. Not only had the law come to establish legal doctrines that maintained the new distribution of economic and political power, but, wherever it could, it actively promoted a legal redistribution of wealth against the weakest groups in the society.

The rise of legal formalism can be fully correlated with the attainment of these substantive legal changes. If a flexible, instrumental conception of law was necessary to promote the transformation of the postrevolutionary American legal system, it was no longer needed once the major beneficiaries of that transformation had obtained the bulk of their objectives. Indeed, once successful, those groups could only benefit if both the recent origins and the foundations in policy and group self-interest of all newly established legal doctrines could be disguised. There were, in short, major advantages in creating an intellectual system which gave common law rules the appearance of being self-contained, apolitical, and inexorable, and which, by making "legal reasoning seem like mathematics," conveyed "an air . . . of . . . inevitability" about legal decisions.[1]

The sources of legal formalism as it developed after 1850 can be traced to a much earlier bifurcation of American legal thought, for in reality two competing ideological tendencies operated on American law after the Revolution. The first, which dominated the law until around 1850, was largely shaped by the efforts of mercantile and industrial groups to capture and transform the system of private law that existed in 1776.

In commercial law, entrepreneurial groups from the beginning sought to change the law inherited from the colonial period and to restrict the power of the state to enforce substantive standards of fair dealing. They also regularly sought the state's aid in creating a more efficient system of debt collection, and after the Panic of 1819, they strongly increased their enthusiasm for bankruptcy legislation, seeking, however, to limit its benefits to "merchants and traders."[2] In general, the course of commercial law during the nineteenth century was to perfect the remedial system while circumscribing the power of courts—and, far more important, legislatures—to intervene substantively.

In property and tort law, the interventionist state was inseparably linked to the earliest goals of low cost development. In this area, it was the constant aspiration of the developers first to seek as much support for change as they could from the courts, while only involving the more politically volatile legislatures when they could not. It is remarkable how well they succeeded. The basic system of tort and property law (other than rules of inheritance and systems of title recordation) was judicially created. And, by and large, it was strongly geared to the aspirations of those who benefited most from low cost economic development.

These efforts to overthrow precommercial and antidevelopmental legal doctrines and institutions powerfully supported the instrumentalist character of private law before the Civil War. A second, seemingly contradictory, tendency however, was also established quite early in postrevolutionary constitutional law. By forging constitutional doctrines under the Contracts Clause barring retroactive laws and giving constitutional status to "vested rights," this line of intellectual development sought basically to limit the ability of the legal system — more specifically, of the legislature — to bring about redistributions of wealth. While the first tendency underlined and acknowledged the malleable — and hence political — character of law, the second sought to depoliticize the law and to insist upon its objective, neutral, and facilitative character.

While commercial and entrepreneurial interests thus saw the private law shaped to their own needs and interests, they managed also to derive the benefits of an antidistributive ideology developed in public law. One cannot but be struck by the sharp contrast between the utilitarian and instrumentalist character of early nineteenth century private law and the equally emphatic antiutilitarian, formalistic cast of public law. If public law during this period was dominated by a conservative fear that legislatures might invade "vested property rights" — that is, that it might be used to redistribute wealth for equalitarian ends — the reality of the private law system was that it invariably tolerated and occasionally encouraged disguised forms of judicially sanctioned economic redistribution that actually increased inequality.

More than any other jurist of the nineteenth century, Joseph Story brought each of these two contradictory tendencies to their highest fulfillment. His private law opinions are, by and large, highly utilitarian and self-consciously attuned to the goals of promoting procommercial and developmental legal doctrines. By contrast, his public law opinions are usually starkly formalistic, often antiquarian, and therefore have frequently puzzled historians by running contrary to the supposed prevailing economic needs of his day. These differences between the character of public and private law, however, can be traced directly to an underlying conviction held by all orthodox nineteenth century legal thinkers that the course of

American legal change should, if possible, be developed by courts and not by legislatures. The persistent formalism of public law, in short, was related to the infinitely greater threat of redistribution that statutory interferences with the economy represented.

Here, then, is the basic ambiguity in early nineteenth century American law: a public law devoted to preventing redistribution, a private law undergoing massive doctrinal change and bringing economic redistribution about, if only by inadvertance. Moreover, those who took the lead in erecting public law doctrines into major constitutional and ideological barriers to the "political" uses of law were also among the foremost advocates of a transformation of the private law system into one that subsidized economic development, if necessary, through the redistribution of wealth. For most of the antebellum period, then, while mercantile and entrepreneurial groups generally tied themselves to an instrumental conception of private law to achieve the goal of low cost economic development, they were simultaneously able to develop a noninstrumentalist conception of public law.

After 1825 or 1830, however, even private law doctrine reveals traces of an emerging formalism as it begins to take on an aggressively apolitical caste. The insistence on a separation between law and morality emphasized by Nathan Dane and most later treatise writers appears together with regular warnings of the dangers of using law for redistributive purposes. The attack on equitable doctrines of substantive justice in contract law was one example of this insistence on a rigid separation between law and morality. Since only the market could supply "neutral" principals of distribution free from all "political" (that is, dangerously equalitarian) influences, it became the task of the law to create legal doctrines that simply mirrored the market. Most of the basic dichotomies in legal thought of the nineteenth century—between law and politics, law and morality, objective and subjective standards, distributional and allocational goals—arose to establish the objective nature of the market and to neutralize and hence defuse the political and redistributional potential of law.

The desire to separate law and politics has always been a central aspiration of the American legal profession. From the time of its earliest incarnation in postrevolutionary constitutional theory, politics in American thought has usually represented power and will, the clash of interests, and the subjectivity of values. Law, by contrast, has been the only plausible claimant to the role of objectivity and political neutrality. The legal profession, in turn, has had every reason to insist on its own autonomy. If law is simply a product of power or will, any special claims of the profession to determine the nature and scope of legal development is undermined. The special power of the legal profession in American society has always been grounded in some theory of the distinctively objective and autonomous nature of law.

Consistent with postrevolutionary theories of sovereignty, until around 1820 legal thinkers sought to justify the special position of lawyers and judges in terms of a "trustee" or delegation theory of power. But as the Codification Movement began to develop momentum in the 1820s, with its ultimate charge that law was inherently a product of power and will, a new and different justification for the common law power of judges rapidly developed, which continued throughout the century. The new and defensive emphasis in orthodox legal theory on the "scientific" nature of the law arose simultaneously as a reaction to the claim of the radical codifiers that the common law was political.

Perry Miller has shown the dominance of the equation of law with science in all antebellum legal theorizing.[4] Except for the identification of "science" with systematization and classification, however, there is no coherent content or methodology to be found in these persistent claims to the scientific character of law. What does seem extremely clear, nevertheless, is that the attempt to place law under the banner of "science" was designed to separate politics from law, subjectivity from objectivity, and laymen's reasoning from professional reasoning.

One of the most important consequences of the profession's effort to depoliticize the law during the codification controversy of the 1820s and 1830s was the establishment of the treatise tradition. It is surely no coincidence that the first volume of Kent's *Commentaries* (1826), which inaugurated that tradition, appears just as the movement to codify law was gaining strength. Nor is Kent's new emphasis on the scientific character of the rules of private law at all surprising. If, for example, we compare his *Commentaries* to the earlier postrevolutionary forms of legal literature, we can begin to appreciate the important change in legal consciousness that the treatise tradition represented. From what we know of the lectures of Chancellor Wythe at William and Mary or those of James Wilson at the Philadelphia College of Law, or indeed of Kent's own first efforts at Columbia in 1794, there is a decidedly proprofessional turn in legal thinking and writing during the 1820s and 1830s.[5] The earlier efforts, modeled after Blackstone and designed for the nonpracticing gentleman lawyer, emphasized jurisprudence, constitutional law, and political science. By contrast, Kent and, even to a greater extent, his successors in the treatise tradition like Joseph Angell and Joseph Story, generally sought to write high quality technical handbooks for practicing lawyers.

That the treatise tradition reflects a growing professional market for the law book is, of course, true. But it is primarily with its underlying legal consciousness that I am here concerned. The legal treatise was regarded by its admirers as above all demonstrating the "scientific" nature of the law. Through classification of subjects, it sought to show that law proceeds not from will but from reason. Through its "black letter" presentation of supposed "general principles" of law it sought to suppress all controversy over policy while promoting the comforting ideal of a logical, symmetrical, and most important, inexorable system of law. Finally, its decided focus upon the technicalities of private law illustrates an increasingly well-organized and self-conscious profession's yearning for an objective, apolitical conception of law.

During the 1820s and 1830s, then, orthodox legal thought was put decisively on the defensive as both law and the profession came to be challenged by the Codification Movement for being fundamentally political. As a result, orthodox legal thought began to retreat from self-conscious policy-making goals, emphasizing more and more the apolitical, deductive, and "scientific" character of legal reasoning. Before the 1840s, however, the underlying formalism of the treatise tradition still stood somewhat separate from every day judicial thought. The postrevolutionary goal of transforming the common law remained dominant, and legal change still continued to serve the interests of newly emerging industrial and financial groups. Only during the 1840s do we finally begin to see a convergence between the growing formalism of the legal writers and the more general desires of the newly powerful economic groups to sharply curtail the ability of the legal system to bring about further redistributions of wealth.

With this background in mind, the rise of legal formalism can be understood as resulting from the convergence and synthesis of three major factors in antebellum America. First, it is a rough measure of the rise in the power of the postrevolutionary legal profession and is a culmination of the Bar's own separate and autonomous professional interest, especially after the Codification Movement, in representing law as an objective, neutral, and apolitical system. Second, it mirrors a convergence of interest between the elite of the legal profession and the newly powerful commercial and entrepreneurial interests, an alliance which, beginning after the Revolution, enabled the Bar to achieve real prestige and power for the first time. Finally, it represents the successful culmination of efforts by mercantile and entrepreneurial interests during the preceding half century to transform the law to serve their interests, leaving them to wish for the first time to "freeze" legal doctrine and to conceive of law not as a malleable instrument of their own desires and interests but as a fixed and inexorable system of logically deducible rules.

The 1840s and even more the 1850s thus represent an attempt to check the overtly political uses of law and to establish a system of legal thought which sharply distinguished between law and policymaking. What came to be certified as purely "legal," of course, were those rules of law that had been established during the previous half century to implement a market regime, primarily new rules of contract, property, and commercial law, now thoroughly exorcised of earlier protective or paternalistic doctrines and reshaped to serve the interests of the wealthy and the powerful.

By 1850 one can identify in a variety of different legal areas a drawing away from interests in substantive ends and a resurgence of formal or procedural concerns. One consequence of the shift from "political" to formal legal criteria was the sharply antilegislative trend that began to take hold in the courts. For example, by the 1840s in New York, where a strikingly instrumentalist vision of law had appeared earliest, judges also led the way in reversing the earlier policy of judicial support for legislative encouragement of economic growth. One symptom was a dramatic upsurge in judicial review of legislation. Until 1820 only three statutes had been declared unconstitutional by the New York courts. During each of the next two decades only three more statutes were invalidated. But in the 1840s, the total jumps to 14 and in the next decade to 25. The numbers, of course, continued to rise after the Civil War.[6]

Three major areas of the law (there were many others) illustrate the reasons for this change of consciousness. One area was the law of eminent domain — the one truly explosive legal "time bomb" in all antebellum law. From the postrevolutionary period on, it was the eminent domain power that had intimately linked the destinies of entrepreneurial groups to that of the state. Whatever the theory of just compensation, in practice the eminent domain power had always underlined the potential of the state to redistribute wealth.

As in the analogous case of the mill acts, there regularly appeared to be considerable "slippage" between actual compensation and the fair market value of appropriated land, especially since "consequential" or "speculative" damages were often left uncompensated. Nevertheless, until around 1840 the rising entrepreneurial groups, though regularly suspicious of this confiscatory potential within their midst, had used the eminent domain power ruthlessly to their own advantage. But after 1840, we can see a shift from the expansive and growth-oriented utilitarian legal categories of private law to the restrictive, formal categories of public law. Before that time, the overwhelming tendency of courts was to define "public purpose" substantively.[7] Any compensated taking of property would be upheld so long as it could be plausibly connected to the promotion of economic growth. In the next decade, however, a major change became apparent, which may be related to the sharp increase in state taxation that followed from the Panic of 1837 and the resulting collapse of the earlier system for financing economic development. In general, a widespread fear of legislatively authorized redistributions of wealth began to overshadow the enthusiasm for eminent domain as an important instrument of cheap economic growth. Thus in 1843 a sharply divided New York Court proclaimed the basically new position that the judiciary itself would ultimately decide which takings were of a public nature.[8] It struck down as unconstitutional a colonial statute that had been employed hundreds of times before to enable landowners to build private roads through another's property. Creating for the first time a sharp distinction between public and private takings—no doubt drawing spiritual guidance from the earlier and analogous public-private corporation doctrine of the *Dartmouth College* case—the court refused to accept what before had been a standard and virtually unchallenged utilitarian justification of these acts.

Like the analogous mill acts,[9] it had always been successfully argued before that these private road statutes promoted a "public purpose" simply because they promoted economic growth. Indeed, the earliest challenges to the eminent domain powers of both privately owned turnpikes and railroads had also been regularly rejected by a similarly utilitarian vision of what constituted a public purpose. Some railroad lawyers and judges in fact had even gone so far as to argue that the existence of constitutional provisions restricting compensated takings to public purposes meant that there was no constitutional barrier at all to uncompensated takings for private purposes.[10] They had thereby underlined the law of eminent domain's extreme redistributive potential. During the 1840s and 1850s judges everywhere began to turn away from the view that a public purpose inhered in any state promoted activity that simply increased the gross national product. The New York judges rejected the orthodox view of "public purpose" when they overturned their state's private road statute. "If the power exists to take the property of one man without his consent and transfer it to another, it may be exercised without any reference to the questions of compensation," the New York Supreme Court explained in a *non sequitur* which nonetheless expressed its most basic fears of redistribution.[11] Similarly, after 1840 the public nature of the railroad's exercise of the eminent domain power was no longer justified simply by its contribution to economic growth but rather in terms of the more limited and less "political" formal criterion of the public right of equal access to its facilities.[12]

One can also see a similar shift from an instrumental to a formalist legal consciousness in Chief Justice Shaw's attempts at rationalizing the mill acts. From the time they were extended to cotton mills, the mill acts represented the most extreme invasion of the eminent domain power into private activities as well as the most blatant instance of the ruthless exercise of that power to bring about redistributions of wealth. After 1814 the Massachusetts Supreme Judicial Court never doubted that this "private" flooding of land could nonetheless be characterized as promoting a public purpose simply because it contributed to national wealth.[13] But in the course of Shaw's own term as Massachusetts Chief Justice (1830-60) one sees a growing reluctance to rationalize the mill acts as an exercise of the power of eminent domain, since any legislatively authorized taking could then be upheld simply because it contributed to economic improvement. By the end of his term, in fact, Shaw had successfully created an entirely separate rationale for the mill acts as an application of historically special principles of riparian law, thus completely detaching it from the more general redistributional potential of the law of eminent domain.[14] So successful was Shaw's effort that it enabled even a suspicious postwar United States Supreme Court to turn back a constitutional challenge to the acts on the grounds that they had nothing to do with the general scope of the state's power to take property.[15] By "formalizing" the inquiry—that is, by reclassifying the problem into a supposedly nonpolitical doctrinal category— Shaw was able to defuse its general redistributive significance.

Contract law provides another excellent illustration of the deep pressure toward formalism in nineteenth century law. During the first half of the century, legal writers and judges were largely successful in banishing fair price conceptions from the law of contract and in shifting the focus of contract law to the determination of whether there had been a "meeting of minds." As a result, one of the central problems of the "new" contract law was the development of supposedly nonpolitical criteria for distinguishing between "free" and "unfree" wills, while at the same time barring any substantive inquiry into the equivalency of an exchange, which now had come to be regarded as "subjective" and "political." This proscription against examining "consequences" or "outcomes" eventually led to the creation of a disembodied conception of "unfree" states of mind which measured the impairment of "will" by criteria entirely detached from concrete social or economic forms of coercion. Moving from the abstract view "that coercion necessarily implies that the party to whom it is applied has no volition," courts deduced "the converse notion that where he has volition, or the ability to make a choice, there is no coercion or duress."[16] As a result, "the search for evidence that the 'will' had been destroyed" buttressed the view "that relief for any type of pressure depends on a showing of complete absence of consent." And, since "the instances of more extreme pressure were precisely those in which the consent expressed was *more* real," courts in practice tended to the view that "the more unpleasant the alternative, the more real the consent to a course which would avoid it."[17] Thus legal thought came to conceive of undue influence and duress, not as reflecting a continuum of unequal bargaining power which depended on a subtle interaction between the actor and the pressures of his environment, but as a set of abstract formal volitional categories that measured whether a person "possessed" sufficient reason so that his will could not be "overborne." The consequence of this conception was to sharply restrict those cases in which courts were likely to find economic duress.

It is important to see that the postulates of a market regime required that individualized equitable inquiries be suppressed through the development of abstract and formal legal categories.

The freedom of the "market" was essentially a freedom of individuals and groups to coerce one another, with the power to coerce reinforced by agencies of the state itself. Even though the larger implications of this idea were by no means understood, one simple and quite obvious deduction had already been made — this is, that if the "market" was to be free, any form of external regulation was objectionable. Regulation by court enforced rules of private law seemed just as unwise and dangerous as regulation by statute or administrative action. From this point of view, where urgent need or special disadvantage compelled agreement to the terms proposed, these circumstances must be disregarded since they differed only in degree from the basic conditions which governed the exchange of goods and services throughout society.[18]

Likewise, in fraud doctrine, the postulates of a market economy required the development of a sharp distinction between "facts" on the one hand and "estimates" and "opinions" on the other, whose effect was to sharply limit the finding of fraud. Since opinions, estimates, or interpretations were regarded as subjective — and part of a legitimate distribution of individual talents and attitudes — the only forms of knowledge which the market system could justly protect against misrepresentations were thought to be bare statements of "facts."[19] There thus developed an elaborate set of formal doctrines distinguishing "fact" from "opinion," which also had the effect of sharply circumscribing judicial intervention into the bargaining system.[20]

These efforts during the first half of the nineteenth century to disengage the contract system from substantive criteria of fairness produced internal pressures to generate "objective" and "nonpolitical" doctrinal measures of "free will" and "meeting of the minds." These pressures were generalized still further beginning in the 1850s into an "objective" theory of contract. What most distinguished the "objective" theory was its insistence upon establishing uniform and general rules of law with reference to which contracting parties would be required to shape their conduct. With this change, contract law was no longer conceived of as simply implementing the parties' "wills," but rather its categories were often treated as existing somehow prior to individual bargains.[21]

I discussed earlier how economic pressures toward uniformity and predictability, as well as efforts to restrict the scope of jury discretion, also ultimately contributed to this shift toward formal and objective legal rules. But it is important to see as well that this trend toward uniformity necessarily required doctrines of greater abstraction and generality, which in turn had the effect of detaching contract doctrine from the particularities of individual cases and of creating internal logical pressure to conceive of the law of contract as a system of disembodied logical interrelationships. This tendency throughout the second half of the nineteenth century to seek higher levels of generality and inclusiveness of legal doctrine is one of the more important characteristics associated with the development of legal formalism. One of the most important results of this trend in thought was to prevent particularized equitable inquiries into the circumstances of individual cases and to destroy more particular legal rules in the name of promoting the "rule of law."

But how was this move toward generality in fact accomplished? The history of insurance law in the nineteenth century provides one prime example. In 1800 most lawyers regarded the marine insurance contract as the most prominent subcategory of Contract. By 1850, however, the law of insurance had become entirely isolated from the main stream of contract law and was by then considered a separate, technical body of law. The source of this crucial change seems to be traceable to the fact that the economic and moral premises that underlay insurance law were becoming increasingly subversive of the newer economically dominant contract of sale. Marine insurance doctrine had developed during the eighteenth century in a homogeneous economic setting of reciprocal business relationships among merchants of relatively equal bargaining power. The clearly proseller contract doctrines developed early in the nineteenth century for the law of sales, by contrast, seem to reflect the fact that this branch of law reached maturity in a period when economic relations between economically sophisticated "seller-insiders" and relatively unsophisticated "buyer-outsiders" were becoming dominant.

The assumptions underlying these two radically different models of business dealings first began to clash in a battle over adopting the rule of caveat emptor early in the nineteenth century. Every time that a court adopted the caveat emptor rule for sales contracts, it was confronted with the argument of counsel that the "true" common law contract rule that had already emerged from insurance law required the full disclosure of all material facts necessary to rationally judge the level of risk in a bargain. As a result, during the first half of the nineteenth century, judges first "distinguish" sales from insurance contracts, then later claim that the "true" paradigm of contract is the sale of goods and not the insurance contract, and finally they end all contradiction by treating the law of insurance as a separate and autonomous area of the law whose "principles" have nothing to do with the general law of contract. Contract law, in short, achieved generality by "purifying" itself through the expulsion of legal doctrines that had become alien to the norms of nonintervention underlying the market economy of the nineteenth century. We have already witnessed an identical pattern in Chief Justice Shaw's efforts to separate the mill acts from the law of eminent domain, thus purifying eminent domain law by expelling this increasingly "alien" precedent for the principle of state intervention and redistribution.

Another powerful symbol of the increasing tendency toward formalism is the merger of Law and Equity first accomplished in the New York Field Code of 1848. The passage of this Code of Civil Procedure represents the final exhaustion of the earlier Codification Movement's most radical goal of codifying and thereby transforming substantive law as well. For just as this last phase of the Codification Movement itself is marked by the conversion of substantive legal conflict into exclusively technical debates over procedure, the Code's merger of Law and Equity also represented the triumph of the formalization of Equity.

While the merger of Law and Equity has usually been portrayed by uncritical legal historians as effecting the "rationalization" of civil procedure, in fact it marks the final and complete emasculation of Equity as an independent source of legal standards. The subjection of Equity to formal rules was a prominent article of faith within the orthodox nineteenth century movement to conceive of law as science. Indeed, Equity was regularly attacked by the treatise writers as inherently discretionary and "political." Justice Story's "scientific" treatise on *Equity Jurisprudence* (1836) marks a major step in the transformation of Equity from an eighteenth century system of substantive rules derived from "natural justice" to a nineteenth century positivist conception of Equity as simply providing a more complete and inclusive set of procedural remedies. Story's treatise, for example, was influential in bringing about the final overthrow of the eighteenth century "just price" doctrine that Courts of Equity would not enforce grossly unequal contracts.[22]

Perhaps the most important subject of equitable jurisdiction involved supervision of mortgages. The law of mortgages carried with it an accumulation of several centuries of received equitable doctrines that had crystallized much earlier under a premarket and paternalistic social order.[23] In the early nineteenth century, by contrast, Chancellor Kent single-handedly undermined the well-established protective doctrine that foreclosure of a mortgage in equity was an exclusive remedy.[24] By providing creditors an additional contractual remedy at law for any deficiency, Kent contributed to the subversion of the long-standing regulatory role of Equity. Another important erosion of traditional equitable powers can be seen in the systematic pattern of statutory reductions of the time period in which a mortgagee could exercise his equity of redemption, which had the effect of undermining equitable supervision of this major area of debtor-creditor relations.[25] This broad tradition of equitable supervision of mortgages for a time had also stood in the way of the exploitative use of the more recently established chattel mortgage. Yet, in one of the great triumphs of form over substance common law judges during the nineteenth century instead began to treat these transactions as conditional sales, thus entirely freeing this economic relationship from regulatory and paternalistic equitable mortgage doctrines.[26]

The movement to merge Law and Equity begun by the Field Code of 1848 thus represents another instance of the subjection of an already internally eroded tradition of substantive justice to an increasingly formal set of legal rules, which were themselves now stridently justified as having nothing to do with morality.

There were, in short, extremely deep and powerful currents which moved American law to formalism after 1850. As the fear of "political" uses of private law began to increase, the strikingly different premises that had earlier characterized public and private law now began to merge. The independent professional ideology of the Bar, once confined and restricted because of the more powerful transforming ambitions of emerging economic groups, now for the first time comes into full bloom. A scientific, objective, professional, and apolitical conception of law, once primarily a rhetorical monopoly of a status-hungry elite of legal thinkers, now comes to extend its domain and to infiltrate into the every day categories of adjudication.

This alliance between intellect and power could only finally come into being after the transforming surge of postrevolutionary legal activity had become uncongenial to its beneficiaries. For the paramount social condition that is necessary for legal formalism to flourish in a society is for the powerful groups in that society to have a great interest in disguising and suppressing the inevitably political and redistributive functions of law.

NOTES
INDEX

Notes

I. THE EMERGENCE OF AN INSTRUMENTAL CONCEPTION OF LAW

1. D. Boorstin, *The Americans: The National Experience* 35 (1958).
2. R. Pound, *The Formative Era of American Law* (1938); C. Haar, ed., *The Golden Age of American Law* (1965).
3. M. Howe, "The Creative Period in the Law of Massachusetts," 69 *Proceedings of the Massachusetts Historical Society* 237 (1947-50).
4. T. Parsons, *Memoir of Theophilus Parsons* 239 (1859).
5. Lindsay v. Commissioners, 2 Bay 38, 45 (S.C. 1796).
6. Liebert v. The Emperor, Bee's Admir. Rep. 339, 343 (Pa. 1785).
7. Silva v. Low, 1 Johns. Cas. 184, 191 (N.Y. 1799).
8. Winston v. Saidler, 3 Johns. Cas. 185, 196 (N.Y. 1802).
9. 3 Cai. R. 307, 314 (1805). For the common law rule, see Merritt v. Parker, 1 Coxe L. Rep. 460 (N.J. 1795).
10. Thurston v. Koch, 4 Dall. 348, app. xxxii (C.C.A. Pa. 1803).
11. Commonwealth v. Pullis (The Philadelphia Cordwainers Case) (1806) in 3 J. Commons *et al.*, eds., *A Documentary History of American Industrial Society* 229-30 (1910).
12. Jackson v. Brownson, 7 Johns. 227 (N.Y. 1810) (Spencer, J.). See also Findlay v. Smith, 6 Munf. 134, 142 (Va. 1818).
13. Conner v. Shepherd, 15 Mass. 164 (1818); Nash v. Boltwood (Mass. 1783) in W. Cushing, "Notes of Cases in the Supreme Judicial Court of Massachusetts, 1772-1789" (ms. in Treasure Room, Harvard Law School Library). In the eighteenth century the Massachusetts legislature had substantially reduced the period during which a mortgagor could recover the equity of redemption in his land so "that the improvement of real estates thus taken might not be delayed or lost to the mortgagee." J. Sullivan, *The History of Land Titles in Massachusetts* 102 (1801).
14. Stetson v. Massachusetts Mutual Fire Ins. Co., 4 Mass. 330 (1808).

15. J. Otis, *A Vindication of the British Colonies,* in 1 B. Bailyn, ed., *Pamphlets of the American Revolution* 563 (1965).

16. 1 *Journals of the Continental Congress* 69 (1904).

17. 9 W. Hening, ed., *The Statutes at Large: Being a Collection of all the Laws of Virginia* 127 (1821); 5 F. Thorpe, ed., *Federal and State Constitutions, Colonial Charters and Other Organic Laws . . .* , 2598 (1809).

18. E. Brown, *British Statutes in American Law, 1776-1836* 24 (1964). Rhode Island enacted a reception provision in 1798, while Connecticut waited until 1818.

19. "Charge to the Grand Jury," Quincy's Mass. Rep. 234-35 (1767).

20. 2 *Legal Papers of John Adams* 199 (L. Wroth and H. Zobel, eds., 1965).

21. 1 *The Law Practice of Alexander Hamilton* 10, 11, 33 (J. Goebel, ed., 1964).

22. The distinction appears in Calvin's Case, 7 Coke Rep. 1, 17 (1608). It is discussed in J. Smith, *Appeals to the Privy Council from the American Plantations* 466-69 (1950) and St. G. Sioussat, *The English Statutes in Maryland* 18 (1903). Coke's categories were expanded considerably by Chief Justice Hold who in 1693 acknowledged that in "uninhabited" lands founded by Englishmen British law prevailed. He held, however, that Jamaica was a "conquered" land. Blankard v. Galdy, 2 Salk. 411.

23. E. Brown, *supra* note 18, at 17; Sioussat, *supra* note 22, at 25-27.

24. Sioussat, *supra* note 22, at 21.

25. This was Blackstone's position. 4 W. Blackstone, *Commentaries on the Laws of England* 107 (1765-69).

26. J. Smith, *supra* note 22, at 482-87.

27. Blackwell v. Wilkinson, Jefferson's Rep. 73, 77 (Va. 1768).

28. Anderson v. Winston, Jefferson's Rep. 24, 27 (Va. 1736).

29. "The Right of the Inhabitants of Maryland to the Benefit of the English Laws" (1728), in Sioussat, *supra* note 22, at 82. Dulany, however, was unusual among his contemporaries in maintaining that protection of liberty required the reception of English statutes as well as the common law. These Acts, he argued, "have always been deemed, as essential a Part of the Security, of the Subject to his Rights and Privileges as the Common Law itself." *Id.,* at 99.

30. Josiah Quincy's argument for the defense in Rex v. Wemms (1770), quoting 4 Blackstone, *Commentaries* 3, in 3 *Legal Papers of John Adams, supra* note 20, at 160; Hutchinson "Charge to the Grand Jury," Quincy's Mass. Rep. 113 (1765).

31. J. Galloway, "A Letter to the People of Pennsylvania" (1760), in Bailyn *supra* note 15, at 266-67; W. Brattle, in 3 *The Works of John Adams* 518 (Charles Francis Adams, ed., 1851). See also G. Wood, *The Creation of the American Republic, 1776-1787* 294-95 (1969).

32. Charge of Justice Edmund Trowbridge in Rex v. Wemms (1770), 3 *Legal Papers of John Adams, supra* note 20, at 288; 3 *Works of John Adams, supra* note 31, at 527, 540, 546.

33. Watts v. Hasey, Quincy's Mass. Rep. 194 (1765); Hooton v. Grout, Quincy's Mass. Rep. 343, 362, (1772); Robinson v. Armistead, Barradall's Rep. 223 (Va. 1737).

34. 2 *Legal Papers of John Adams, supra* note 20, at 127; Goebel, *supra* note 21, at 50n; West v. Stigar, 1 H. and McH. 247, 254 (Md. 1767); Chisholm v. Georgia, 2 Dall. 419, 448 (1793).

35. United States v. Hudson and Goodwin, 7 Cranch 32 (1812). This view has prevailed even though several members of the Supreme Court were disposed to reconsider the case four years later. At that time, however, the United States Attorney General refused to argue that the *Hudson* case should be overturned, and as a result, the Supreme Court did not reconsider its decision. United States v. Coolidge, I Wheat. 415 (1816). For a learned discussion of the subject, see P. DuPonceau, *A Dissertation on the Nature and Extent of the Jurisdiction of the Courts of the United States* (1824).

36. For the political background of the battle over federal common law crimes, see 1 C. Warren, *The Supreme Court in United States History* 157-64 (1922).

37. 7 *The Writings of Thomas Jefferson* 451 (Paul Leister Ford, ed., 1892-1899). A year earlier Jefferson had written that "Of all the doctrines which have been broached by the federal government, the novel one, of the common law being in force & cognizable as an existing law in their courts, is to me the most formidable. . . . If this assumption be yielded to, the state courts may be shut up." *Id.*, at 383-84.

38. [J. Sullivan], *A Dissertation upon the Constitutional Freedom of the Press* 31, 40-41, 48-54 (1801). In his opinion in United States v. Coolidge, 1 Gall. 488 (C.C.A. Mass. 1813), Mr. Justice Story denied that there was a threat of federal usurpation in the argument in favor of common law crimes. "I admit in the most explicit terms," he wrote, "that the Courts of the United States are Courts of limited jurisdiction, and cannot exercise any authorities which are not confided to them by the Constitution and laws made in pursuance thereof."

39. In United States v. Hudson & Goodwin, 7 Cranch 32 (1812), the Supreme Court declared that it was "unnecessary" to decide whether Congress could confer common law jurisdiction on the federal courts, since it was "enough that such jurisdiction has not been conferred by any legislative act." But one year later, the author of that opinion, Mr. Justice Johnson wrote: "We do not deny, nor we suppose was it ever denied — that if . . . Congress had by Law vested in this Court jurisdiction over all cases to which the punishing power of the United States might . . . be extended, it would then rest with this Court to decide (wild and devious as the track assigned them would be) to what cases that jurisdiction extended." *Trial of William Butler for Piracy* 34-35 (n.d., n.p.) [1813?]. (The text of this unreported circuit court opinion appears in a pamphlet in the Treasure Room, Harvard Law School Library.) See also 1 Blackstone, *Commentaries: With Notes of Reference to the Constitution and Laws of the Federal Govern-*

ment of the United States and of the Commonwealth of Virginia 429-30 (St. George Tucker, ed., 1803).

40. Modern scholarship has suggested that the Act did confer such jurisdiction on the federal courts. See C. Warren "The History of the Federal Judiciary Act of 1789," 37 *Harv. L. Rev.* 73 (1923).

41. Tucker's Blackstone, *supra* note 39, at 405.

42. *Id.*, at 438.

43. United States v. Worrall, 2 Dall. 384 (1798).

44. 2 *The Miscellaneous Essays and Occasional Writings of Francis Hopkinson* 97 (1792).

45. *An Essay upon the Government of the English Plantations on the Continent of America* 23, 39 (Louis B. Wright, ed., 1945).

46. *The Superior Court Diary of William Samuel Johnson* xviii (John T. Farrell, ed., 1942). In another Connecticut criminal case in 1743, the legislature told the court that, since there was no statute, the judge should prescribe a penalty "according to their best skill and judgment." *Id.*, at xviii n. 3.

47. N. Chipman's Rep. 61, 67. 2 Z. Swift, *A System of the Laws of the State of Connecticut* 365-66 (1796).

48. *The Trial of William Butler for Piracy, supra* note 39, at 28, 21-25, 32.

49. J. Goodenow, *Historical Sketches of the Principles and Maxims of American Jurisprudence in Contrast with the Doctrines of the English Common Law on the Subject of Crimes and Punishments* 3-4, 6, 33 (1819). For a discussion of Goodenow's influence on Ohio law, see A. Hadden, *Why Are There No Common Law Crimes in Ohio?* (1919). It is important to note, however, that despite his far-reaching attack on common law crimes, Goodenow expressly exempted common law civil actions from his analysis. "Civil actions," he wrote, "are founded in the private rights and private wrongs of the individuals; in which the legislative power of the civil state has nothing to do. Natural justice and right reason, are foundation of all our private rights." Goodenow, *supra* at 36.

50. S. Thorne, *A discourse upon the . . . Statutes . . .* (1942); J. Gough, *Fundamental Law in English Constitutional History* 117-22 (1955).

51. 1 W. Blackstone, *Commentaries* 108-109.

52. Quoted in Wood, *supra* note 31, at 301-02.

53. 4 *The Writings of Thomas Jefferson* 115; *Notes on the State of Virginia* 148 (1853). On prerevolutionary efforts to find a "pure" common law system in ancient English law, see Bailyn *supra* note 15, at 26-27; J. Pocock, *The Ancient Constitution and the Feudal Law* (1957). One of Jefferson's earliest efforts to put forth a theory of an ancient common law appears in "Whether Christianity is Part of the Common Law," 1 *The Writings of Thomas Jefferson* 360 [1764?].

54. "Oration at Scituate" (1836) in *Memoirs, Speeches and Writings of Robert Rantoul, Jr.*, 278 (Luther Hamilton, ed., 1854).

55. 1 *The Works of James Wilson* 180 (Robert G. McCloskey, ed., 1967); *Id.*, II, 506; I, 121-122; I, 353 ff. Wood, *supra* note 31, at 292-96.

56. Blackstone, *supra* note 39, at 53, n. 10.

57. 1 Root's Conn. Rep. iv, ix-xiii.

58. J. Sullivan, *supra* note 13, at 337-40.

59. Commonwealth v. Pullis in Commons, *supra* note 11, at 233.

60. H. Brackenridge, *Law Miscellanies* 84 (1814); E. Whittlesey, "Reeve & Gould's Lectures," I, 1 (1813) (ms. 4024, Treasure Room, Harvard Law School Library); "Reeve's Lectures," I, 4-5 (n.d.) (ms. 2013, Treasure Room, Harvard Law School). There is a similar, but much less detailed, discussion of the same point in the Whittlesey manuscript, above, I, 6-7, which suggests that these remarks were made around 1813. In another undated manuscript, reprinted in J. Smith, *Development of Legal Institutions* 479-81 (1965), Reeve declares that in new cases courts "to all intents make the law." On the basis of the latest English report cited in the manuscript, Smith dates it at 1802.

61. Brackenridge, *supra* note 60, at 54, 52; Z. Swift, *A Digest of the Law of Evidence* v-vi (1810).

62. For the report of the Judges declaring which English statutes were binding in Pennsylvania, see 3 Bin. 595 (1808). [H. Brackenridge], *Considerations on the Jurisprudence of the State of Pennsylvania* (1808). Brackenridge, *supra* note 60, at 382, 469, 75.

63. N. Chipman's Rep. 63-65.

64. *Id.*, at 66n; 1 Z. Swift, *A System of the Laws of the State of Connecticut*, 46, 41 (1795).

65. Silva v. Low, 1 Johns. Cas. 184, 190 (N.Y. 1799).

66. "Charge to the Grand Jury by the Chief Justice," Quincy's Mass. Rep. 232, 234 (1767); Swift, *supra* note 61, at vi; Lee v. Boardman, 3 Mass. 238, 247-48 (1807); Thurston v. Koch, 4 Dall. 348, app. xxxiv (C.C.A.Pa. 1805); Silva v. Low, 1 Johns. Cas. 190 (N.Y. 1799).

67. W. Duane, *The Law of Nations, Investigated in a Popular Manner* 3 (1809); Vandenheuvel v. United Ins. Co., 2 Cai. Cas. 217, 285, 289 (N.Y. 1805); Ludlow v. Bowne, 1 Johns. 1, 7 (N.Y. 1806); Brackenridge, *supra* note 60, at 344; DuPonceau, *supra* note 35, at 124-25.

68. Jefferson, *Notes on Virginia* 140; 2 *The Works of James Wilson*, *supra* note 55, at 540; "Powers and Rights of Juries," Quincy's Mass. Rep. 558-72.

69. A. Addison, *Charges to Grand Juries of the Counties of the Fifth Circuit in the State of Pennsylvania* 53 (1800). (This volume often appears as an appendix to Addison's Reports.) Kirby's Conn. Rep. iii.

70. J. Morison, *Life of Jeremiah Smith*, 165-66, 173-74; [H. Binney], *Bushrod Washington* 21-22 (1858); Bentaloe v. Pratt, Wallace Cir. Rep. 58, 60 (E.D. Pa. 1801); The Schooner Lively, 15 Fed. Cas. 634-35 (1812); "Report of Phillip Schuyler for Board of Directors of the Western Inland Lock Navigation Co. to the New York Legislature," (1798) in A. Gallatin,

"Report on Roads and Canals," *American State Papers*, Class I, Misc. I, 779; N.Y. Stat. ch. 101 (1798); Silva v. Low, 1 Johns. Cas. 184, 199 (N.Y. 1799).

71. DuPonceau, *supra* note 35, at 115-16; Kline v. Husted, 3 Cai. R. 275, 278 (N.Y. 1805); Spear v. Bicknell, 5 Mass. 125, 133 (1809).

72. Riddle v. Proprietors of the Locks and Canals of Merrimack River, 7 Mass. 169, 180-81 (1810); "A Dissertation on the Negotiability of Notes," N. Chipman's Rep. 89, 95. Compare M'Cullough v. Houston, 1 Dall. 441 (Pa. 1789); Mackie's Exec. v. Davis, 2 Wash. 281 (Va. 1796).

II. THE TRANSFORMATION IN THE CONCEPTION OF PROPERTY

1. For Blackstone, the right of property consisted of "that sole and despotic dominion which one man claims and exercises over the external things of the world, in total exclusion of the right of any other individual in the universe." 2 W. Blackstone, *Commentaries* *2.

2. 3 *id.* at *217-18.

3. See F. Bohlen, "The Rule in Rylands v. Fletcher," 59 *U. Pa. L. Rev.* 298 (1911).

4. Use your own [property] so as not to harm another's.

5. Though there had been controversies involving diversion of water for irrigation or saw and grist mills in the colonial period, the rise of large New England cotton mills after 1815 intensified the conflict.

6. See pp. 47-53 *infra*.

7. Merritt v. Parker, 1 Coxe L. Rep. 460, 463 (N.J. 1795). See also Beissell v. Sholl, 4 Dall. 211 (Pa. 1800); Livezay v. Gorgas (Pa. 1811), in H. Brackenridge, *Law Miscellanies* 454-56 (1814).

8. Perkins v. Dow, 1 Root 535 (1793), citing Howard v. Mason (1783). "If I can dispose of, and absorb upon my land, the whole of the stream excepting a sufficiency for necessary purposes," a legal commentator explained, "I have the prior right, because I am above him on the stream and have the first opportunity." 2 Z. Swift, *A System of the Laws of the State of Connecticut* 86 (1796). This rule did not, of course, express a theory of prior appropriation, since, regardless of who first occupied the stream, the upper proprietor was granted "that artificial advantage which the situation of his ground will admit." *Id.* at 87. Indeed, it appears to be the only instance in American law of adoption of the precivil code French rule. See 3 J. Kent, *Commentaries* 439 n.(a) (4th ed. 1840).

9. Ingraham v. Hutchinson, 2 Conn. 584, 591 (1818). Interestingly, the judge whose opinion overruled these eighteenth century decisions was Zephaniah Swift, whose *System of Laws* (1796) had justified those decisions. See note 8 *supra*.

10. W. Cushing, "Notes of Cases Decided in the Superior & Supreme Judicial Courts of Massachusetts from 1772 to 1789" (ms. Treasure Room, Harvard Law School).

11. Weston v. Alden, 8 Mass. 136 (1811); Bent v. Wheeler (Mass. 1800), in J. Sullivan. *The History of Land Titles in Massachusetts* 273-74 (1801); 3 N. Dane, *A General Abridgement and Digest of American Law* 16 (1824).

12. See note 40 *infra*.

13. Palmer v. Mulligan, 3 Cai. R. 307, 312 (N.Y. Sup. Ct. 1805) (Spencer, J.); Ingraham v. Hutchinson, 2 Conn. 584, 595 (1818) (Gould, J., dissenting). In Sherwood v. Burr, 4 Day 244 (1810), the Connecticut Supreme Court resorted to the doctrine of prescription to deal with the novel problem of downstream obstruction, thereby implying that without long use the upstream owner would have no right of action for obstruction.

14. 3 Cai. R. 307, 313-314 (1805).

15. *Id.* at 314.

16. 15 Johns. 213, 218 (N.Y. 1818).

17. *Id.* at 218.

18. Outside New York, I have found only one other case before 1825 that followed the principle that diversion or obstruction of water might be justified by a doctrine of reasonable use which took account of a right to equal exploitation of water. Runnels v. Bullen, 2 N.H. 532, 537 (1823). See also Merritt v. Brinkerhoff, 17 Johns. 306 (1820).

19. *Watercourses* 37 (1st ed. 1824).

20. *Id.* at 41.

21. 24 F. Cas. 472 (No. 14,312) (C.C.D.R.I. 1827).

22. Professor Lauer correctly sees the "almost schizophrenic . . . nature" of these early transitional water cases, which wavered between "both the pre-existent law and the need for just apportionment of the water." Lauer, "Reflections on Riparianism," 35 *Mo. L. Rev.* 1, 8 (1970).

23. Story continued: "The consequence of this principle is, that no proprietor has a right to use the water to the prejudice of another. It is wholly immaterial, whether the party be a proprietor above or below, in the course of the river; the right being common to all the proprietors on the river, no one has a right to diminish the quantity which will, according to the natural current, flow to the proprietor below, or to throw it back upon a proprietor above." 24 F. Cas. at 474.

24. *Id.*

25. Though Tyler v. Wilkinson has often mistakenly been understood to expound a doctrine of proportionate use, I have found no subsequent case in which Story upheld any interference with the flow, unless based on prescriptive right as in Tyler itself. See Lauer, *supra* note 22, at 8. Indeed, Story often seemed routinely to apply a "natural flow" rule in granting injunctive relief,, Farnum v. Blackstone Canal Corp., 8 F. Cas. 1059 (No. 4675) (C.C.D.R.I. 1830); Mann v. Wilkinson, 2 Sumner 273 (C.C.D.R.I. 1835). More illuminating still, in Webb v. Portland Mfg. Co., 29 F. Cas. 506 (No. 17,322) (C.C.D. Me. 1838), he found against a defendant who had extracted a proportionate part of the water from a stream for mill purposes even though there was no proof of actual damage. And in Whipple v. Cumberland Mfg. Co., 29 F. Cas 934 (No. 17,516) (C.C.D. Me. 1843), he

upheld an action against a downstream mill owner for flowing back water, again without proof of actual injury to the plaintiff's mills. There was no need to prove damage, he wrote, since "the principle of law goes much further; for every riparian proprietor is entitled to have the stream flow in its natural channel, as it has been accustomed to flow, without any obstruction by any mill or riparian proprietor below on the same stream. . . . And if any mill or riparian proprietor below on the same stream does . . . undertake to obstruct or change the natural stream, then, although the riparian proprietor above cannot establish in proof, that he has suffered any substantial damage thereby, still he is entitled to recover nominal damages, as it is an invasion of his rights." *Id.* at 935-36. By allowing a damage action — and presumably an injunction — without proof of actual injury, Story went beyond the common law in restraining exploitation of water resources. It was one thing to hold that economic development could not take place at another's expense, and quite another to allow an existing riparian owner to prevent exploitation of surplus water on the ground that he might use it at some future time. Not only was Story's formulation contrary to any supposed right of equal use; it also resulted in a rule still more monopolistic and exclusionary than anything the common law had required.

26. Contrast Story's views with those put forth by the Vermont court in the same year: "The common law of England seems to be that each land owner, through whose land a stream of water flows, has a right to the water in its natural course, and any diversion of the same to his injury, gives him a right of action . . . Should this principle be adopted here, its effect would be to let the man who should first erect mills upon a small river or brook, control the whole and defeat all the mill privileges from his mills to the source." In the absence of "wanton waste" or "obstruction," it allowed the upstream proprietor to build a mill dam that would interfere with the natural flow of water. Martin v. Bigelow, 2 Aiken 184, 187 (Vt. 1827).

27. See, *e.g.,* Omelvany v. Jaggers, 2 S.C. (Hill) 634 (1835); Buddington v. Bradley, 10 Conn. 213 (1834); Arnold v. Foot, 12 Wend. 330 (N.Y. Sup. Ct. 1834).

28. Elliot v. Fitchburg R.R., 64 Mass. (10 Cush.) 191, 195 (1852).

29. Snow v. Parsons, 28 Vt. 459, 462 (1856). For a much earlier recognition of this principle, see Hoy v. Sterrett, 2 Watts 327, 332 (Pa. 1834), in which the court, though citing *Tyler v. Wilkinson* for the purpose of rejecting the plaintiff's claim to a right derived from prior occupancy, nevertheless held that if "the water was no longer detained than was necessary for a proper enjoyment of it . . . for the use of [the defendant's] mill, it is damage to which the plaintiff must submit."

30. R. Fogel, *Railroads and American Economic Growth* 123 (1964).

31. J. Angell, *Watercourses* vii (2d ed. 1833).

32. Damage without legal injury.

33. "The Law of Water Privileges," 2 *Am. Jurist* 25, 27 (1829).

34. 49 Mass. (8 Met.) 466 (1844).

35. *Id.* at 476-77.

36. *Id.* at 477. In more conventional cases as well, Shaw's court often insisted that any use of water for manufacturing purposes was prima facie reasonable, regardless of proportionality. See *e.g.,* Pitts v. Lancaster Mills, 54 Mass. (13 Met.) 156 (1847), where a manufacturing company was held justified in obstructing a stream so long as "they detained the water no longer than was necessary to raise their own head of water and fill their own pond." *Id.* at 158. See also Hoy v. Sterrett, 2 Watts 327 (Pa. 1834).

37. 49 Mass. (8 Met.) at 477.

38. The latter system, which prevails in many western states today, confers on the earlier occupant of a stream the right to continue using water to the extent that he had before the arrival of a later occupant. Compare Wiel, "Waters: American Law and French Authority," 33 *Harv. L. Rev.* 133 (1919) with Maass and Zobel, "Anglo-American Water Law: Who Appropriated the Riparian Doctrine?," 10 *Public Policy* 109 (1960).

39. The first modern English decision on water rights, Bealey v. Shaw, 6 East 208 (K.B. 1805), is typical. The court held that an upstream proprietor could not divert a greater proportion of a stream than he had a right to on the basis of long use. Of the four judges who sat, two seemed to think that the plaintiff should prevail because he had previously appropriated the portion of the stream at issue, while the other two arrived at the same result on the ground that the plaintiff was entitled to the flow as a natural right regardless of priority. See also Merritt v. Parker, 1 Coxe L. Rep. 460 (N.J. 1795), which prevents upstream development by citing both theories.

40. In Massachusetts, however, an early departure from the common law tradition led to a rather complicated evolution of legal doctrine, resulting in the earliest consideration of a separate theory of prior appropriation. Most eighteenth century Massachusetts water cases proceeded from the theory that the sole basis for a right to the flow of water was a claim of prescription founded on immemorial usage. 3 N. Dane, *supra* note 11, at 14-17; J. Sullivan, *supra* note 11, at 272-74; 1 *Legal Papers of John Adams* 68 n.90, 82 n.108, (L. Wroth and H. Zobel eds. 1965). But see Clark v. McCarney (1772), *id.* at 68, in which both parties appear to have been apportioned water privileges by a common grantor, and Pearson v. Tenny (1802), 3 N. Dane, *supra* note 11, at 14, in which a contractual relationship governed. I have also found one Massachusetts action for diversion in which the plaintiff prevailed though he asserted neither a prescriptive right nor a contractual agreement. Clark v. Billings (1783), cited in W. Cushing, *supra* note 10.

The theory that in the absence of agreement only prescription created a property interest in water was derived from a much older English rule that had been all but superseded by the first quarter of the seventeenth century, when English courts declared that the right to water was a "natural right" not founded on long usage. Shury v. Piggot, 3 Bulst. 339 (K.B. 1625). Before that decision the English cases had evolved from the theory that only ancient mills could prescribe for water to the doctrine that even a recently erected mill located on an ancient stream was entitled to an uninterrupted flow. See Lauer, "The Common Law Background of the Riparian Doctrine," 28 *Mo. L. Rev.* 60, 83-85 (1963). Thus, even before Shury v. Piggot, the English courts had arrived at a "natural flow" doctrine that allowed any mill owner to bring an action for diversion, though it was still couched in the language of ancient use. While English law moved on to an explicit statement of a modern "natural right" doctrine, eighteenth century Massachusetts law remained fixed at the earliest stage in the English evolution.

In Shorey v. Gorrell (1783), cited in W. Cushing, *supra* note 10, the Massachusetts Supreme Court misconstrued English authority to hold that the only basis for preventing interference with the flow of water was a right gained by prescription. In rejecting the plaintiff's claim to a right derived from mere priority of appropriation, the court also observed that, since his mill was in existence for only eighteen years, it was not old enough for a prescriptive right to accrue. After this case, Massachusetts pleadings always sound in prescription when a contract is not involved, but a crucial change takes place that imitates the earlier English developments discussed above. By 1801 James Sullivan's statement of Massachusetts water law, *supra* note 11, at 272-73, included only the requirement that a mill be located on an ancient stream, which again amounted to no more than a disguised version of the natural right doctrine. See also Perham, *American Precedents of Declarations* 196, #1, note (1802), which explicitly states the fictionalized result of this change: "In a declaration for turning a watercourse, it is good to state it as *an ancient watercourse* which has been accustomed to run to the Plf.'s mill, without setting forth any prescription; for these words are tantamount." And although there is some evidence around the turn of the century that courts were beginning to look skeptically at these prescriptive claims, see note 52 *infra,* the theory was not finally abandoned in Massachusetts until 1827. See Anthony v. Lapham, 22 Mass. (5 Pick.) 175 (1827). See also Weston v. Alden, 8 Mass. 136 (1811); Colburn v. Richards, 13 Mass. 420 (1816); Cook v. Hull, 20 Mass. (3 Pick.) 269 (1825).

The original Massachusetts decision consciously to limit the action for diversion to prescriptive claims was based on far-reaching economic considerations, analogous to those that later induced courts to modify the natural flow rule. In Shorey v. Gorrell, cited in W. Cushing, *supra* note 10, the Massachusetts court equated the right to prevent diversion with a right to recover for competitive injury, which it regarded as pernicious. "For it is too much opposed to the general good," Chief Justice Cushing declared, "to say that a man shall not erect a mill or institute a school merely because there are others in the neighbourhood." Indeed, the claim — and rejection — of a rule of prior appropriation in this early case illustrates again that it was only when the natural flow rule was rejected that there was any need for a separate theory of prior appropriation. By the turn of the century, however, these considerations seem to have been forgotten, and the continuing persistence of prescriptive claims apparently reflects only an attempt to satisfy the technical rule that title and not merely possession had to be alleged in order to bring an action for diversion. See Pearson v. Tenny (1802), 3 N. Dane, *supra* note 11, at 14. See also 2 W. Blackstone *Commentaries* *195-99; Jones v. Waples, 2 Del. Cas. 159, 162-63 (1802); Twiss v. Baldwin, 9 Conn. 291, 301-02 (1832).

41. In 1821 the Massachusetts Supreme Court intimated adoption of a rule of priority, Hatch v. Dwight, 17 Mass. 289 (1821), to which Chancellor Kent, in a thoroughly confused statement of water law, offered the prestigious approbation of his *Commentaries.* "If the right of prior occupancy . . . did not go thus far," Kent wrote, acknowledging the basic dilemma of proponents of the old law, "the water privilege would seem to be rendered wholly useless for mill purposes to all parties." 3 *Commentaries* 358 n.(b) (1st ed. 1828).

42. 3 *Commentaries* 355 n.(b) (1st ed. 1828).

43. Palmer v. Mulligan, 3 Cai. R. 307, 314 (N.Y. Sup. Ct. 1805).

44. See 7 W. Holdsworth, *A History of English Law* 343-50 (1926); "Acquisition of Title by Prescription," 19 *Am. Jurist* 96, 98-101 (1838).

45. 3 W. Blackstone, *Commentaries* *218-19; Mosley v. Chadwick, 7 Barn. & Cres. 47 n. (a) (K.B. 1782); Chadwick v. Proprietors of Haverhill Bridge (1787) in 2 N. Dane, *Abridgement* 686-87 (1823); Tripp v. Frank, 4 T.R. 668 (1792).

46. Indeed, even after the attack on royal monopolies in England, custom legitimized long-standing monopolies that would otherwise have been illegal. 10 W. Holdsworth, *supra* note 44, at 402.

47. 3 W. Blackstone, *Commentaries* *217. See also J. Sullivan, *supra* note 11, at 268; Story v. Odin, 12 Mass. 157 (1815).

48. See Ingraham v. Hutchinson, 2 Conn. 584 (1818); Sherwood v. Burr, 4 Day 244 (Conn. 1810). For the doctrine of prescription in Massachusetts water cases, see note 40 *supra*.

49. See Cort v. Birkbeck, 1 Doug. 218 (K.B. 1779); White v. Porter, Hard. 177, 145 Eng. Rep. 439 (Exch. 1672); 3 W. Blackstone, *Commentaries* *235.

50. Mosley v. Walker, 7 Barn. and Cres. 40 (K.B. 1827). In a related sphere, the doctrine of prescription operated in the American colonies so as to transform long established private associations into municipal corporations, thereby conferring upon them the ancient privileges of suing, levying taxes, and making by-laws that were legally binding on all their members. After 1780 Massachusetts no longer recognized corporations by prescription. O. Handlin and M. Handlin, *Commonwealth: Massachusetts, 1774-1861,* at 162. (1947); *cf.* 1 N. Dane, *Abridgement* 459 (1823). Similarly, in Virginia after the Revolution all corporations were created by legislative act. 1 W. Blackstone, *Commentaries* 472 n.2 (St. G. Tucker ed. 1803). Nevertheless, long use was admitted as evidence of a presumably destroyed corporate charter, Dillingham v. Snow, 5 Mass. 547, 551-52 (1809), and some states even went further to allow proof of long use when

there was no reason to suppose that an original charter ever existed. See Greene v. Dennis, 6 Conn. 292, 302-04 (1826).

51. See 2 W. Blackstone, *Commentaries* 31 n.2, 36 n.7, 266 n.1 (St. G. Tucker ed. 1803). Similarly, Zephaniah Swift of Connecticut, who must have been familiar with the multitude of prescriptive actions in neighboring Massachusetts, nevertheless remarked in 1796 that "this country has been so lately settled, that the right of prescription, had hardly had time to operate." 1 Z. Swift, *A System of the Laws of the State of Connecticut* 442 (1795). Later, as Chief Justice of Connecticut, Swift recognized prescription with a vengeance in Ingraham v. Hutchinson, 2 Conn. 584 (1818). In Cortelyou v. Van Brundt, 2 Johns. 357, 361 (N.Y. Sup. Ct. 1807), counsel for the prevailing side argued that "Prescriptive rights are not to be favoured; in *England* they are considered as rights, the origin of which cannot be traced; but in this country almost every right can be traced to its origin." But see 1 N. Dane, *supra* note 50, at 459, which asserts that "though the country is young, yet it is old enough for prescription." Nevertheless, Dane himself was greatly skeptical of the entire doctrine of prescription and called forth a major legal controversy over its legitimacy. See p. 45 *infra.*

52. In Palmer v. Mulligan, 3 Cai. R. 307 (N.Y. Sup. Ct. 1805), the court ignored the plaintiff's prescriptive claim to water, treating it as if it were no more than a claim of prior appropriation. See *id.* at 315 (Thompson, J., dissenting). By the end of the eighteenth century, Massachusetts courts seem to have treated allegations of prescription as nontraversable, Perham *supra* note 40, perhaps producing a reaction. Writing in 1801, Sullivan cited a 1798 water rights decision based on prescription "which called for more investigation than had been usual in cases of this kind," J. Sullivan, *supra* note 11, at 273, an indication that the Massachusetts courts were beginning to examine prescriptive claims substantively. Indeed, in 1803 the Massachusetts court stated that even sixty years use was not sufficient to establish title by prescription. Devereux v. Elkins, 5 N. Dane, *Abridgement* 568 (1824).

In Connecticut, two water cases in 1783 and 1793 also rejected the prescriptive claim of a downstream landowner to an undiminished flow of water. Perkins v. Dow, 1 Root 535 (1793), citing Howard v. Mason (unreported 1783). In two other important early nineteenth century cases, the Pennsylvania and Connecticut supreme courts rejected prescriptive claims to exclusive fishing rights, Carson v. Blazer, 2 Binn. 475 (Pa. 1810), and to an exclusive stagecoach franchise, Nichols v. Gates, 1 Conn. 318 (1815). In a much narrower opinion, the New York Supreme Court sidestepped the question of whether the right to fish could be acquired by prescription, holding that even if it could, prescription could not justify building fishing huts on another's land. Cortelyou v. Van Brundt, 2 Johns. 357 (1807).

By this time, however, the English courts had already changed the theory of prescription, having begun in 1761 to analogize the prescriptive period to that required for the statute of limitations to run. J. Angell, *An*

Inquiry Into the Rule of Law which Creates a Right to an Incorporeal Hereditament by an Adverse Enjoyment for Twenty Years 23-31 (1827); 7 W. Holdsworth, *supra* note 44, at 343-50. Thus, by 1815, the attack on prescription had abated as the American courts followed the English judges in reducing the prescriptive period, first to sixty years, Thurston v. Hancock, 12 Mass. 220, 225 (1815), and eventually to twenty years, Bolivar Mfg. Co. v. Neponset Mfg. Co., 33 Mass (16 Pick.) 241 (1834), and in one state to fifteen years, Ingraham v. Hutchinson, 2 Conn. 584 (1818).

53. 2 Conn. 584 (1818).

54. *Id.* at 595.

55. *Id.*

56. Stokes v. Upper Appomatox Co., 30 Va. 343, 360, 3 Leigh 318, 334 (1831); Hoy v. Sterrett, 2 Watts 327, 330-31 (Pa. 1834); Thurber v. Martin, 68 Mass. (2 Gray) 394 (1854); Pratt v. Lamson, 84 Mass. (2 Allen) 275, 287-89 (1861); see Norton v. Volentine, 14 Vt. 239, 245-46 (1842), applying the same doctrine to upstream users; *cf.* Holsman v. Boiling Spring Bleaching Co., 14 N.J. Eq. 335, 344-46 (1862). See also "Acquisition of Title by Prescription," 19 *Am. Jurist* 96 (1838). Modern water law confirms this trend. "Prescriptive rights cannot be acquired against potential upstream users and can be acquired against downstream users only where their 'reasonable' use has been interfered with for the full statutory period." Haber, "Introductory Essay" to *The Law of Water Allocation,* at xxix (D. Haber and S. Bergen, eds. 1958). On the other hand, the intimately related issue of whether a nonmill owner could bring a cause of action in order to prevent the prescriptive period from running, *e.g.,* Whipple v. Cumberland Mfg. Co., 29 F. Cas. 934, 935-36 (No. 17,516) (C.C.D. Me. 1843), did not undergo any comparable change in spite of Chief Justice Shaw's efforts in that direction. Compare Elliot v. Fitchburg R.R., 64 Mass. (10 Cush.) 191, 196-97 (1852), with Lund v. New Bedford, 121 Mass. 286 (1876), and Parker v. Griswold, 17 Conn. 288, 301-08 (1845). See also Haar and Gordon, "Legislative Change of Water Law in Massachusetts," in *The Law of Water Allocation, supra* at 23, n.98.

57. 1 N. Dane, *supra* note 11, at 459.

58. 3 *id.* at 54-55. It is unclear from the passage whether Dane is attacking prescription in principle or only various applications of the doctrine. Nevertheless, he was understood as challenging the doctrine itself.

59. J. Angell, *supra* note 52, at 63.

60. *Id.* at 102-03.

61. 19 Wend. 309 (N.Y. Sup. Ct. 1838).

62. The decision had the effect of overthrowing, as one contemporary put it, "a fatal enemy to modern improvements in building." Noyes, "The Legal Rules Governing Enjoyment of Light," 23 *Am. Jurist* 52, 57-58 (1840).

63. 19 Wend. at 318. In his *Commentaries,* Chancellor Kent had entirely misunderstood the question. Having incorrectly assumed that the right to light could be based on mere priority and not prescription, he pro-

posed that one could lose the right by nonuse for a prescriptive period. "It is a wholesome and wise qualification of the rule," he concluded, "considering the extensive and rapid improvements that are every where making upon real property." 3 J. Kent, *Commentaries* 450 (2d ed. 1832). Because he started from an erroneous major premise, the "qualification" he proposed was entirely consistent with traditional legal conceptions.

64. See "The Natural Right of Support from Neighbouring Soil," 1 *Am. L. Rev.* 1, 9-11 (1866). The first case to reject the doctrine was Mitchell v. Rome, 49 Ga. 19 (1873).

65. Charles River Bridge v. Warren Bridge, 24 Mass. (7 Pick.) 344 (1829), *aff'd,* 36 U.S. (11 Pet.) 420 (1837).

66. 24 Mass. (7 Pick.) at 449-50. The prevailing technical objection to prescription at the turn of the century was that the country was too young for prescription to operate. See note 51 and text at note 58 *supra.* As Morton's opinion demonstrates, by the 1830s the technical challenge had shifted to an exposure of the fictional character of the "lost grant" theory of prescription, see *e.g.,* Parker v. Foote, 19 Wend. 309 (N.Y. Sup. Ct. 1838); Spear v. Bicknell, 5 Mass. 125, 130 n.(a) (Rand. ed. 1835); 3 N. Dane, *supra* note 11, at 54-55, which had been only recently developed by the English courts. See 7 W. Holdsworth, *supra* note 44, at 347-49.

67. Tyler v. Wilkinson, 24 Fed. Cas. 472, 474 (No. 14,312) (C.C.D.R.I. 1827). The entire question of whether prescription created a conclusive presumption of right was in a state of total confusion in this period. This confusion, of course, reflected the courts' uneasiness with the doctrine of prescription itself. For Angell's attempt to create a coherent rule, see J. Angell, *supra* note 52, at 36-68.

68. The same doctrine is advanced in Parker v. Foote, 19 Wend. 309 (N.Y. Sup. Ct. 1838).

69. H. Baldwin, *A General View of the Origin and Nature of the Constitution of the United States* 157 (1837). Baldwin intended this work to be included as an appendix to volume 11 of Peters Reports, but it was not prepared in time for publication there. It does appear in 9 *L. Ed.* 868, 949 (1837).

70. 36 U.S. (11 Pet.) at 562-63.

71. F. Hilliard, *Elements of Law* viii (1835).

72. Province Laws 1713, ch. 15.

73. Dench v. Jones (1783), in W. Cushing, *supra* note 10; T. Parsons, "Precedents" 51-52 (1775) (ms. Treasure Room, Harvard Law School). Published common law actions for flooding between 1768 and 1778 do not mention the mill act. Keen v. Turner (1768), in 1 *Legal Papers of John Adams* 242 (L. Wroth and H. Zobel, eds. 1965); Wilkins v. Fuller (1770), in *id.* at 274.

74. J. Sullivan, *supra* note 11, at 277-78. The frequency of mill act cases after 1798 reflects the growing conviction that the statutory remedy was finally intended to preclude any common law action. Buckman v. Tufts (1800), in J. Sullivan, *supra* note 11, at 278-82; 3 N. Dane, *supra* note 11, at 16; Batchelder v. Peabody (1800), in J. Story, *A Selection of*

Pleadings in Civil Actions 457-58 (1st ed. 1805); Lowell v. Spring, 6 Mass. 398 (1810). As late as 1813, the Supreme Judicial Court still reserved decision on the question whether the statutory remedy was exclusive, Staple v. Spring, 10 Mass. 72, 74-75 (1813), and so held only in 1814. Stowell v. Flagg, 11 Mass. 364, 365 (1814). In his note to Johnson v. Kittredge, 17 Mass. 76, 79 n.5, (B. Rand ed. 1832), Benjamin Rand nevertheless argued that the statute "should have been regarded as only giving an additional remedy," which, of course, would have made it nearly worthless.

75. A complete list of all such legislation to 1884 appears in Head v. Amoskeag Mfg. Co., 113 U.S. 9, 17 n.* (1884).

76. In England, by contrast, the construction of mills was encouraged by the typically feudal device of allowing the proprietor to recover damages against a customer who took his trade elsewhere. 3 W. Blackstone, *Commentaries* *235; J. Angell, *supra* note 31 at 119 n.19.

77. Act of Feb. 27, 1795, ch. 74, [1794-96] Mass. Acts & Resolves 443.

78. See Wroe v. Harris, 2 Va. (2 Wash.) 126 (1795).

79. Kentucky and, before 1798, Rhode Island also required prior authority. In 1798 Rhode Island modeled its statute after that of Massachusetts. "The Law of Water Privileges," 2 *Am. Jurist* 25, 31-32 (1829). See P. Coleman, *The Transformation of Rhode Island 1790-1860*, at 76-77 (1963).

80. For an example of a common law trespass action for flooding, see Wright v. Cooper, 1 Tyler 425 (Vt. 1802). *Cf.* Merritt v. Parker, 1 Coxe L. Rep. 460 (N.J. 1795). See also King v. Tarlton, 2 Har. & McH. 473 (Md. 1790).

81. See pp. 50-51.

82. See Bradley v. Amis, 3 N.C. 399, 400 (1806) (note by reporter): "Sometimes the profits of . . . merchant mills . . . are of much greater value in one year, than the fee simple of the annoyed property. . . . The object of the law cannot be obtained but by damages equivalent to the profits gained by the erection, or by damages to such an amount as will render those profits not worth pursuing." See ___ v. Deberry, 2 N.C. 248 (1795). The identical view prevailed as to trespasses. In trespass, "the value of the property, or the amount of the injury done to it, is not the only ground of damages. . . . Were it otherwise, a person so disposed might forcibly dispossess another of any articles of property at his pleasure, and compel the owner, however unwilling, to accept of the value in its stead." Edwards v. Beach, 3 Day 447, 450 (Conn. 1809).

83. Merritt v. Parker, 1 Coxe L. Rep. 460, 465 (N.J. 1795).

84. Gleason v. Gary, 4 Conn. 418 (1822); Dimmett v. Eskridge, 20 Va. (6 Munf.) 308 (Va. 1819); Hodges v. Raymond, 9 Mass. 316 (1812).

85. Not all states had courts of equity. There was no explicit grant of equity jurisdiction to restrain a nuisance in Massachusetts, for example, until 1827, although many contended that a general statute of 1817 conferred such authority. See Charles River Bridge v. Warren Bridge, 23 Mass. (6 Pick.) 376, 383, 394 (1828); Woodruff, "Chancery in Massachusetts," 9 *B.U.L. Rev.* 168 (1929).

86. 11 Mass. 364 (1814).

87. *Id.* at 366.

88. *Id.* at 368.

89. In his Litchfield Law School lectures, Judge Tapping Reeve questioned whether "other people have a right to build so near [an existing millowner] as to lessen his profits." "Henry H. Fuller's Notes of Lectures on Tapping Reeve and James Gould at the Litchfield Law School," vol. III, at 467-68 (1812-13) (ms. Treasure Room, Harvard Law School).

90. Skipwith v. Young, 19 Va. (5 Munf.) 276, 278 (1816).

91. J. Angell, *supra* note 19, at 62.

92. Act of Feb. 26, 1825, ch. 153, § 1 [1822-25] Mass. Laws 658.

93. Wolcott Woollen Mfg. Co. v. Upham, 22 Mass. (5 Pick.) 292, 294 (1827).

94. Avery v. Van Deusen, 22 Mass. (5 Pick.) 182 (1827).

95. Act of Feb. 26, 1825, ch. 153, § 3 [1822-25] Mass. Laws 658.

96. Even without a specific statutory provision, however, the Supreme Judicial Court had long assumed that a defendant did not have to pay for flooding in the absence of damages, Lowell v. Spring, 6 Mass. 398 (1810). Indeed, in a 1769 common law action for flooding the court apparently had allowed the defendant to show that his action had improved the plaintiff's land. Keen v. Turner, in 1 *Legal Papers of John Adams, supra* note 73, at 242, 244-45.

97. Merritt v. Parker, 1 Coxe L. Rep. 460, 466 (N.J. 1795).

98. 23 Mass. (6 Pick.) 94 (1828).

99. *Id.* at 96.

100. "The Law of Water Privileges," 2 *Am. Jurist* 25, 30-31, 34 (1829).

101. Stowell v. Flagg, 11 Mass 364, 366 n.a (Rand ed. 1832).

102. Maine Citizens Memorial To the Legislature (c. 1833), broadside J 38 (ms. Treasure Room, Harvard Law School). Maine, which separated from Massachusetts in 1820, continued to enforce the Massachusetts mill act. For another attack on the act, see "The Requisites to Dower and Who are Capable of It," 20 *Am. Jurist* 47, 60-63 (1838).

103. Act of March 22, 1830, ch. 122, § 2 [1828-31] Mass. Laws 474.

104. At the same time, however, the legislature began to play a more active role in the process of industrialization by specifically including in the charters of manufacturing corporations powers of flooding even more extravagant than the mill act would support. See *e.g.,* Boston & Roxbury Mill Dam Corp. v. Newman, 29 Mass. (12 Pick.) 467 (1832).

105. Fiske v. Framingham Mfg. Co., 29 Mass. (12 Pick.) 68 (1832). But *cf.* Chase v. Sutton Mfg. Co., 58 Mass. (4 Cush.) 152 (1849).

106. *Id.* at 69 (argument of plaintiff's counsel).

107. 29 Mass. (12 Pick.) at 70.

108. *Id.* at 70-71.

109. *Id.* at 71-72. At a later date in a strictly common law action dealing with the right of riparian owners to the use of water, he extended this concept still further in the interest of promoting economic development, by suggesting that where a stream could satisfy the full power needs of only

one manufacturer, the law would encourage this monopoly by recognizing an exclusive right in the first appropriator — this time, without compensation. Cary v. Daniels, 49 Mass. (8 Met.) 466 (1844).

110. He was not successful in carrying the court with him. In the same year, when Shaw was not sitting, the Massachusetts Supreme Court rested the entire case for the mill acts on an expansive conception of the power of eminent domain. Boston & Roxbury Mill Dam Corp. v. Newman, 29 Mass. (12 Pick.) 467 (1832). Shaw himself on occasion reverted to a "pure" eminent domain theory. Hazen v. Essex Co., 66 Mass. (12 Cush.) 475, 477-78 (1853); cf. Chase v. Sutton Mfg. Co., 58 Mass. (4 Cush.) 152, 169-70 (1849). Yet, clearly troubled by this rationale, he declared in 1851 that "the principle on which the law is founded is not, as has sometimes been supposed, the right of eminent domain. . . . It is not in any proper sense a taking of property of an owner of the land flowed, nor is any compensation awarded by the public." Murdock v. Stickney, 62 Mass. (8 Cush.) 113, 116 (1851).

111. J. Sullivan, supra note 11 at 334-35.

112. 2 Z. Swift, supra note 8 at 82. Similarly, in 1801 the Pennsylvania Supreme Court refused to allow an action of waste against a tenant in dower. "It would be an outrage on common sense," the court stated, "to suppose that what would be deemed waste in England, could receive that appellation here. Lands in general with us are enhanced by being cleared, provided a proper proportion of woodland is preserved for the maintenance of the place." Hastings v. Crunckleton, 3 Yeates 261, 262 (Pa. 1801).

113. Jackson v. Brownson, 7 Johns. 227, 232 (N.Y. 1810).

114. Id. at 236. Some may regard this conclusion as too extreme. Jackson v. Brownson was an action for forfeiture of the tenancy due to breach of the covenant against waste. "If this was an action of covenant to recover damages," Spencer conceded, "then, indeed, we should have a right to give the covenant not to commit waste a greater latitude of construction." Id. While it is true that he would have relaxed his reluctance to find waste were the consequences less grave than forfeiture, still this conclusion was based on contract principles and not on a theory of a right to prevent waste inherent in the ownership of land.

115. See, e.g., Findlay v. Smith, 20 Va. (6 Munf.) 134, 142 (1818), in which the court refused to find waste on the ground that "the law of waste, in its application here, varies and accommodates itself to the situation of our new and unsettled country." "England is an old country," wrote Dane in 1824, "where trees are raised even for fuel by planting and industry, almost as much as corn or grain is. This is a new country, where, except in some parts, of late years, the great object has been to destroy trees, to clear up the land, and to bring the wilderness or natural forests into cleared lands for cultivation and pasturing." 3 N. Dane, supra note 11, at 214. See also 1 W. Cruise, A Digest of the Law of Real Property 116 n. 2 (Greenleaf ed. 1849).

116. See Elwes v. Maw, 3 East 38 (K.B. 1802).

117. 27 U.S. (2 Pet.) 137 (1829).

118. The most important prior decisions are Whiting v. Brastow, 21 Mass. (4 Pick.) 310 (1826), and Holmes v. Tremper, 20 Johns. 29, 30 (N.Y. 1822), which distinguishes Elwes v. Maw, 3 East 38 (K.B. 1802), as inapplicable to less than major constructions.

119. 27. U.S. (2 Pet.) at 145.

120. 2 J. Kent, *Commentaries* 343 (3d ed. 1836). In the first two editions of the *Commentaries* Kent had merely repeated the conventional English view. The author of "Fixtures," 10 *Am. Jurist* 53, 56 (1833), states that the distinction between trade and agricultural fixtures "does not seem to have been generally admitted to prevail" in America.

121. 2 J. Kent, *Commentaries* 346 (3d ed. 1836).

122. Nash v. Boltwood (1783) in W. Cushing *supra* note 10. A report of the decision was first published in J. Story, *supra* note 74 at 366 (1805).

123. 4 Mass. 533 (1808).

124. 15 Mass. 164 (1818).

125. *Id.* at 167 (emphasis added).

126. *Id.*

127. See 3 N. Dane, *supra* note 11, at 219.

128. There is, however, another more general hypothesis that could explain Parker's support for the view that immediate development would make land less valuable. Land speculation was one of the most important businesses in postrevolutionary America. "Judges, land officers, government officials and their families, and military officers, jumped in at the very beginning of their residencies in new territories to acquire claims." P. Gates, *History of Public Land Law* 92 (1968). Many territorial judges were involved in land speculation, and there were many cases of judicially sanctioned land frauds. A large part of the legal profession was also involved in speculating in land. A. Chandler, *Land Title Origins* 484 (1945); M. Rohrbough, *The Land Office Business* 21 (1968).

The historical literature on government land policies emphasizes the disastrous consequences of governmental encouragement of speculation. "Men coming into a new community were regarded as intruders who would seriously retard the growth of the area for years by withholding land from development while they waited for its value to rise." P. Gates, *supra*, at 149. In short, Parker's opinion may reflect a more general prospeculative and antidevelopmental legal ideology that flourished under the doctrine of free transferability of land. See, *e.g.*, Claiborne v. Henderson, 13 Va. (3 Hen. & Munf.) 322, 331 (1809).

129. Libby v. Swett (1804) in Francis Dana Papers, Mass. Historical Society (Box 15 "Supreme Judicial Court Cases"). Published reports of this case appear in J. Story, *supra* note 74, at 365 and 4 N. Dane, *Abridgement* 675 (1824). See also Webb v. Townsend, 18 Mass. (1 Pick.) 21 (1822); Ayer v. Spring, 9 Mass. 8 (1812).

130. *E.g.*, Braxton v. Coleman, 9 Va. (5 Call) 433 (1805).

131. Humphrey v. Phinney, 2 Johns. 484 (1807) (improvements); Dorchester v. Coventry, 11 Johns. 510 (1814) (increased land value).

132. Winder v. Little, 1 Yeates 152, 154 (1792).

133. Thompson v. Morrow, 5 S. & R. 289, 291 (1819).

134. Horsford v. Wright, 1 Kirby 3 (Conn. 1786).

135. Guerard v. Rivers, 1 Bay 265 (1792); Liber v. Parsons, 1 Bay 19 (1785).

136. Horsford v. Wright, 1 Kirby 3 (1786).

137. Humphrey's Adm'r v. M'Clenachan's Adm'r, 15 Va. (1 Munf.) 493, 500 (1810); Mills v. Bell, 7 Va. (3 Call) 320, 326-27 (1802). In Lowther v. Commonwealth, 11 Va. (1 Hen. & Munf.) 202 (1806), a purchaser had recovered the full value of the land at the time of eviction from the seller. The seller then petitioned the state, his grantor, for compensation. The court held that he could only recover his sale price from the state, since in the original suit against him, the measure of damages should have been limited to that sum. Though clearly overruled by Humphrey's Adm'r v. M'Clenachan's Adm'r, *supra*, this case was never mentioned.

138. Gore v. Brazier, 3 Mass. 523, 545 (1807). Massachusetts distinguished between breach of covenants of seisin, for which only the purchase price was recoverable, Marston v. Hobbs, 2 Mass. 433, 440 (1807), and breach of covenants of quiet enjoyment for which the value at the time of eviction was recoverable. The theory was that the breach of the former took place before the eviction, while eviction was the cause of the latter breach.

139. Bender v. Fromberger, 4 Dall. 436, 441-46 (Pa. 1806). Although the decision does not mention whether the plaintiff was entitled to interest on the purchase price, the jury had returned damages of $2,979 and the purchase price was $2,390. *Id.* at 436. The court affirmed the verdict. Two Pennsylvania trial court rulings in 1804 and 1806 had previously split on the question of damages. See *The Institutes of Justinian* 619-20 (T. Cooper trans. 1812) (translator's notes). Finding the issue unsettled, a third trial court instruction apparently requested that the jury return a special verdict, specifying separate damages for the purchase money and the value of improvements. 4 Dall. at 445 (opinion of Smith, J.).

140. Staats v. Ten Eyck, 3 Cai. 111 (1805) (opinion of Livingston, J.) (increased land values); Pitcher v. Livingston, 4 Johns. 1 (1809) (improvements).

141. Liber v. Parsons, 1 Bay 19, 20 (1785).

142. Pitcher v. Livingston, 4 Johns. 1, 11-12 (1809).

143. *Id.* at 14.

144. Bender v. Fromberger, 4 Dall. 436, 444 (Pa. 1806).

145. Talbot v. Bedford's Heirs, 3 Tenn. 447, 455 (1813).

146. Cox's Heirs v. Strode, 5 Ky. 273, 278-79 (1811).

147. Phillips v. Smith, 1 *Carolina Law Repository* 475 (N.C. 1814); Furman v. Elmore, 2 Nott & McC. 189, 198-204 (S.C. 1812).

148. In Stout v. Jackson, 23 Va. (2 Ran.) 132 (1823), a divided court limited recovery to the purchase price plus interest, but it was only in Threlkeld's Adm'r v. Fitzhugh, 29 Va. (2 Leigh) 451 (1830) that the earlier rule was finally abandoned.

149. See note 138 *supra*.

150. *The Institutes of Justinian* 610, 620 (Cooper trans. 1812) (translator's notes).

151. *W. Sampson's Discourse and Correspondence with Various Learned Jurists upon the History of the Law* 78 (Thompson ed. 1826). The date of the letter is not given, but it was clearly written between 1823 and 1826.

152. See, *e.g.*, the contractarian reasoning of the Tennessee and Kentucky courts in Talbot v. Bedford's Heirs, 3 Tenn. 447, 454-55 (1813) and Cox's Heirs v. Strode, 5 Ky. 273, 278-79 (1811).

153. See note 128 *supra*.

154. 2 J. Kent, *Commentaries* 336-37 (4th ed. 1840).

155. For the Massachusetts statute of 1807, see "Improvements on Land," 2 *Am. Jurist* 294 (1829), which is largely devoted to establishing the civil law pedigree of the statute.

156. Message of the Governor of Kentucky Transmitting to the General Assembly the Report of Henry Clay and John Rowan Together With the Petition Presented by Them to the Supreme Court of the United States for a Re-Hearing of the Case Involving the Validity of the Occupying Claimant Laws 12 (1823).

157. Green v. Biddle, 21 U.S. (8 Wheat.) 1 (1823). The court had first decided against the statute in 1821, Justice Story writing a short, unsatisfactory opinion. The case was reargued in 1822, with Justice Washington the next year writing a more elaborate opinion striking the law down. Justice Johnson wrote a brilliant opinion upholding the constitutionality of the statute, though on the basis of a procedural point he formally concurred with the majority. 2 C. Warren, *The Supreme Court in United States History* 96-102 (1926).

158. 2 J. Kent, *Commentaries* *337-38 (4th ed. 1840).

159. Nelson v. Allen, 9 Tenn. 360, 379 (1830). In the third edition of his *Commentaries* (1836), Kent cited Nelson v. Allen for this proposition. 2 *Commentaries* 335 n.(d). However, Judge Catron, who also participated in the decision, added a note to the report of the case, which stated that he regarded "the examination of the constitutional question" as "gratuitous in this cause." 9 Tenn. at 386. Even Kent acknowledged that the case allowed the good faith occupant to recover the value of his improvements in a separate suit in equity. 2 *Commentaries* 336 n.(c). The full opinion in Nelson seems to deny the validity of even that procedure, 9 Tenn. at 380, 383-84, although Judge Catron's note clearly permits it. *Id.* at 386.

160. 2 J. Kent, *Commentaries* 336 (4th ed. 1840) lists eight states as having a statute. I have also added the limited New York statute discussed by Kent, *id.* at 335, and the Tennessee statute, see note 159 *supra*. In the twelfth edition of Kent's *Commentaries,* Oliver Wendell Holmes, Jr. stated that the principle of protecting good faith purchases had "been adopted very generally by the courts or legislatures of the several States." 2 J. Kent, *Commentaries,* 336 n.(1) (O. Holmes ed. 1873).

III. SUBSIDIZATION OF ECONOMIC GROWTH
THROUGH THE LEGAL SYSTEM

1. An Act for Highways, Mass. Colonial Statutes (ch. 23, 1693).

2. See, *e.g.*, An Act for the better Clearing Regulating and further Laying out Public High Roads in the City and County of Albany, N.Y. (1742), (3 *Colonial Laws of New York* 262 [1894]).

3. T. Jefferson, *Notes on the State of Virginia* 148 (1853 ed.). When Virginia actually settled on the principle of compensation is difficult to determine. Although a Virginia law passed in 1785 seems to have provided for compensation for land taken for roads, Va. Stat. ch. 75 (1785), it was asserted by a judge of the Virginia high court in 1831 that there was no payment for rights of way "until a very late period." Stokes v. Upper Appomatox Co., 3 Leigh 318, 337 (1831).

4. In Pennsylvania from the time of William Penn there was a consistent policy of giving landowners 6 percent more land than had been purchased so that roads could later be built. For this reason, the Pennsylvania Supreme Court in 1802 held that it was not a violation of the state constitution's "just compensation" clause to transfer land to a turnpike company without providing compensation. M'Clenachan v. Curwin, 3 Yeates 362, 6 Binn. 509 (Pa. 1802). It appears that a new statute was enacted in response to this decision, and two years later the Court interpreted it to require compensation. New Market and Budd Street, 4 Yeates 133 (Pa. 1804). It soon became "usual" for the state to provide compensation in turnpike acts. Stokely v. Robbstown Bridge Co., 5 Watts 546, 547 (Pa. 1836).

In New Jersey, under both the proprietary and state governments, no compensation was made for rights of way on the same theory as in Pennsylvania. "Compensation was first allowed, when companies were incorporated to make roads." Bonaparte v. Camden & Amboy R.R., 3 Fed. Cas. 821, 824 (1830) (argument of defendant's counsel).

5. Grant, "The 'Higher Law' Background of the Law of Eminent Domain," 6 *Wisc. L. Rev.* 67, 70 (1931).

6. 1 W. Blackstone, *Commentaries* 139 (Christian ed., 1855) [hereinafter cited as *Commentaries*].

7. See Van Horne's Lessee v. Dorrance, 2 Dall. 304 (Pa. 1795). C. G. Haines, *The Revival of Natural Law Concepts* (1935); B. F. Wright, *American Interpretations of Natural Law* 7, 50, 237 (1962).

8. State v. Dawson, 3 Hill 100, 103 (S.C. 1836); Stark v. M'Gowen, 2 Nott & M'Cord 387 (S.C. 1818); Lindsay v. Comm'rs, 2 Bay 38 (S.C. 1796). Compensation, however, was provided by South Carolina for damages in canal building. D. Kohn & B. Glenn, eds., *Internal Improvement in South Carolina, 1817-1828,* 327, 516 (1938).

9. Gardner v. Village of Newburgh, 2 Johns. Ch. 162, 168 (N.Y. 1816). See also Raleigh & Gaston R.R. v. Davis, 2 D & B 451, 459-61 (N.C. 1837).

10. Gardner v. Village of Newburgh, 2 Johns. Ch. at 166. However, in Rogers v. Bradshaw, 20 Johns. 735, 744-45 (Ct. of Err. 1823) and Jerome v. Ross, 7 Johns. Ch. 315 (N.Y. 1823), Kent reversed his earlier position that advance compensation was required. Along with his contemporaries, he was prepared to relax his earlier views for the sake of promoting the Erie Canal. "If there was ever a case," he declared, ". . . in which all petty private interests should be made subservient to the interest of an entire people, this is one." 7 Johns. Ch. at 342. But his last word published in his *Commentaries* held that "the better opinion" was that advance compensation should be required, emphasizing that anything to the contrary in the earlier opinions was mere dictum. 2 J. Kent, *Commentaries* 339 n. (4th ed. 1840). Kent's earlier strict view was applied to a private company in Bonaparte v. Camden & Amboy R.R., 3 Fed. Cas. 821 (1830) and in Bloodgood v. Mohawk & Hudson R.R., 18 Wend. 9 (N.Y. 1837).

11. M'Clenachan v. Curwin, 3 Yeates 362, 6 Binn. 509 (Pa. 1802).

12. Compare "An Act to continue 'An Act for Opening and Better a Mending, and Keeping in Repair, the Public Roads and Highways within this province,' " Pa. Stats. Ch. 1309 (1787), which continues the policy of a 1772 highway law allowing compensation only for improvements with "An Act to appoint commissions to regulate the streets, lanes and alleys in the District of Southwork," Ch. 1310 (1787), which for the first time allows damages (and a jury trial) for the taking of unimproved land.

The first canal incorporation statute establishing the Schuylkill and Susquehanna Navigation Co., Pa. Stat. 1577 (1791), contained an elaborate provision for recovering all damages resulting from eminent domain and became the model for subsequent Pennsylvania canal statutes. By contrast, the first turnpike incorporation statute establishing the Philadelphia and Lancaster Turnpike Co., Ch. 1629 (1792), allowed damages only for improvements. It was this provision that was upheld against constitutional challenge in M'Clenachan v. Curwin, 3 Yeates 362, 6 Binn. 509 (Pa. 1802).

13. Commonwealth v. Fisher, 1 Pen. & W. 462, 465 (Pa. 1830).

14. State v. Dawson, 3 Hill 100, 103 (S.C. 1835); Stark v. M'Gowen, 1 Nott & M'Cord 387 (S.C. 1818); Lindsay v. Comm'rs, 2 Bay 38 (S.C. 1796).

15. Beekman v. Saratoga & Schenectady R.R., 3 Paige 45, 57 (N.Y. 1831) (argument of counsel); Harvey v. Thomas, 10 Watts 63, 66-67 (Pa. 1840) (Gibson, C.J.).

16. Beekman v. Saratoga & Schenectady R.R., 3 Paige 45, 57-58 (N.Y. 1831) (argument of counsel).

17. Harvey v. Thomas, 10 Watts 63, 67 (Pa. 1840).

18. The Case of "The Philadelphia & Trenton R.R.," 6 Wharton 25, 46 (Pa. 1840); "On the Liability of the Grantee of a Franchise To an Action at Law for Consequential Damages, From its Exercise," 1 *Am. L. Mag.* 52 (1843).

19. "Restrictions upon State Power in Relation to Private Property," 1 *U.S. Law Intell.* 4, 5, 4 (1829).

20. Report of the Directors of the Western & Northern Inland Lock

Navigation Cos. to the New York Legislature (1795), U.S. Senate, 10th Congress, Rep. 250 (1808) in Albert Gallatin, "Report on Roads and Canals," *American State Papers* (W. Lowrie & W. Franklin, eds.), Class X, Miscellaneous, I, 724, 770, 772 [1834].

21. Report of the Board of Directors of the Western Inland Lock Navigation Co. to the New York Legislature (1798) in Gallatin, *id.* at 779.

22. 1797 N.Y. Stat. ch. 26.

23. 1798 N.Y. Stat. ch. 101.

24. Steele v. Western Inland Lock Navigation Co., 2 Johns. 283 (N.Y. 1807).

25. Ex Parte Jennings, 6 Cow. 518 (S. Ct., N.Y. 1826) *reversed sub nom* Canal Commissioners v. The People, 5 Wend. 423 (Ct. of Errors 1830); People v. Canal Appraisers, 13 Wend. 355 (S. Ct. 1835) reversed Canal Appraisers v. People, 17 Wend. 571 (Ct. of Errors 1836); Commissioners of the Canal Fund v. Kempshall, 26 Wend. 404 (Ct. of Errors 1841); Starr v. Child, 20 Wend. 149 (S. Ct. 1838) reversed 4 Hill 369 (Ct. of Errors 1842).

26. See Bronson, J., dissenting in Starr v. Child, 20 Wend. 149, 158-59 (N.Y. 1838).

27. Ex Parte Jennings, 6 Cow. 518, 523 (N.Y. 1826) (argument of commissioner's counsel).

28. Note, *id.* at 550.

29. Rogers v. Bradshaw, 20 Johns. 735, 740 (N.Y. 1823).

30. Carson v. Blazer, 2 Binn. 475 (Pa. 1810).

31. Cates v. Wadlington, 1 McCord 580 (S.C. 1822).

32. Canal Appraisers v. The People, 17 Wend. 571 (N.Y. 1836).

33. Commissioners of the Canal Fund v. Kempshall, 26 Wend. 404 (N.Y. 1841).

34. Child v. Starr, 4 Hill 369 (N.Y. 1842).

35. Letter of Charles G. Paleske in Gallatin, *supra* note 20, at 828, 829.

36. Letter of Alexander Wolcott in Gallatin, *supra* note 20, at 869.

37. Gallatin, *supra* note 20, at 828.

38. Stevens v. Proprietors of the Middlesex Canal, 12 Mass. 466, 468 (1815).

39. C. Roberts, *The Middlesex Canal, 1793-1860* 178 (1938). Many years later, in 1860, after the canal itself was relegated to the scrap heap of history without compensation, the Massachusetts legislature proposed to reimburse those property owners who had been deprived of compensation. O. Handlin and M. Handlin, *Commonwealth: Massachusetts 1774-1861* 222 (1947).

40. The Annual Report of the Canal Commissioners of the State of New York 25 (1825); N. Miller, *The Enterprise of a Free People: Aspects of Economic Development in New York State during the Canal Period, 1792-1838* 57-58 (1967).

41. "Consequential Damages," *supra* note 18, at 66, 60.

42. Report of the New York State Assembly, Doc. No. 284 (1841).

43. First Annual Report of the Boston & Maine R.R. Extension Co. 16 (1844) in Mass. Gen. Ct. Comm. on Railways and Canals, *Annual Reports of Railroad Corps. in Mass. for 1844* (1845). Expenditure for land and damages constituted almost 35 percent of the budget of the Old Colony Railroad. First Annual Report of the Old Colony Rail-Road. *Id.* at 86. In other cases, land damages were relatively small, owing to the "liberal desire" of landowners "of promoting so great a public improvement." Boston & Worcester R.R. Corp., Report of the Directors 8 (1832).

44. Eighth Annual Report of the Directors of the Western Railroad Corp. to the Massachusetts Legislature (January, 1844) in Mass. Gen. Ct. Comm. on Railways and Canals, *Annual Rpts. of Railroad Corps. in Mass. for 1843* (1844).

45. Thirteenth Annual Report of the Boston & Worcester Railroad (1844) in Mass. Gen. Ct., *supra* note 43, at 26.

46. "Consequential Damages," *supra* note 18. T. Sedgwick, *A Treatise on the Measure of Damages* 110-11 (1847).

47. See Hay v. Cohoes Co., 2 N.Y. 159 (1849); Tremain v. Cohoes Co., 2 N.Y. 163 (1849); Fish v. Dodge, 4 Denio 311 (N.Y. 1847).

48. Indeed, courts had held that for an intentional trespass, the amount of actual injury was not the limit of damages. "Were it otherwise, a person so disposed might forcibly dispossess another of any article of property at his pleasure, and compel the owner, however unwillingly, to accept of the value in its stead." Edwards v. Beach, 3 Day 447, 450 (Conn. 1809).

49. "Between 1816 and 1840, about $125,000,000 was spent on canal building, and at least three states had so strained their credit as to be brought to the verge of bankruptcy." G. R. Taylor, *The Transportation Revolution, 1815-1860* 52 (1951).

50. L. Hartz, *Economic Policy and Democratic Thought: Pennsylvania, 1776-1860* 159 (1948).

51. Losee v. Buchanan, 51 N.Y. 476, 484 (1873).

52. Palmer v. Mulligan, 3 Caines 307 (N.Y. 1805).

53. Steele v. Western Inland Lock Navigation Co., 2 Johns. 283 (N.Y. 1807). Compare Coleman v. Moody, 4 H & M 1 (Va. 1809), where the court only allows damages actually foreseen as a bar to further damages.

54. Thurston v. Hancock, 12 Mass. 220 (1815); Panton v. Holland, 17 Johns. 92 (N.Y. 1819).

55. 1 Pick. 418 (Mass. 1823).

56. Lansing v. Smith, 8 Cow. 146 (N.Y. 1828). See also Hollister v. Union Co., 9 Conn. 436 (1833).

57. 1 Pick. at 431. This theory, in one form or another, reappears in later cases. See Lexington & Ohio R.R. v. Applegate, 8 Dana 289, 298 (Ky. 1839); Radcliff's Executors v. Mayor, 4 N.Y. 195, 207 (1850).

58. M'Clenachan v. Curwin, 3 Yeates 362, 6 Binn. 509 (Pa. 1802).

59. Lansing v. Smith, 8 Cow. at 149.

60. 2 J. Kent, *Commentaries* 340 n. (4th ed. 1840); Charles River

Bridge v. Warren Bridge, 36 U.S. (11 Pet.) 420, 638 (Story, J., dissenting).

61. See, *e.g.,* Hollister v. Union Co., 9 Conn. 436, 446 (1833).

62. Barron v. Baltimore, 2 *Am. Jur.* 203, 206-07 (Md. 1828), reversed 7 Pet. 243 (1833). [The decision of the Maryland Court of Appeals, reversing the lower court, is unreported, but a summary of that decision based on court records appears in Cumberland v. Willison, 50 Md. 138, 150-55 (1878). The United States Supreme Court then dismissed the appeal from the decision of the Maryland Court of Appeals, holding that it had no jurisdiction, since the "just compensation"clause of the Fifth Amendment applied only to the national government.]

63. See T. Sedgwick, *supra* note 46, at 63-112; "Consequential Damages," *supra* note 18, at 72.

64. Radcliffe v. Mayor of Brooklyn, 4 N.Y. 195, 206 (1850).

65. *Id.* at 207.

66. However, in The Case of "The Philadelphia & Trenton R.R.," 6 Whar. 25, 46 (Pa. 1840), Chief Justice Gibson applied the immunity to a railroad corporation, holding that the constitutional provision requiring just compensation did not extend to consequential damages. The fact that the state had usually provided compensation, he declared, "was of favor, not of right."

67. 3 *Commentaries* 217-18. See Fish v. Dodge, 4 Denio 311 (N.Y. 1847).

68. Henry H. Fuller, "Notes on Lectures of Tapping Reeve and James Gould at the Litchfield Law School" (1812-13) Vol. III, at 465-66 (ms. LMS 2014, Treasure Room, Harvard Law School).

69. 8 Dana 289 (Ky. 1839).

70. *Id.* at 305.

71. *Id.* at 309.

72. Compare Pennsylvania v. Wheeling Bridge Co., 54 U.S. (13 How.) 518, 577 (1851) with 54 U.S. (13 How.) at 602, 605, 608 (Daniel, J., dissenting).

73. St. Helen's Smelting Co. v. Tipping, 11 H.L.C. 642, 11 Eng. Rep. 1485 (1865). Even as late as 1890 the Maryland Court of Appeals, in a widely followed opinion, refused to balance social utilities in a nuisance case. "It may be convenient to the defendant," the court declared, "and it may be convenient to the public, but, in the eye of the law, no place can be convenient for the carrying out of a business which is a nuisance, and which causes substantial injury to the property of another." Susquehanna Fertilizer Co. v. Malone, 73 Md. 268, 277 (1890). Only in the twentieth century did official and formal nuisance doctrine incorporate a balancing test. See, *e.g.,* Rose v. Socony-Vacuum Corp., 54 R.I. 411 (1934), though similar results had been reached much earlier under the various technical exceptions that have been discussed.

74. See, *e.g.,* Hudson & Delaware Canal Co. v. N.Y. & Erie R.R., 9 Paige 323 (N.Y. 1841); Lexington & Ohio R.R. v. Applegate, 8 Dana 289, 298 (Ky. 1839).

75. Steele v. Western Inland Lock Navigation Co., 2 Johns. 283 (1807).

76. Van Schoick v. Delaware & Raritan Canal Co., Spencer's Rep. 249, 254 (N.J. 1843).

77. 3 *Commentaries* 219.

78. *Ibid.* See Nichols v. Pixly, 1 Root 129 (Conn. 1789); Harrison v. Sterett, 4 H & McH. 540 (Md. 1774).

79. Lansing v. Smith, 8 Cow. 146, 156-68 (N.Y. 1828) *aff'd* 4 Wend. 9 (1829); Barron v. Baltimore, 2 *Am. Jur.* 203 (Md. 1828), 32 U.S. (7 Pet.) 243, 244 (1833).

80. 8 Cow. at 157-58.

81. Hart v. Mayor of Albany, 9 Wend. 571 (N.Y. 1832).

82. Smith v. Boston, 7 Cush. 254, 255 (Mass. 1851).

83. Blood v. Nashua & Lowell R.R., 2 Gray 137 (Mass. 1854); Brightman v. Fairhaven, 7 Gray 271 (Mass. 1856).

84. Governor & Co. of the British Cast Plate Manuf. v. Meredith, 4 T.R. 794, 100 Eng. Rep. 1306 (1792); Sutton v. Clarke, 6 Taunt 29, 128 Eng. Rep. 942 (1815).

85. Steele v. Western Inland Lock Navigation Co., 2 Johns. 283 (N.Y. 1807); Callender v. Marsh, 1 Pick. 418 (Mass. 1823); Lansing v. Smith, 8 Cow. 146 (N.Y. 1828). See also J. Angell, *Watercourses* 66 (1st ed. 1824).

86. One of the first cases openly to recognize an immunity limited to public officials was Sayre v. Northwestern Turnpike Rd., 10 Leigh 454 (Va. 1839). But by this time the doctrine already had been extended to cover private companies as well.

87. Bloodgood v. Mohawk & Hudson R.R., 18 Wend. 9, 30 (N.Y. 1837); Barron v. Baltimore, 2 *Am. Jur.* 203, 212-13 (Md. 1828).

88. Shrunk v. President of the Schuylkill Navigation Co., 14 Serg. & Rawl. 71, 83 (Pa. 1826).

89. L. Hartz, *supra* note 50, at 85.

90. Spring v. Russell, 7 Greenl. 273, 289-90 (Me. 1831).

91. "Consequential Damages," *supra* note 18, at 66.

92. T. Sedgwick, *supra* note 46, at 110. In the same year, even Joseph Angell accepted this doctrine, although his earlier, more conservative work on *Watercourses* had conceded only the limited immunity conferred on public officials. Compare *Treatise on Tidewaters* 93-97 (2d ed. 1847) with *Watercourses* 108-09 (2d ed. 1833).

93. The one clear exception is Hollister v. Union Co., 9 Conn. 436, 445-46 (1833).

94. Ten Eyck v. Delaware & Raritan Canal Co., 3 Harr. 200 (N.J. 1841); Hooker v. New-Haven & Northampton Co., 14 Conn. 146 (1841) (3-2 decision).

95. People v. Kerr, 27 N.Y. 188, 190 (1863).

96. Bellinger v. N.Y. Central R.R., 23 N.Y. 42 (1861).

97. See Scott v. Bay, 3 Md. 431 (1853); Fish v. Dodge, 4 Denio 311 (N.Y. 1847).

98. By this time, incidentally, English railroads already had been made liable by statute for activity which "injuriously affected" land, I.

Redfield, *Law of Railways* 116 (2d ed. 1858), a factor undoubtedly contributing to "the immense cost in the construction of English railroads . . . mainly derived from the extravagant prices which are demanded, and have to be paid at the outset for the land." H. M. Flint, *The Railroads of the United States* 26 (1868).

99. See pp. 98-99 *infra*.

100. Commonwealth v. Reed, 34 Pa. 275, 281-82 (1859).

101. "A Reading On Damages in Actions Ex Delicto," 3 *Am. Jur.* 287, 288 (1830). Although the piece was unsigned, authorship is attributed to Metcalf in T. Sedgwick, *supra* note 46, at 45n.

102. *Id.* at 292.

103. *Id.* at 305.

104. 2 *Treatise on the Law of Evidence* 209 (1st ed. 1846). Greenleaf's views were also presented in "The Rule of Damages in Actions Ex Delicto," 9 *Law Rep.* 529 (1847). This unsigned piece is attributed to Greenleaf in 10 *Law Rep.* 49 (1847).

105. T. Sedgwick, *supra* note 46, at 46n. See also *id.* at 38-44. Sedgwick's views were also presented in "The Rules of Damages in Actions Ex Delicto," 10 *Law Rep.* 49 (1847). The article is signed "T.S."

106. 2 *Treatise on the Law of Evidence* 242 n.2 (2d ed. 1848).

107. Sedgwick, *supra* note 46 at 214.

108. *Id.* at 63-112.

109. See p. 67 *infra*.

110. C. Roberts, *supra* note 39, at 177.

111. L. Hartz, *supra* note 50 at 159.

112. *Id.* at 160.

113. See Beekman v. Saratoga & Schenectady R.R., 3 Paige 45, 68-69 (N.Y. 1831) (Argument of Counsel).

114. Bonaparte v. Camden & Amboy R.R. 3 Fed. Cas. 821 (1830).

115. Beekman v. Saratoga & Schenectady R.R., 3 Paige 45 (N.Y. 1831).

116. Willyard v. Hamilton, 7 Ohio 111 (pt. 2) (1836).

117. Raleigh & Gaston R.R. v. Davis, 2 D & B 451 (N.C. 1837).

118. See Bonaparte v. Camden & Amboy R.R., 3 Fed. Cas. 821 (1830). One of the earliest of such statutes was passed in New York in connection with the Erie Canal. 1817 Stat. ch. 262, sec. 3. But the canal commissioners had been in the habit of offsetting benefits as early as 1810. See W. W. Campbell, *Life and Writings of DeWitt Clinton* 54 (1849). In 1829 Massachusetts amended its mill act to allow for consideration of benefits from flooding. *Gen. Laws* 1829, ch. 122.

119. *The New Constitution* 167 (1849).

120. *Sic utere tuo, ut alienum non laedas* (Use your own [property] so as not to harm another's).

121. 3 *Commentaries* 217-18.

122. C. H. S. Fifoot, *History and Sources of the Common Law* 164 (1949).

123. Scovel v. Chapman, 2 Root 315 (Conn. 1795); — v. Jackson, 2 N.C. 14 (1792); M'Clures v. Hammond, 1 Bay 99 (S.C. 1790). For two early eighteenth century carrier cases in which negligence was alleged, see Gassarett v. Bogardus (1701) and Smith v. Bill (1710-11) in R. B. Morris, ed., *Select Cases of the Mayor's Court of New York City, 1674-1784* 361, 395 (1935).

124. Stephens v. White, 2 Wash. 203 (Va. 1796); Cross v. Guthery, 2 Root 90 (Conn. 1794); Coker v. Wickes (R.I. 1742), reported in Chafee, *Records of the Rhode Island Court of Equity, 1741-1743,* 35 *Publications of the Colonial Society of Mass.* 91, 105-07 (1944).

125. Sparhawk v. Bartlet, 2 Mass. 188 (1806); Brown v. Lord, Kirby's Rep. 209 (Conn. 1787).

126. Jones v. Abbee, 1 Root 106 (Conn. 1787); Staphorse v. New Haven, 1 Root 126 (Conn. 1789); Abel v. Bennet, 1 Root 127 (Conn. 1789). See also 2 *Dane Abr.* 649-57.

127. 3 *Commentaries* 415-16.

128. *Id.* at 165.

129. Lobdell v. New Bedford, 1 Mass. 153 (1804); Bill v. Lyme, 2 Root 213 (Conn. 1795); Swift v. Berry, 1 Root 448 (Conn. 1792); Harris v. Moore, 1 Coxe 44 (N.J. 1790).

130. Lord v. Fifth Mass. Turnpike Corp., 16 Mass. 106 (1819); Riddle v. Prop. of Locks, 7 Mass. 169 (1810); Townsend v. Susquehannah Turnpike Rd., 6 Johns. 90 (N.Y. 1809).

131. 3 W. Blackstone, *Commentaries* 415-16. See also Lansing v. Fleet, 2 Johns. Cas. 3, 5 (N.Y. 1800).

132. Patten v. Halsted, 1 Coxe 277, 279 (N.J. 1795).

133. R. Pound, *An Introduction to the Philosophy of Law* 91 (rev. ed. 1954). In Johnson v. Macon, 1 Wash. 4 (Va. 1790), for example, the Virginia high court held that the lower court erred in requiring proof of negligence in order to hold a sheriff liable for an escape. Though the statute expressly required that negligence be shown, the court declared that it "ought to be presumed." *Id.* at 5.

134. 1 J. Comyns, *Digest of English Law* 202 (1785 ed.).

135. All the above quotations appear in 3 *Dane Abr.* 31-33.

136. *Id.* at 33.

137. *Id.* at 35.

138. Bussy v. Donaldson, 4 Dall. 206 (Pa. 1800), cited *id.* at 35.

139. Clark v. Foot, 8 Johns. 421 (N.Y. 1811), cited in *ibid.*

140. Fifoot, *supra* note 122 at 164.

141. See, *e.g.,* Bussy v. Donaldson, 4 Dall. 206 (Pa. 1800); Waldron v. Hopper, 1 Coxe 339 (N.J. 1795); Van Cott v. Negus, 2 Caines 235 (N.Y. 1804). Cases involving injury to pedestrians did not appear until after 1825. See M'Allister v. Hammond, 6 Cow. 342 (N.Y. 1826); Lane v. Crombie, 12 Pick. 177 (Mass. 1831). The earliest American collision case of which I am aware is Waterman v. Gillings (1770) (unpublished records of Plymouth, Mass. Court of Common Pleas). This case, involving a ship collision, is

founded on an allegation of negligence. It is apparently the source of the one declaration for collision that appears in Theophilus Parsons' "Precedents" (1775) (ms. LMS 1118, Treasure Room, Harvard Law School) under the heading "Case for Carelessly Managing a Vessel" p. 53. "Cases of [ship] collision were relatively infrequent [during the colonial period]. . . . The seas were wide and sinkings at sea were rare. Collision was more common in harbors and channels, and because the damage was slight the cases did not get into the courts." C. M. Andrews, "Introduction" to *Records of the Vice-Admiralty Court of Rhode Island, 1716-1752* 31 (D. Towle, ed. 1936). Professor Andrews does, however, cite two instances of collisions in colonial admiralty court records. *Ibid.* In a study of colonial Massachusetts admiralty records, Professor Wroth also found that "actions for collisions were unaccountably few." Wroth, "The Massachusetts Vice Admiralty Court" in G. Billias, ed., *Law and Authority in Colonial America* 32, 42 (1965).

142. For a surprisingly and atypically late example of the unwillingness of common law judges to enforce legal duties between strangers, see the English case of Heaven v. Pender, 11 Q.B.D. 503 (1883).

143. C. Gregory, "Trespass to Nuisance to Absolute Liability," 37 *Va. L. Rev.* 359, 365-70 (1951).

144. See Leame v. Bray, 3 East. 593, 102 Eng. Rep. 724 (1803); Scott v. Shepherd, 2 W. Blackstone 892, 96 Eng. Rep. 525 (1773) (opinion of Blackstone, J.); Reynolds v. Clarke, 2 Ld. Raymond 1399, 92 Eng. Rep. 410 (1726). For an excellent discussion of these cases, see E.F. Roberts, "Negligence: Blackstone to Shaw to ?" 50 *Cornell L. Rev.* 191 (1965).

145. See Winfield and Goodhart, "Trespass and Negligence," 49 *L.Q.R.* 359 (1933).

146. An examination of a number of nineteenth century cases bears this out.

In Taylor v. Rainbow, 2 H&M 423 (Va. 1808) we find the court reporter observing of the difference between trespass and case that "the *law* says there is a nice distinction, but the *reason* . . . is often difficult to discover." *Id.* at 423. And the defendant's counsel, in arguing that the plaintiff brought the wrong action, declared: "It is unnecessary to *reason* on the propriety of keeping up the boundaries of action: it is a *settled rule of law* that they *must* be preserved." *Id.* at 430. Of the three judges who wrote opinions in the case, none puts the distinction on substantive law. Judge Fleming, for example, emphasized that he saw no substantive difference between the two writs, since the ends of justice would be served by either, yet he felt "tied down, and bound by precedents" establishing the direct-indirect distinction. *Id.* at 444.

Likewise, in Purdy v. Delavan, 1 Caines 304, 312 (N.Y. 1803), there was an elaborate argument concerning the distinction between trespass and case, which nevertheless failed to mention any distinction in the level of liability between the writs. One judge simply concluded that "the boundary between case and trespass is faintly delineated, and not easily discerned."

Id. at 322. In Gates v. Miles, 3 Conn. 64, 67 (1819), the Connecticut Supreme Court also spelled out only procedural reasons for preserving the difference between the actions: "As no suit can be maintained for trespass *vi et armis* after three years, and as in trespass on the case there is no limitation, it becomes highly important to preserve the established boundaries between these actions."

An 1817 case in New York, Foot v. Wiswall, 14 Johns. 304, marks a significant turning point because it emphasizes how late it was before lawyers came to regard even the allegation of negligence in an action on the case as limiting liability. In a ship collision case, the plaintiff's counsel still argued the strict liability doctrine that the defendant had "acted at his peril." The action was brought in case, he pointed out, only because that form of action was required when a servant brought about an injury. "If the defendant had been at the helm of his boat at the time," he concluded, "there is no doubt that the plaintiffs could have recovered in an action of *trespass*; and there is no reason why they should not be equally entitled to recover in an action of *trespass on the case*, or for negligence; the distinction between the two actions being purely technical." *Id.* at 306. Nevertheless, the New York Supreme Court upheld the verdict for the defendant, clearly indicating for the first time it was for the plaintiff to prove and for the jury to determine whether the defendant had violated some standard of care. Three years later the court also upheld a trespass action for a collision only after minutely examining the evidence for proof of carelessness. Percival v. Hickey, 18 Johns. 257, 289-90 (1820). Thus, it is not surprising that by 1826, when the New York court elaborately explained why "it is still important to preserve the distinction between the actions" it failed to discuss any differences in substantive law, mentioning only technical differences in costs and pleadings. M'Allister v. Hammond, 6 Cow. 342, 344 (N.Y. 1826). For within a few short years, actions in both trespass and case had been simultaneously put to the test of negligence.

Outside New York, Benjamin L. Oliver, Jr., of Massachusetts was the first clearly to state that "without any negligence or fault whatever, it seems no action can be maintained" in either trespass or case. B. Oliver, *Forms of Practice; or American Precedents* 619 (1828).

147. 6 Cush. 292 (Mass. 1850).

148. *The Common Law* 84-85 (Howe ed. 1968).

149. Gregory, *supra* note 143, at 365.

150. See note 146 *supra*.

151. Cole v. Fisher, 11 Mass. 137 (1814). The court stated that if the parties agreed on the damages "a contest about the form of action will be of little avail to the defendant." For even if the defendant were successful in invoking the direct-indirect distinction to show that the plaintiff had brought the wrong action, "the expenses of [the first action] might be properly urged as a ground for further damages" in a second action. *Id.* at 139. More important, the court's impatience is associated with the beginning of the disintegration of strict liability under the pressure of the negligence

idea. The defendant had stepped out of his shop in order to discharge a gun for the purpose of drying it. The shot caused a horse standing across the highway to flee in fright and thereby to destroy a carriage to which it was harnessed. In an action in trespass, the court's analysis of whether the plaintiff had brought the proper writ already shows the transformation of the direct-indirect distinction. For the purpose of determining whether the injury was sufficiently immediate to allow trespass to lie, the court looked to whether "the horse and chaise were in plain sight, and near enough to be supposed to excite any attention or caution on the part of the defendant." *Id.* at 138. In short, the court had unconsciously shifted the criterion of "directness" to the negligence test of whether the defendant had adverted to the danger. It was groping for the modern distinction between intentional and negligent injuries.

152. Agry v. Young, 11 Mass 220 (1814). 2 *Dane Abr.* 487. The report of the case says nothing about amendment. Dane presumably learned of the procedure from one of the lawyers in the case.

153. Fales v. Dearborn, 1 Pick. 344 (Mass. 1823). See also Parsons v. Holbrook (1816) (unpublished records of Norfolk, Mass. Court of Common Pleas), in which plaintiff alleges in a trespass action that defendant "with force and arms negligently caused a carriage collision," suggesting the simultaneous conversion of the writ of trespass to a negligence standard.

154. Sproul v. Hemmingway, 14 Pick. 1 (Mass. 1833).

155. See note 202 *infra*.

156. On the basis of a diametrically opposed major premise, Professor E. F. Roberts has also argued that the significance of *Brown v. Kendall* had been exaggerated. "In fact," he maintains, "*Brown v. Kendall* did not remove strict liability from the law: it was not then there." Roberts, *supra* note 144, at 204. First, Roberts places too much emphasis on the tort excuse of "inevitable accident," finding in it a much earlier defense of no negligence. *Id.* at 203. Compare note 176 *infra*. There is no indication that in America "inevitable accident" was an earlier surrogate for the negligence principle. Second, Roberts simply "assume[s] that English and American tort law were pretty much parallel in their development" through the first half of the nineteenth century. *Id.* at 201. In fact, they were substantially different. See note 146 *supra,* note 177 *infra*.

In his *Elements of Law* (1835), Francis Hilliard, the Boston lawyer who was later to write the first Anglo-American treatise on the law of torts, shows no trace of recognition of a modern action for negligence. His only extended discussion of negligence is in connection with bailment. *Id.* at 101-02. He makes no distinction between actions in trespass and case in terms of whether negligence need be alleged, *id.* at 242-43, and his conception of negligence in the master-servant area is limited to the eighteenth century contractual notion of neglect. *Id.* at 30.

157. Hill & Denio 193 (N.Y. 1843) [Lalor supp. 1857]. Although Holmes acknowledged this case in *The Common Law* (1881), the romantic

role that has been assigned to *Brown v. Kendall* has fostered a tendency to ignore the significance of prior cases. It is true, however, that because of a changeover in court reporters, *Harvey v. Dunlop* was not published until 1857, seven years after Shaw's decision. In any event, that decision was every bit as self-confident as the Shaw opinion, and, more important, study of the prior New York cases underlines how this doctrine was merely the conclusion of a twenty-five year trend directed toward triumph of the negligence doctrine in that state.

158. See note 146 *supra*.

159. Lehigh Bridge v. Lehigh Coal & Navig. Co., 4 Rawle 8 (Pa. 1833).

160. Sullivan v. Murphy, 2 *Law Rep.* 247 (1839).

161. 1 *Commentaries* 431. Compare *id.* at 431 n.24 (Christian ed., New York 1832).

162. See note 142 *supra*.

163. In his lectures at the Litchfield Law School Judge Tapping Reeve elaborated upon Blackstone's example. "If the servant of a Blacksmith, while shoeing a horse, designedly lame the horse, both master and servant are liable. If he had lamed the horse, thro carelessness or want of skill, the master only would be liable. But in each of these cases if the master had done the act himself he would have been liable, and ought therefore to be liable when it is done by the servant. The Blacksmith is liable on the ground of an implied contract which he, and all other mechanics are under to perform the business of their trade in a workmanlike manner" (citing cases) Vol. I., p. 146 (ms. 2013, Treasure Room, Harvard Law School) (18??).

164. 1 *Laws of Connecticut* 223 (1795). *Cf.* M'Manus v. Crickett, 1 East. 106, 102 Eng. Rep. 43 (1800).

165. The only cases in which Swift writes of the "negligence" or "misconduct" of the servant are cases in which there is a status or contractual relationship between the master and the plaintiff. *Id.* at 223-24.

166. M'Manus v. Crickett, 1 East. 106, 102 Eng. Rep. 43 (1800); Morley v. Gaisford, 2 H. Bl. 441, 126 Eng. Rep. 639. But see Grinnell v. Phillips, 1 Mass. 530 (1805).

167. See J. Wigmore, "Responsibility for Tortious Acts: Its History II," 7 *Harv. L. Rev.* 382 (1894), 3 *Select Essays in Anglo-American Legal History* 520-33 (1909). 2 *Dane Abr.* 511-13. *Cf.* R. Morris, *Studies in the History of American Law* 238-39 (1963).

168. Surprisingly, this limited function of negligence reappears as late as 1848 in Simon Greenleaf's *Evidence*. The defendant is liable for a trespass, he wrote, "if it appear that the act was done by his direction or command, or by his *servant* in the course of his master's business, or while executing his orders with ordinary care" II, 579 (2d ed.). Writing in 1835, Francis Hilliard uses negligence in the master-servant context interchangeably with "neglect" or unintentional activity, not with carelessness. *Supra* note 156, at 30.

169. Bussy v. Donaldson, 4 Dall, 206, 208 (Opinion of Smith, J.) (Pa. 1800). See also Snell v. Rich, 1 Johns. 305 (N.Y. 1806).

170. 2 Bay 345, 1 Brev. 178 (S.C. 1802).

171. 2 Bay at 350.

172. 1 Brev. at 179-80.

173. Although both reporters were judges at the time, the two reports of the case have somewhat different emphases. The report of Bay suggests that the court slightly changed (or believed it was changing) the English rule, so that the master would not be liable "for any unauthorized or casual act committed without the knowledge or approbation of the master." 2 Bay at 350-51. The shorter report of Brevard does not indicate that there was any change in the English rule. 1 Brev. at 180.

174. 2 Bay at 350.

175. Beaulieu v. Fingham, Y.B. 2 H. IV, fo. 18, pl. 6 (1401). The plaintiff declared that the servant had "so negligently cared for his fire."

176. Tubervil v. Stamp, 1 Salk. 13, 91 Eng. Rep. 13 (1698); Beaulieu v. Fingham, Y.B. 2 H. IV. fo. 18, pl. 6 (1401). A. Ogus, "Vagaries in Liability for the Escape of Fire," 1969 Cambr. L. Jo. 104, 105-06. Compare J. Wigmore, "Responsibility for Tortious Acts: Its History III," 7 Harv. L. Rev. 441, 448-49 (1894); 3 Select Essays 511-12 (1909), and 11 W. Holdsworth, A History of English Law 606 (1938) with P. Winfield, "The Myth of Absolute Liability," 42 L.Q.R. 37, 46-50 (1926).

177. For a discussion of the extent to which the statute, 6 Anne c. 31, R.C. c. 58 (1707), affected the common law in England, see 11 Holdsworth supra note 176 at 607-08. St. George Tucker stated that the statute had no force in Virginia, 1 W. Blackstone, Commentaries 431 n.15 (Tucker ed. 1803) and I have been able to find no evidence that it was received in any state. For the colonial rule of absolute liability for fire, see R. Morris, supra note 167, at 242-44. For a nineteenth century American reiteration of the common law rule of strict liability, see Opinion of Atty. Gen. William Wirt, "Claim For Damage by Fire" (1819) in Opinions of Attorneys General 163-64 (1851).

178. See Wigmore, supra note 176.

179. 1 Commentaries 431.

180. 8 W. Holdsworth, A History of English Law 469 n.3 (1926).

181. The case can be read as merely applying one of the narrow excuses available under the old common law rule. Even in Turbervil v. Stamp, 1 Salk 13, 91 Eng. Rep. 13 (1698), the court stated that it would take evidence that the fire spread because of a sudden wind. On the other hand, the court in Snee v. Trice not only distinguishes between "accident" and "negligence" and seems clearly to be thinking in terms of a standard of care, but plaintiff's counsel argued primarily in terms of whether the slaves had adverted to the risk of danger. 2 Bay at 346-47.

182. Wingis v. Smith, 3 McCord 400 (S.C. 1825).

183. See, e.g., Jewett v. Brown (Mass. 1797) (unpublished records of Essex Mass. Common Pleas); Washburn v. Tracy, 2 Chip. 128 (Vt. 1824).

184. Blin v. Campbell, 14 Johns. 432 (N.Y. 1817); M'Allister v. Hammond, 6 Cow. 342 (N.Y. 1826); Wilson v. Smith, 10 Wend. 324 (N.Y. 1833); Hartfield v. Roper, 21 Wend. 615 (N.Y. 1839).

185. Williams v. Holland, 10 Bing. 112, 131 Eng. Rep. 848 (1833).

186. Winfield & Goodhart, *supra* note 145, at 359.

187. I have found only one difference that may have made an action on the case more attractive to plaintiffs than trespass. In Connecticut, there was a three year limitation on bringing trespass actions, but no limitation for case, Gates v. Miles, 3 Conn. 64 (1819), but I have found no other decision in any state that indicates this was a general advantage. Another difference might also have made case more attractive, if it had been followed. In *Adams v. Hemmenway,* 1 Mass. 145 (1804), the court held that case and not trespass lay against a defendant who fired at and wounded the plaintiff's ship master, so that the vessel was forced to turn back from its voyage. The plaintiffs sued for the expected value of the voyage. Even though there was also direct injury, the court held that the plaintiffs could not recover in trespass for the consequential injuries. However, there is no indication that this rule was ever observed, even in Massachusetts. 2 *Dane Abr.* 487-89. I have been able to find no other American case in which the doctrine of *Adams v. Hemmenway* was applied, and it was often rejected. See, *e.g.,* Johnson v. Courts, 3 H & McH. 510 (Md. 1796); Wilson v. Smith, 10 Wend. 324, 328 (N.Y. 1833). See also Taylor v. Rainbow, 2 H & M 423, 428, 441-43 (Va. 1808), in which the court rejects the argument by plaintiff's counsel that he should be allowed to sue in case for a direct injury because otherwise the plaintiff could not recover for consequential damages in trespass. Accord Riddle v. Prop. of Locks, 7 Mass. 169, 172 (1810). In the only other case I have been able to find that resembles *Adams v. Hemmenway,* the court, while nonsuiting the plaintiff for suing in trespass for consequential injury arising from a direct injury, stated that the defect would have been cured if the plaintiff had made a special plea for consequential damages. Robinson v. Stokely, 3 Watts 270 (Pa. 1834). On the other hand, there were definite secondary advantages for plaintiffs who sued in trespass. They gained certain pleading advantages in trespass, since the defendant often was required to answer with a special plea. Indeed, the books are filled with cases of defendants who stumbled amid the mysteries of special pleading to a declaration in trespass. See M'Allister v. Hammond, 6 Cow. 342, 346 (N.Y. 1826), which concluded that the action on the case "is altogether the most favorable to the defendant." A second advantage is that the plaintiff often needed a smaller damage recovery in trespass to receive an award of the costs of litigation than if he sued in case. *Ibid.* Another difference is that some courts required proof of actual damage in case, while they allowed the jury to estimate the damage in trespass. Cole v. Fisher, 11 Mass. 137, 139 (1814).

188. Leame v. Bray, 3 East. 593, 102 Eng. Rep. 724 (1803); Scott v. Shepherd, 2 W. Blackstone 892, 96 Eng. Rep. 525 (1773) (Opinion of Blackstone, J.).

189. See, *e.g.,* Taylor v. Rainbow, 2 H & M 423, 440 (Va. 1808) (Opinion of Tucker, J.); Gates v. Miles, 3 Conn. 64, 75 (dissenting opinion).

190. Dalton v. Favour, 3 N.H. 465 (1826).

191. Formally, the "not guilty" plea to a trespass action only denied that the trespass had occurred. Yet, it seems obvious that the real issue for the jury in virtually all cases was not whether the defendant had actually collided with the plaintiff, but whether, in fact, the collision was due to the defendant's carelessness. Thus, for example, in Barber v. Backus (1824) (unpublished records of Berkshire (Mass.) Court of Common Pleas), the defendant pleaded to a trespass action that "if any hurt or damage happened to . . . [plaintiff] or his wagon, the same was occasioned by the wrongful act of [the plaintiff]." And the jury found that the "defendant is not guilty of the trespass alleged," even though he quite obviously had admitted that a collision had occurred. See also Oomen v. Wellington (1828) (unpublished records of Suffolk, Mass. Court of Common Pleas); Dunn v. Bernard (Sept. 1823) (unpublished records of Norfolk, Mass. Court of Common Pleas) and Wilbore v. Pickins (March 1816) (unpublished records of Bristol, Mass. Common Pleas.), in which it seems equally unlikely that the jury merely found that no collision had taken place. Thus, it is no surprise that in 1823, the Massachusetts Supreme Court should casually refer to a trespass action for "negligently driving" a carriage. Fales v. Dearborn, 1 Pick. 344 (Mass. 1823). Similarly, in Percival v. Hickey, 18 Johns. 257 (N.Y. 1820), the New York court took it for granted that proof of carelessness was necessary if the plaintiff was to prevail in a ship collision case. In short, the negligence standard had already developed an "underground" existence. The first legal writer to recognize the underground consensus that had developed concerning negligence was Benjamin Oliver, Jr., who wrote, in 1828: "It seems reasonable, that, if two vessels run foul of each other in a dark night, one can maintain no action against the other; for, if otherwise, then if both were injured, each might maintain an action against the other; which would be absurd; and so for the same reason, if both parties are to blame, neither should be allowed to bring an action against the other." B. Oliver, *supra* note 146, at 619.

192. The first English contributory negligence case is Butterfield v. Forrester, 11 East. 60, 103 Eng. Rep. 926 (1809). The first American case that might be described as involving contributory negligence as a matter of law is Bush v. Brainard, 1 Cow. 78 (N.Y. 1823), but the most influential decision is Smith v. Smith, 2 Pick. 621 Mass. 1824). There is a dictum recognizing a doctrine of contributory negligence in Farnum v. Concord, 2 N.H. 392 (1821). There is also language suggestive of contributory negligence in Wood v. Waterville, 4 Mass. 422 (1808) and Gorden v. Butts, 1 Penn. 334 (N.J. 1807), but I believe these cases can best be understood in other terms. There was a defense of contributory negligence in Bindon v. Robinson, 1 Johns. 516 (N.Y. 1806), but the case was decided on a technicality of pleading.

193. Washburn v. Tracy, 2 Chip. 128 (Vt. 1824). Lane v. Crombie, 12 Pick. 177 (Mass. 1831) involved the running down of a pedestrian.

194. Smith v. Smith, 2 Pick. 621 (Mass. 1824); Thompson v. Bridgewater, 7 Pick. 188 (Mass. 1829); Howard v. N. Bridgewater, 16 Pick. 189

(Mass. 1834); Adams v. Carlisle, 21 Pick. 146 (Mass. 1838). For other states, see Farnum v. Concord, 2 N.H. 392, 393 (1821) (dictum); Harlow v. Humiston, 6 Cow. 189 (N.Y. 1826).

195. Not only did the overwhelming number of early American contributory negligence cases arise out of highway obstructions, but so did the first English contributory negligence decision, Butterfield v. Forrester, 11 East 60, 103 Eng. Rep. 926 (1809).

196. This explains why the burden of proving the absence of contributory negligence was originally placed on the plaintiff. Smith v. Smith, 2 Pick. 621 (Mass. 1824); Lane v. Crombie, 12 Pick. 177 (Mass. 1831). It was regarded as an essential part of a good cause of action that the plaintiff show he had not caused the injury.

197. 2 S. Greenleaf, *Treatise on the Law of Evidence* 451-52, 580, 583 (2d ed., 1848); T. Sedgwick, *supra* note 46, at 146.

198. W. Malone, "The Formative Era of Contributory Negligence," 41 *Ill. L. Rev.* 151 (1946).

199. The one possible exception is the South Carolina fire case, *Snee v. Trice,* discussed pp. 92-93 *supra.* Since that case stands so completely alone, however, it is perhaps more sensible to view it as limited to the special problem of liability of masters for the injuries of their slaves.

200. 14 Pick. 1 (Mass. 1833).

201. The primary question in the case was whether a nonnegligent defendant was liable for a ship collision caused by the negligence of a steamboat he had hired to bring him into shore. Although the problem of contributory negligence of the plaintiff was absent, the court was still able to view the problem of determining carelessness as involving only a question of causation. Hence, the decision is hardly a ringing statement of modern negligence principles.

202. See Worster v. Prop. of Canal Bridge, 16 Pick. 541 (Mass. 1835) (one count for negligence not founded on a statute); Barnard v. Poor, 21 Pick. 378 (Mass. 1838); Howland v. Vincent, 10 Met. 371 (Mass. 1845); Tourtellot v. Rosebrook, 11 Met. 460 (Mass. 1846). Note the time lag between these decisions and Benjamin Oliver's 1828 statement insisting that proof of negligence need be shown. B. Oliver, *supra* note 146, at 619.

203. Lehigh Bridge v. Lehigh Coal & Navig. Co., 4 Rawle 8 (Pa. 1833).

204. The earlier cases, all in New York, were Clark v. Foot, 8 Johns. 421 (1811); Panton v. Holland, 17 Johns. 92 (1819); Livingston v. Adams, 8 Cow. 175 (1828).

205. Ten Eyck v. Delaware & Raritan Canal Co., 3 Harr. 200, 203 (N.J. 1841); Beekman v. Saratoga & Schenectady R.R., 3 Paige 45 (N.Y. 1831).

206. Stowell v. Flagg, 11 Mass. 364 (1814). See also Varick v. Smith, 5 Paige 137, 143-47 (N.Y. 1835).

207. 14 Conn. 146.

208. *Id.* at 156-57.

209. Clark v. Foot, 8 Johns. 421, 422 (1811). I have put to one side Snee v. Trice, 2 Bay 345 (S.C. 1802) as primarily involving the liability of mas-

ters for acts of their slaves. In any case, it had little influence on the course of development of American law.

210. See p. 93 *supra.*

211. Barnard v. Poor, 21 Pick. 378 (Mass. 1838), is a decision on the measure of damages for injury from fire, although all parties seem to agree that the defendant's carelessness must be shown for him to be liable. Wilson v. Peverly, 2 N.H. 548 (1823) also assumes that a servant's negligence is necessary to establish the master's liability for fire, although the case denies recovery on the basis of a very narrow construction of the master's liability.

212. Jordan v. Wyatt, 4 Gratt. 151 (Va. 1847); Tourtellot v. Rosebrook, 11 Met. 460 (Mass. 1846); Ellis v. Portsmouth & Roanoke R.R., 24 N.C. 138 (1841).

213. The only possible exception is Vermont, which established contributory negligence as a bar to recovery in 1824, Washburn v. Tracy, 2 Chip. 128 (1824), and denied recovery in trespass for personal injuries caused without negligence in 1835, Vincent v. Stinehour, 7 Vt. 62 (1835). Yet, neither of these cases involved economically important events, and the subject did not become of economic significance until Claflin v. Wilcox, 18 Vt. 605 (1846).

214. 1 Beven, *Principles of the Law of Negligence* 679 (2d ed. 1895).

215. A. Hirschman, *The Strategy of Economic Development* 71 (1958). For an excellent discussion of the application of welfare economics to economic history, see H. Scheiber, *The Ohio Canal Era: A Case Study of Government and Economy 1820-1861* 391-97 (1969).

216. R. Fogel, *Railroads and American Economic Growth* 208-24 (1964). Fogel's conclusions have been questioned in A. Fishlow, *American Railroads and the Transformation of the Ante-Bellum Economy* 57-62 (1965).

217. O. Handlin and M. Handlin, *supra* note 39, at 65, 81-85.

218. D. Sowers, *The Financial History of New York State from 1789 to 1912* 114 (1914). Between 1815 and 1826 a small property tax was enacted to pay off the state debt. *Id.* at 115. The general property tax was instituted in 1843, *id.* at 123, and it accounted for $1.26 million dollars or 40 percent of the state budget in 1854. *Id.* at 326-27. For a discussion of an aborted plan to impose a property tax in areas to be benefited by the Erie Canal, see N. Miller, *supra* note 40, at 68-70.

219. L. Hartz, *supra* note 50, at 17-18, 299. The revived interest in the history of state intervention by New Deal historians such as the Handlins, Hartz, and George Rogers Taylor has skewed the literature of antebellum economic history. These writers tended indiscriminately to lump together all forms of governmental financing of enterprise, since they do illustrate the legitimacy of state intervention. But they did not give equal attention to the effects that different forms of financing had upon the distribution of wealth. See, *e.g.,* G. Taylor, *supra* note 49, at 48-52, 376.

220. For the general increase in taxation, see G. Taylor, *supra* note 49, at 375-76.

221. The better-known attack on state financial aid to enterprise is discussed by L. Hartz, *supra* note 50, at 113-26.

222. The one partial exception is the law of nuisance, which even in the twentieth century has continued to be infused with strict liability conceptions. See W. Prosser, "Nuisance Without Fault," 20 *Tex. L. Rev.* 399 (1942). But see P. Keeton, "Trespass, Nuisance, and Strict Liability," 59 *Col. L. Rev.* 457 (1959). The vitality of the strict liability tradition in nuisance explains the blasting cases that began to appear around 1850. In Hays v. Cohoes, 2 N.Y. 159 (1849), the New York Supreme Court held that the defendant was liable even in the absence of negligence for a blasting operation on his own land which caused damage by throwing stone and debris onto an adjoining building. The court did not rely on the fact that the action involved a physical invasion of the plaintiff's land, and I think it a mistake to place any emphasis on the trespass. See C. Gregory, "Trespass to Negligence to Absolute Liability," 37 *Va. L. Rev.* 359, 370-72 (1951). All the cases the court cited involved nuisance actions. Indeed, only four years later the Maryland high court also held a defendant liable in a blasting case in which there was direct injury, and the express theory was that he had committed a nuisance. Scott v. Bay, 3 Md. 431 (1853).

223. Wilde v. Minsterly, 2 *Rolle Abr.* 564 (1639) overruling Slingsby v. Barnard, 1 Rolle Rep. 430, 81 Eng. Rep. 586 (1617).

224. Brown v. Robins, 4 H & N 186, 157 Eng. Rep. 809 (1859); Stroyan v. Knowles, 6 H & N 454, 158 Eng. Rep. 186 (1861).

225. 12 Mass. 220, 224 (1815).

226. *Id.* at 224-25.

227. See Lasala v. Holbrook, 4 Paige 169 (N.Y. 1833).

228. 12 Mass. at 228.

229. *Id.* at 230.

230. F. Bohlen, "The Rule in Rylands v. Fletcher," 59 *U. Pa. L. Rev.* 298, 373, 423 (1911). For an American water case reflecting these views, see Evans v. Merriweather, 4 Ill. 492 (1842).

231. See Lasala v. Holbrook, 4 Paige 169, 171 (N.Y. 1833), where the distinction between "natural" and "artificial" activity is made explicit.

232. Panton v. Holland, 17 Johns 92 (N.Y. 1819); Lasala v. Holbrook, 4 Paige 169 (N.Y. 1833).

233. Not only was this extreme view later reaffirmed in Massachusetts, Foley v. Wyeth, 2 Allen 131 (1861), but the case was understood by contemporaries as going this distance. In his note to *Thurston v. Hancock*, first published in 1833, Benjamin Rand criticized the decision for denying liability even for unreasonable activity. "Why might not the plaintiff as lawfully build on the confines of his own lands in a populous and crowded city, as the defendant could dig on the confines of his land?" he asked. "If the plaintiff, then, had lawfully erected a building on his own lands, on a safe and proper foundation, so as not to require any extraordinary support from the adjoining soil, and to allow the defendant without prejudice to use his lands for ordinary purposes, was it lawful for the defendant, by digging a pit in an unusual manner, and to an extraordinary depth, and for

no ordinary purpose, in his own soil, to undermine or loosen the foundation of the building erected by the plaintiff? It is said, the defendants had a right to make what advantage they could of their own property. Had not the plaintiff the same right? Having exercised this right in a reasonable way, had the defendants any right to use their property in an unusual manner, so as to injure him?" 12 Mass. at 228 n. (1865 ed.).

234. Radcliff's Executors v. Mayor of Brooklyn, 4 N.Y. 195, 203 (1850).

235. Balston v. Bensted, 1 Camp. 463, 170 Eng. Rep. 1022 (1808); Dexter v. Providence Aqueduct Co., 1 Story 387 (U.S.C.C.A. 1832); Smith v. Adams, 6 Paige 435 (N.Y. 1837).

236. The term is used by Perloff and Wingo, "Natural Resource Endowment and Regional Economic Growth" in J. Friedmann and Alonso, eds., *Regional Development and Planning* 215, 239 (1964).

237. 18 Pick. 117, 121 (Mass. 1836) quoting 1 Domat's *Civ. Law* Tit. 12 §2.

238. See the brilliant discussion in the leading English case of Acton v. Blundell, 12 M & W 324, 152 Eng. Rep. 1223 (1843).

239. 25 Penn. St. 528, 532 (1855).

240. 12 Ohio St. 294, 311 (1861).

241. Ellis v. Duncan, 21 Barb. 230, 235 (N.Y. 1855).

242. Meeker v. East Orange, 77 N.J.L. 623, 637-38 (1909) (emphasis in original).

243. "The Rights and Obligations of Riparian Proprietors," 7 *Am L. Reg.* 705, 716 (1859).

244. See Pickard v. Collins, 23 Barb. 444, 459 (N.Y. 1856); Chatfield v. Wilson, 28 Vt. 49, 56-58 (1856).

245. Pickard v. Collins, 23 Barb. 444, 459 (N.Y. 1856).

246. J. W. Hurst, *Law and the Conditions of Freedom in the Nineteenth Century United States,* 7-10 (1956).

247. Haar and Gordon, "Legislative Change of Water Law in Massachusetts" in D. Haber and S. Bergen, eds., *The Law of Water Allocation in the Eastern United States* 1, 25 (1958); Ellis, "Some Legal Aspects of Water Use in North Carolina," *id.* at 189, 292; Arens, "Michigan Law of Water Allocation," *id.* at 377, 393.

248. See, *e.g.,* Martin v. Riddle, 26 Penn. St. 415 n. (1848); Kauffman v. Griesemer, 26 Penn. St. 407 (1856); Butler v. Peck, 16 Ohio St. 334 (1865).

249. See, *e.g.,* Buffum v. Harris, 5 R.I. 243 (1858); Goodale v. Tuttle, 29 N.Y. 459 (1864).

250. Luther v. Winnisimmet, 9 Cush. 171 (Mass. 1851); Flagg v. Worcester, 13 Gray 601 (Mass. 1859).

251. Gannon v. Hargadon, 10 Allen 106, 109 (Mass. 1865).

IV. COMPETITION AND ECONOMIC DEVELOPMENT

1. Penn. Stat. ch. 980 (1781) *repealing* Penn. stat. ch. 472 (1761).

2. Ellis v. Marshall, 2 Mass. 269 (1807).

3. 17 U.S. (4 Wheat.) 518 (1819).

4. 4 J. S. Davis, *Essays in the Earlier History of American Corporations* 24 (1917). By 1801 only eight manufacturing corporations had received charters. *Id.* at 269.

5. Currie's Admin. v. Mutual Ass. Soc., 4 H & M 315, 347-48 (Va. 1809).

6. Wales v. Stetson, 2 Mass. 143, 146 (1806). In the same year, the New York Council of Revision, possessing veto power over acts of the legislature, also recognized the private nature of turnpike corporations for the first time. In rejecting a bill giving turnpike commissioners certain powers over toll roads, the council declared that "the rights vested in the stockholders of a turnpike company, incorporated by law, are as sacred and as much entitled to protection as any other private rights." However, the bill was passed over the Council's veto. A. B. Street, *The Council of Revision of the State of New York* 339 (1859).

7. 4 H & M at 350.

8. Ellis v. Marshall, 2 Mass. 269 (1807). Despite his distinction between public and private corporations, Parker still could not manage to think consistently about the corporation out of its eighteenth century context. Thus, he surprisingly continued to rely on English cases involving towns, deriving support from a decision which held that a town could not be incorporated without the consent of a *majority* of its inhabitants. *Id.* at 277. In fact, this case thoroughly undermined his position, since the whole point of *Ellis v. Marshall* was that even a majority vote would not bind the defendant without his consent.

9. Brewer v. New Gloucester, 14 Mass. 216 (1817); Chase v. Merrimack Bank, 19 Pick. 564, 568-69 (Mass. 1837).

10. Andover and Medford Turnpike v. Gould, 6 Mass. 40 (1809). See also Handlin and Handlin, "Origins of the American Business Corporation," 5 *Journal of Economic History* 1 (1945); E. M. Dodd, *American Business Corporations Until 1860* 75-93 (1954).

11. Riddle v. Prop. of Locks, 7 Mass. 169, 186-87 (1810).

12. 3 W. Blackstone, *Commentaries* 219 (Christian ed., 1855), hereinafter cited as *Commentaries*. It was not until the nineteenth century that judges came to see that the rule concerning interference with water led to almost equally anticompetitive results.

13. *Ibid.* Some have mistakenly assumed that Blackstone developed this rule in order to promote eighteenth century economic liberalism. See D. Boorstin, *The Mysterious Science of the Law* 184-86 (1941). In fact, it was recognized in the middle of the seventeenth century, *Rolle Abr.* 107, and was based on fifteenth century yearbook cases. See Y.B. 11 Henry IV 47 (1408); Y.B. 22 Henry VI 14 (1444). The latter case, however, which is translated in C. Haar, *Land Use Planning* 105 (1971), underlines the uncertainty with which English judges confronted the question of competitive injury and indicates that these cases by no means settled the issue in English law.

14. Compare Cort v. Birbeck, 1 Doug. 218, 99 Eng. Rep. 143 (K.B.

1779) (Mansfield, C.J.) with Richardson v. Walker, 2 B & C 827, 107 Eng. Rep. 590 (K.B. 1824). We can trace the miller's monopoly at least as far back as Fermor v. Brooke, Cro. Eliz. 203, 78 Eng. Rep. 459 (1590) and Hix v. Gardiner, 2 Bulst. 195, 80 Eng. Rep. 1062 (K.B. 1614). After the Restoration, Chancery also intervened and began to decree destruction of mills that competed either with the King's mill or that of his grantee. Otherwise the chancellor simply enjoined the new miller from taking away custom from an existing mill. Currier v. Cryer, Hard. 21, 145 Eng. Rep. 360 (Ch. 1655); Green v. Robinson, Hard. 174, 145 Eng. Rep. 438 (Ch. 1660); White v. Porter, Hard. 177, 145 Eng. Rep. 439 (Ch. 1660); Mayor of Scarborough v. Skelton, Hard. 184, 145 Eng. Rep. 443 (Ch. 1661). The House of Lords confirmed this practice in Smallman v. Brayne, Colles 49, 1 Eng. Rep. 174 (1698). In the eighteenth century the doctrine was upheld in Drake v. Wigglesworth, Willes 654, 125 Eng. Rep. 1369 (C.P. 1752) and Cort v. Birbeck, *supra*, as well as in The Duke of Norfolk's Case, 4 Madd. 83, 112, 56 Eng. Rep. 639, 650 (Ch. 1819). Finally, only in the nineteenth century, in Richardson v. Walker, 2 B & C at 839, 107 Eng. Rep. at 594, did a court overturn this line of decisions, citing "the increase of population and the alteration of manners [that] would make the mischief of such a restriction in these times incalculable."

15. 3 *Commentaries* 235.

16. "Declaration and Plea Book" 32, 33 (1735?) (ms. LMS 1046, Treasure Room, Harvard Law School Library). See note 13 *supra*. See also J. Angell, *Watercourses* 119n. (2d ed. 1833).

17. Mosley v. Walker, 7 Barn. and Cres. 40, 108 Eng. Rep. 640 (1827); Mosley v. Chadwick, 7 Barn. and Cres. 47 n. 1, 108 Eng. Rep 642 n. (a) (1782).

18. 3 *Commentaries* 219.

19. 10 W. Holdsworth, *A History of English Law* 402 (1938).

20. See, *e.g.,* Tripp v. Frank, 4 T.R. 666, 100 Eng. Rep. 1234 (1792); Huzzy v. Field, 2 C.M. & R. 432, 150 Eng. Rep. 186 (1835).

21. 3 *Commentaries* 219.

22. Marston's Petition (1794) (unpublished records of Essex, Mass. Court of Common Pleas). The court held that the plaintiff was "not entitled to compensation for any damage he may have sustained."

23. Chadwick v. Prop. of Haverhill Bridge, 2 *Dane's Abr.* 686-87 (1798). Dane incorrectly cites the date as 1787.

24. Commonwealth v. Knowlton, 2 Mass. 530 (1807); Inhabitants of Arundel v. M'Culloch, 10 Mass. 70 (1813); Davidson v. Fowler, 1 Root 358 (Conn. 1792); Burrows v. Pixley, 1 Root 362 (Conn. 1792).

25. Wales v. Stetson, 2 Mass. 143 (1806).

26. See, *e.g.,* Spring v. Russell, 7 Greenl. 273, 291-94 (Me. 1831); Beekman v. Saratoga & Schenectady R.R., 3 Paige 45, 73 (N.Y. 1831). See also [Josiah Quincy], "Remarks on the Constitutionality of the Memorial of the City Council For an Extension of Faneuil Hall Market" (1823?); Nathan Dane, "The Rights and Obligations of the Proprietors of the Salem Turnpike & Chelsea Bridge Corporation Briefly Examined in Relation to the Proposed Rail Road from Boston to Salem" (1832) (ms. 1032, Treasure

Room, Harvard Law School Library).

27. J. Sullivan, *The Path to Riches* 57 (1792 ed.).

28. Livingston v. Van Ingen, 9 Johns. 507, 573 (N.Y. 1812); see also Charles River Bridge v. Warren Bridge, 7 Pick. 344, 456 (Mass. 1829) (opinion of Morton, J.).

29. Livingston v. Van Ingen, 9 Johns. 507 (N.Y. 1812).

30. B. Hammond, *Banks and Politics in America from the Revolution to the Civil War* 67 (1957).

31. Croton Turnpike Co. v. Ryder, 1 Johns. Ch. 611, 616 (N.Y. 1815).

32. Charles River Bridge v. Warren Bridge, 36 U.S. (11 Pet.) 420, 608 (1837) (dissenting opinion).

33. *Id.* at 639.

34. See Stowell v. Flagg, 11 Mass. 364 (1814); Skipwith v. Young, 5 Munf. 276 (Va. 1816). In England, we have seen, courts had acted on this theory throughout the eighteenth century.

35. See "The Law of Water Privileges," 2 *Am. Jur.* 25, 30-31 (1829).

36. Conn. Pub. Stats. 1824, tit. 69; 1 T. C. Amory, *Life of James Sullivan* 374 (1859).

37. William Cushing, "Notes of Cases Decided in the Superior and Supreme Judicial Courts of Massachusetts from 1772 to 1789" at 37 (ms. LMS 2141, Treasure Room, Harvard Law School Library).

38. Taft v. Sargeants (Mass. 1784) (unpublished records of Supreme Judicial Court, Hampshire, April term).

39. See, *e.g.,* Shorey v. Gorrell (Mass. 1783), in W. Cushing, *supra* note 37, at 39; Palmer v. Mulligan, 3 Caines 307, 313 (N.Y. 1805) (Opinion of Spencer, J.).

40. "Henry H. Fuller's Notes on Lectures of Tapping Reeve and James Gould at the Litchfield Law School" III, 467-68 (1812-13) (ms. LMS 2014, Treasure Room, Harvard Law School Library).

41. Indeed, Asahel Stearns, Story's predecessor at Harvard Law School, seemed inclined to carry the scope of franchises still one step further than Reeve. Stearns, while deploring the inequalities resulting from the English system of prerogative grants, expressed approval in 1824 of franchises given "for the purpose of affecting some object of public or general concern, or even some private concern, which is supposed to be connected with the public good." His list of such franchises included "our insurance and banking institutions, our toll bridges and turnpikes, and almost innumerable manufacturing and other associations." "Lectures on Law" p. 31 (1824?) (ms. LMS 1024, Treasure Room, Harvard Law School Library). Although he made no attempt to say explicitly whether all or none of these franchises were protected from competition, it seems difficult to believe that Stearns was not aware that he was extending the well established English doctrine to these newer franchises.

42. B. Hammond, *supra* note 30, at 67.

43. Attorney General v. Utica Ins. Co., 2 Johns. Ch. 371, 377 (N.Y. 1817).

44. *Supra* note 17.

45. Yard v. Carman, 2 Penn. 681, 687 (N.J. 1812). Monopolistic fishing rights had begun to come under attack in the second half of the eighteenth century. See Opinion of Daniel Dulany, 1 H & McH. 564 (Md. 1768); Freary v. Cooke, 14 Mass. 488 (1779). See also "Caroline County Legislative Petition to Confirm and Ascertain the Common Right of Fishing in Rivers" (Va. 1777) in 1 D. Mays, ed., *The Letters and Papers of Edmund Pendelton 1734-1803* 237 (1967).

46. Carson v. Blazer 2 Binn. 475 (Pa. 1810).

47. Peck v. Lockwood, 5 Day 22 (Conn. 1811). See also Lay v. King, 5 Day 72 (Conn. 1811).

48. See the extensive list of exclusive legislative grants for ferries, bridges and stage-coaches in Gibbons v. Ogden, 22 U.S. (9 Wheat.) 1, 97 n. a (1824) (argument of counsel). See also 1 C. Warren, *History of the Harvard Law School* 399 n. 4 (1908) with reference to state steamboat monopolies.

49. 1 Day 19 (Conn. 1802).

50. *Id.* at 21 (argument of plaintiff).

51. Nichols v. Gates, 1 Conn. 318 (1815).

52. 3 Johns. 27, 37, 39 (N.Y. 1808).

53. 5 Johns. 175 (N.Y. 1809).

54. Livingston v. Van Ingen, 9 Johns. 507 (N.Y. 1812).

55. The New York Council of Revision vetoed the legislature's revocation of the franchise, but it was passed over the council's veto. A. B. Street, *supra* note 6, at 315-16. It is interesting in light of the subsequent development of the contract clause that the council's objection was that the forfeiture did not take place according to due process, which suggests that it already believed that there were at least procedural restraints on the legislative power to affect a revocation. Indeed, intimation of the view that a grant is a contract appeared in the council's deliberations as early as 1780 and 1790. *Id.* at 234, 416.

56. Livingston v. Van Ingen, 9 Johns. at 519.

57. *Id.* at 520-21.

58. *Id.* at 568.

59. Gibbons v. Ogden, 17 Johns. 488, 508 (N.Y. 1820) *reversed* 22 U.S. (9 Wheat.) 1 (1824).

60. *Infra* at pp. 134-136.

61. 11 W. Holdsworth, *A History of English Law* 424 n. 12 (1938). In fact, the intervention of the English Court of Chancery in nuisances cases was "very confined and rare" until 1811. "Since that period the use of the injunction in all of such cases has been increasingly frequent." G. Watrous, "Torts, 1701-1901" in *Two Centuries' Growth of American Law* 113 (1901).

62. 1 J. Dorfman, *The Economic Mind in American Civilization* 481 (1946).

63. 9 Johns. at 547. See the opinion of Yates, J., who adopted this position. *Id.* at 558.

64. Malament, "The 'Economic Liberalism' of Sir Edward Coke," 76

Yale L.J. 1321, 1347-58 (1967).

65. 1 J. Story, *Equity Jurisprudence* 62-63 (1st ed. 1836).

66. J. Kent, Letter to Thomas Washington, 1 *Select Essays in Anglo-American Legal History* 844-45 (1907).

67. 2 J. Story, *Equity Jurisprudence* 156 (1st ed. 1836).

68. Croton Turnpike Co. v. Ryder, 1 Johns. Ch. 611 (N.Y. 1815).

69. 1807 N.Y. Stat. ch. 38.

70. 1808 N.Y. Private Stat. ch. 55.

71. 1 Johns. Ch. at 616.

72. Perhaps Chadwick v. Proprietor of Haverhill Bridge, 2 *Dane's Abr.* 686-87 (1798) is another, but that decision appeared to be shrouded in the mysteries of the doctrine of prescription. From the time toll roads began to be built in New York, the question of competitive injury also began to surface in the New York reports. Two cases in 1805 and 1806 involved lawsuits for competitive injury, but they were reported only because of preliminary procedural disputes. New Windsor Turnpike v. Wilson, Coleman & Caines 467 (N.Y. 1805); New Windsor Turnpike v. Ellison, 1 Johns. 141 (N.Y. 1806).

73. 5 Johns. Ch. 101 (N.Y. 1821).

74. *Id.* at 111-12. It is interesting that the New York Council of Revision in the same year refused to go as far as Kent, at least with respect to the legislature's power to establish a rival grant. While the council successfully vetoed an act that would have interfered with the franchise of a prior toll road, it did so without challenging the legislature's inherent power to make such rival grants. A. B. Street, *supra* note 6, at 394-95.

75. 2 *Dane's Abr.* 686-87.

76. See Morey v. Proprietors of Orford Bridge, Smith's Rep. 91 (N.H. 1804).

77. Anonymous, 2 N.C. 457.

78. Beard v. Long, 4 N.C. 167.

79. Stark v. M'Gowen, 1 Nott & M'Cord 387 (S.C. 1818).

80. Charles River Bridge v. Warren Bridge, 7 Pick. 344, 413 (Mass. 1829) (argument of counsel).

81. J. W. Cadman, Jr., *The Corporation in New Jersey: Business and Politics, 1791-1875* 226-28 (1949).

82. 9 Johns. at 541-42.

83. 7 Conn. 28 (1828).

84. *Id.* at 52.

85. Technically, all this was dicta, since the court ultimately held that many years earlier the first company had asked to be relieved, and was relieved, of its obligation to build the locks. *Id.* at 53-54.

86. "Equity Jurisdiction in the State of New York," 1 *U.S. Law J.* 32, 40, 35 (1822).

87. Sprague v. Birdsall, 2 Cow. 419, 420 (N.Y. 1823).

88. Cayuga Bridge Co. v. Magee, 2 Paige 116 *aff'd* 6 Wend. 85 (N.Y. 1830).

89. *Id.* at 119.

90. For an excellent discussion of the case, see S. Kutler, *Privilege and Creative Destruction: The Charles River Bridge Case* (1971).

91. Charles River Bridge v. Warren Bridge, 7 Pick. 344, 352 (Mass. 1829). The Charles River Bridge, the first to connect Boston with the mainland, was regarded as "one of the marvels of the United States, attracting many persons from other parts of the United States to view it." 1 C. Warren, *supra* note 48, at 509.

92. 7 Pick. at 352.

93. *Id.* at 532.

94. *Id.* at 379-80.

95. *Id.* at 386-87.

96. *Id.* at 385.

97. *Id.* at 390.

98. In fact, the earnings of the Warren Bridge turned out to be so large that it paid for itself two years after construction. But the bridge did not become toll free until 1836, as the legislature realized that an unfavorable Supreme Court decision would subject the company to a crushing liability for damages. 2 C. Warren, *supra* note 48, at 529-30.

99. J. Sullivan, *The History of Land Titles in Massachusetts* 102 (1801).

100. See, *e.g.,* Platt v. Johnson, 15 Johns. 213 (N.Y. 1818); Palmer v. Mulligan, 3 Caines 307 (N.Y. 1805).

101. See Tyler v. Wilkinson, 24 Fed. Cas. 472 (1827); Merritt v. Brinkerhoff, 17 Johns. 306 (N.Y. 1820).

102. Palmer v. Mulligan, 3 Caines 307, 315 (N.Y. 1805).

103. Findlay v. Smith, 6 Munf. 134, 142 (Va. 1818); see also Jackson v. Brownson, 7 Johns. 227 (N.Y. 1810). "The doctrine of waste, as understood in England, is inapplicable to a new, unsettled country." *Id.* at 237 (dissenting opinion of Spencer, J.).

104. 27 U.S. (2 Pet.) 137, 145 (1829).

105. A. Amos and J. Ferard, quoted in 2 *U.S. Law Intell.* 289, 290 (1830); see also "Fixtures," 10 *Am. Jur.* 53 (1833) where the English rule is denounced as "a restraint upon the improvement of real property" having "a general tendency to lessen the amount of its productions and profits."

106. See the discussion of the Massachusetts acts in "Improvements on Land," 2 *Am. Jur.* 294 (1829).

107. 2 J. Kent, *Commentaries* 337-38 (3d ed. 1836).

108. Lasala v. Holbrook, 4 Paige 169, 171 (N.Y. 1833); see also Panton v. Holland, 17 Johns. 92 (N.Y. 1819); Thurston v. Hancock, 12 Mass. 220 (1815).

109. See Callender v. Marsh, 1 Pick. 418 (Mass. 1823); Lansing v. Smith, 8 Cow. 146 (N.Y. 1828) *aff'd* 4 Wend. 9 (1829).

110. 7 Pick. at 514-15.

111. *Id.* at 519.

112. *Id.* at 515.

113. *Id.* at 456.

114. *Id.* at 457.

115. *Id.* at 458.

116. *Id.* at 503.

117. 36 U.S. (11 Pet.) at 608.

118. 7 Pick. at 462.

119. Upon the resignation of Justice Duval in January of 1835, Jackson nominated Taney to the post of associate justice, but the Senate failed to confirm. However, when Chief Justice Marshall died in July of the same year, Taney was again nominated, and because of a change in the composition of the Senate in the recent election, the nomination was confirmed in March of 1836.

120. For the story of this controversy see J. W. Cadman, Jr., *supra* note 81 at 55-59.

121. In this period, attornies general usually maintained private law practices.

122. See newspaper clipping of his Sept. 5, 1833, opinion in "Notebook of Simon Greenleaf" on the *Charles River Bridge* case (Treasure Room, Harvard Law School). For other excerpts from Taney's opinion, see C. Haar, ed., *The Golden Age of American Law* 348 (1965).

123. See L. Hartz, *Economic Policy and Democratic Thought: Pennsylvania 1776-1860* 73-74 (1948).

124. Many arguments similar to Taney's were made after 1833. In his pamphlet "A View of River Rights" (1835), the leading Philadelphia lawyer Charles J. Ingersoll argued that the state was barred from granting a canal company the power to decide whether the city of Philadelphia could extract water from the Schuylkill River. Reminiscent of Chancellor Lansing's argument in the Livingston case, he declared: "If [the state] can so impart its legislative faculty for the purposes of navigation . . . can it go much further, and surrender also the water of these rivers to individuals?" at 17. Again, in 1850 Simon Greenleaf, who had argued and won the *Charles River Bridge* case, reiterated Taney's most radical position, asserting that "any act of the legislature, disabling itself from the future exercise of powers entrusted to it for the public good must be void." 3 S. Greenleaf, ed., *Cruise's Digest of the Law of Real Property* 66-67 n. 1 (1849-50).

125. 36 U.S. (11 Pet.) at 546.

126. See [Theodore Sedgwick] *What is a Monopoly?* (1835).

127. *American Almanac* 189ff (1837); E. M. Dodd, *supra* note 10, at 258-61, 326.

128. As early as 1829, the Middlesex Canal Company recognized the connection between the *Charles River Bridge* case and its ability to resist the competitive threat of railroads. C. Roberts, *The Middlesex Canal, 1793-1860* 154 (1938); see also "The Remonstrance of the Middlesex Canal Company concerning construction of the Boston & Lowell Railroad," Mass. Sen. Rep. No. 21 (1830); Nathan Dane, *supra* note 26; see also Dyer v. Tuskaloosa Bridge Co., 2 Porter 296 (Ala. 1835).

129. See Hudson & Delaware Canal Co. v. N.Y. & Erie R.R., 9 Paige 323 (N.Y. 1841); Thompson v. The New York & Harlem R.R., 3 Sandf. Ch. 625 (N.Y. 1846). Due to the success of the Chesapeake & Ohio Canal in enjoining the rival Baltimore & Ohio R.R., 4 G & J 1 (Md. 1832), the latter's stock "was quoted at very low figures." 1 L. H. Haney, *A Congressional History of Railways in the United States to 1850* 126 (1908). Compare Mohawk Bridge Co. v. Utica & Schenectady R.R., 6 Paige 554 (N.Y. 1837); Tuckahoe Canal Co. v. Tuckahoe R.R., 11 Leigh 42 (Va. 1840); White River Turnpike Co. v. Vt. Central R.R., 21 Vt. 590 (1849) with Chesapeake & Ohio Canal Co. v. Baltimore & Ohio R.R., *supra,* and Enfield Toll Bridge Co. v. Hartford & New Haven R.R., 17 Conn. 40 (1845). See also H. H. Pierce, *Railroads of New York* 11 (1953); M. Reizenstein, *The Economic History of the Baltimore & Ohio Railroad, 1827-53* 29-31 (1897).

130. 36 U.S. (11 Pet.) at 552-53.

131. Letter from Kent to John R. Bleecker, April 7, 1831, 3 *U.S. Law Intell.* 168, 169 (1831).

132. *Id.* at 170, 172-173.

V. THE RELATION BETWEEN THE BAR AND COMMERCIAL INTERESTS

1. P. Miller, *The Life of the Mind in America: From the Revolution to the Civil War* 109 (1965).

2. T. Parsons, *Memoir of Theophilus Parsons* 141-42 (1859).

3. 2 *The Law Practice of Alexander Hamilton* 47, 44 (J. Goebel, ed., 1969), hereinafter cited as *Law Practice of Hamilton.*

4. C. C. Binney, *The Life of Horace Binney* 60 (1903).

5. 1 *Am. L. J.* vi (1808).

6. Letter from Daniel Webster to Justice Story, Dec. 18, 1822, in 1 *The Papers of Daniel Webster* 318 (C. Wiltse, ed. 1974). The case is Tappan v. United States, 2 Mason 393 (1822).

7. 7 Fed. Cas. 418 (No. 3,776) (C.C.D. Mass. 1815).

8. 1 *Life and Letters of Joseph Story* 270 (W. W. Story, ed. 1851).

9. See C. H. S. Fifoot, *Lord Mansfield* 53-54, 108-10 (1936).

10. 2 *Law Practice of Hamilton, supra* note 3, at 20.

11. *Id.* at 18.

12. This statement is made after reading manuscript reports of several hundred Massachusetts cases decided between 1782 and 1805 in the Francis Dana Papers, Massachusetts Historical Society. For an early published case recognizing this procedure, see Livermore v. Newburyport Marine Ins. Co., 1 Mass. 264 (1804).

13. Hague v. Stratton, 8 Va. (4 Call) 84, 85 (1786). A controversy over the judicial power to grant new trials also arose in the New York courts in 1785. After winning a verdict in the trial court, Chancellor Robert Living-

ston, arguing as advocate in his own case, opposed a motion for a new trial. He "denounced with . . . vehemence the judicial authority of lord Mansfield" and opposed "the new-fangled doctrine of lord Mansfield [which] had enlarged and refined upon the power of awarding new trials, so as at last to resolve that mode of trial into the arbitrary discretion of the court." James Kent, *An Address Delivered before the Law Association of the City of New York* 22-23 (1836).

14. Silva v. Low, 1 Johns. Cas. 184, 199 (N.Y. Sup. Ct. 1799) (Lewis, J. dissenting).

15. Silva v. Low, 1 Johns. Cas. 184 (N.Y. Sup. Ct. 1799); Barnewall v. Church, 1 Cai. R. 217 (N.Y. Sup. Ct. 1803); Dow v. Smith, 1 Cai. R. 32 (N.Y. Sup. Ct. 1803); Mumford v. Smith, 1 Cai. R. 520 (N.Y. Sup. Ct. 1804). The only early noninsurance case is Jackson v. Sternbergh, 1 Cai. R. 162 (N.Y. Sup. Ct. 1803).

16. Compare Fuller v. Alexander, 3 S.C.L. (1 Brev.) 149 (1802) with Byrnes v. Alexander, 3 S.C.L. (1 Brev.) 213 (1803) and Wallace & Co. v. DePau, 3 S.C.L. (1 Brev.) 252, 2 Bay 503 (1803).

17. Steinmetz v. Currey, 1 Dall. 234 (Pa. 1788). *Cf.* Swearingen v. Birch, 4 Yeates 322 (Pa. 1806).

18. 1 Z. Swift, *A System of the Laws of the State of Connecticut* 410 (1795).

19. W. Wyche, *Treatise on the Practice of the Supreme Court of Judicature of the State of New York in Civil Actions* 168 (1794) [hereinafter cited as *Treatise on Practice*]. See also Alexander Hamilton's *Practical Proceedings in the State of New York* 139 (1782?), in 1 *Law Practice of Hamilton, supra* note 3, at 118, hereinafter cited as Hamilton's *Practical Proceedings;* Georgia v. Brailsford, 3 Dall. 1, 5 (1794).

20. *The Judicial and Civil History of Connecticut* 163 (D. Loomis and J. G. Calhoun, eds., 1895); Baldwin, "Zephaniah Swift," in 2 W. D. Lewis, ed., *Great American Lawyers* 133-34; East Windsor v. Weatherfield (Conn. 1772) in J. T. Farrell, ed., *The Superior Court Diary of William Samuel Johnson, 1772-1773* 1.

21. Coffin v. Coffin, 4 Mass. 1 (1808); Cook, "Theophilus Parsons" in 2 W. D. Lewis, ed., *Great American Lawyers* 88 (1907-09).

22. W. Nelson, *Americanization of the Common Law: The Impact of Legal Change on Massachusetts Society, 1760-1830* 169 (1975). This discussion of Massachusetts procedure relies entirely upon Nelson, *id.* at 165-72.

23. 1 W. Crosskey, *Politics and the Constitution* 563-77 (1953).

24. 1 *The Works of James Wilson* 279 (R. McCloskey, ed. 1967), hereinafter cited as *Works*.

25. 2 *Works, supra* note 24, at 488.

26. 1 *Works, supra* note 24, at 279.

27. J. Sullivan, *The History of Land Titles in Massachusetts* 338 (1801).

28. Kent to DuPonceau, December 29, 1826, DuPonceau Papers, Historical Society of Pennsylvania, cited in M. Bloomfield, *American Lawyers in a Changing Society, 1776-1876* 361 (1976).

29. 4 Coke's Rep. pt. 8, 81a & 81b (K.B. 1609).

30. 9 Will. III, c. 15 (1698); P. Sayre, "Development of Commercial Arbitration Law," 37 *Yale L. J.* 595, 598-608 (1928).

31. Lord Parker of Waddington [Hubert Lister Parker], *The History and Development of Commercial Arbitration* 15 (Magnes Press, Jerusalem, 1959).

32. W. C. Jones, "Three Centuries of Commercial Arbitration in New York: A Brief Survey," 1956 *Wash. U. L. Q.* 193 (1956). R. Morris, ed., *Select Cases of the Mayor's Court of New York City, 1674-1784,* 44-45, 551-65 (1935).

33. Quoted in Jones, *supra* note 32, at 202.

34. *Letter Book of John Watts* (Collections of the N.Y. His'l. Soc. for 1928) 108, 285 (D. Barck, ed. 1928).

35. Quoted in *The Colonial Merchant* 84 (S. Bruchey, ed., 1966).

36. 1 A. Chroust, *The Rise of the Legal Profession in America* 267-68 (1965).

37. *Id.* at 149.

38. P. Bonomi, *A Factious People: Politics and Society in Colonial New York* 242 (1971).

39. Quoted in *id.* at 210-11.

40. *Id.* at 227.

41. D. Dillon, *The New York Triumvirate* 100, 103, 116-17 (1949).

42. *Letter Book of John Watts, supra* note 34, at 62.

43. *Letter Book of John Watts, supra* note 34, at 331; "Of Abuses in the Practice of the Law" (1753) in *The Independent Reflector* 302 (M. Klein, ed., 1963).

44. 3 J. Campbell, *Lives of the Chief Justices* 275; (3d ed. 1874); W. Vance, "The Early History of Insurance Law," in 3 *Select Essays in Anglo-American Legal History* 98, 113 (A.A.L.S. ed., 1909).

45. 2 T. Clarkson, *A Portraiture of Quakerism* 59-60 (Phila. 1808).

46. The minutes have been published as *Earliest Arbitration Records of the Chamber of Commerce of the State of New York, Founded in 1768 —Committee Minutes, 1779-1792* (New York, 1928?). Indeed, the Chamber established rules regulating damages for the nonpayment of bills of exchange. *A Summary Review of the Laws of the United States* 8 n. (T. Edinburgh and J. Ruddiman, eds., 1788). For a breakdown of the subject matter of these disputes in the years between 1779 and 1784, see Jones *supra* note 32, at 220, n. 114.

47. *Rules and Regulations of the Boston Chamber of Commerce,* Rule VIII, (Boston, 1794). In the same year a standing committee of arbitration was founded in the New Haven Chamber of Commerce. "Early American Arbitration," 1 *Arb. J.* (N.S.) 51, 54 (1946).

48. E. Wolaver, "The Historical Background of Commercial Arbitration," 83 *U. Pa. L. Rev.* 132, 132-38 (1934). On the difficulties of enforcing judgments of merchants courts in actions arising under insurance policies, see Vance, "Insurance Law," *supra* note 44, at 111.

49. B. Austin, *Observations on the Pernicious Practice of the Law, as*

Published Occasionally in the Independent Chronicle, in the Year 1786 8 (1819); reprinted in 13 *A. J. Legal Hist.* 244, 250 (1969).

50. 2 *Works, supra* note 24, at 488-92.

51. [J. Higgins], *Sampson against the Philistines or the Reformation of Lawsuits* 23, 32-34 (2d ed. 1805), hereinafter cited as *Sampson Against the Philistines.* The attacks by both Higgins and Austin, it should be emphasized, were far more populist in tone and content than the quoted passages suggest. Both echoed a strong and widespread popular opposition to the delay and technicality of legal proceedings. Another example, a resolution adopted by "a respectable number of the inhabitants" of Germantown, Pennsylvania, in 1787, denounced "the rapacity of the law, which, in the increase of costs, and delay of justice in our courts, has become such an enormous and oppressive evil, that it is the duty of every real friend to the community, to prevent the people from wasting their property by the chicane of the law, or corruption of our courts of justice." The resolution proposed "that in all cases of altercation, or dispute among ourselves, or our neighbours, we will use our utmost influence to have the same settled by an amicable reference to men, equally selected by each party." 2 *American Museum* 166-67 (2d ed. 1789).

52. Ch. 20 [1791] N. Y. Laws, 14 Sess. 14. During the eighteenth century "the New York judicial system [made] frequent use of the Dutch institution of arbitration which obviated court proceedings by referring the case to the decision of an impartial third party." M. Klein, "The Rise of the New York Bar: The Legal Career of William Livingston," 15 *Wm. & Mary Quar.* 334, 342 (3rd ser. 1958). Connecticut had passed a similar statute in 1753, Virginia in 1789. "Early American Arbitration," 1 *Arb. J.* (N.S.) 53-54, 174 (1946).

53. See Ch. 25, § 1, [1781] N.Y. Laws, 4 Sess. 174, which reenacted a 1768 statute. Ch. 1363, §1 [1768], 28th G.A., 1st Sess., *Laws of N.Y.* 517 (Gaine, ed. 1774). See also W. Wyche, *Treatise on Practice, supra* note 19 at 239. The 1791 New York statute thus had the same effect as the English statute of 9 and 10 Will. III, c. 15 (1698) which also extended arbitration beyond the existing common law power of courts to submit pending suits to arbitrators and allowed parties to a dispute to seek a rule of court before bringing suit. See Shriver v. Maryland, 9 G & J 1 (1837).

54. 2 *Law Practice of Hamilton, supra* note 3, at 379-80. This, of course, excludes informal arbitration by agreement of the parties of which there are no records.

55. Jones, *supra* note 32, at 213.

56. The traditional view was that even where the losing party alleged "that the arbitrators had mistaken their evidence and plain principles of law" a court would not take the "unprecedented" course of looking into the merits of the award. "The reasonableness, or unreasonableness of an award, does not affect its validity [unless] there be . . . misbehavior or corruption in the arbitrators." This was the view of the Connecticut court even

under a statutory scheme of arbitration. Parker v. Avery, Kirby 353 (Conn. 1787). 2 Z. Swift, *supra,* note 18, at 7-8, 17.

57. Allard v. Mouchon, 1 Johns. Cas. 280, 281 (N.Y. 1800).

58. 1 G. Caines *Enquiry into the Law Merchant of the United States; or Lex Mercatoria Americana* 347 (1802).

59. 2 Johns. Cas. 402 (N.Y. 1801).

60. See also Adams v. Bayles, 2 Johns. 374 (N.Y. 1807); Low v. Hallett, 3 Cai. R. 82 (N.Y. 1805).

61. C. Edwards, *The Law and Practice of Referees* 17-18 (1860).

62. *Letter Book of John Watts, supra* note 34, at 285.

63. See 1 Dane *Abridgment* 267. At common law, the winning party in an arbitration would bring a contract action to enforce his award. The importance of a statutory arbitration scheme was that it required the parties to seek a rule of court in advance of arbitration, which thereafter enabled the winning party to levy execution on the loser's property before enforcing the award. 2 Z. Swift, *supra* note 18, at 8.

64. Durell v. Merrill, 1 Mass. 411, 413 (1805) (argument of counsel); Whitney v. Cook, 5 Mass. 139, 143 (1809) (Parsons, C. J.).

65. Mansfield v. Doughty, 3 Mass. 398 (1807).

66. Monosiet v. Post, 4 Mass. 532 (1808). See also Short v. Pratt, 6 Mass. 496 (1810).

67. Fowler v. Bigelow, 8 Mass. 1 (1811); See note by Benjamin Rand in Rand's edition of 8 Mass. at 2, Note (a).

68. 2 Clarkson, *supra* note 45, at 56. Clarkson continued: "Law-suits are at best tedious. They often destroy brotherly love in the individuals, while they continue. They excite also, during this time, not infrequently, a vindictive spirit, and lead to family-feuds and quarrels. They agitate the mind also, hurt the temper, and disqualify a man for the proper exercise of his devotion." *Id.* at 57.

For an excellent discussion of the antilegalism of the evangelical religious tradition in America, see Alan Heimert, *Religion and the American Mind* 179-82 (1966).

69. 2 Clarkson, *supra* note 45, at 56.

70. 1 A. Chroust, *supra* note 36, at 210-11; Odiorne, "Arbitration and Mediation among the Early Quakers," 9 *Arb. Jo.* 161 (1954).

71. 1 Dall. iv. (Preface to reports).

72. Williams v. Craig, 1 Dall. 313, 315 (Pa. 1788).

73. Geyer v. Smith, 1 Dall. 347 (Pa. 1788); Hamilton v. Callender's Ex'rs, 1 Dall. 420 (Pa. 1789).

74. Pringle v. M'Clenachan, 1 Dall. 486, 488 (Pa. 1789).

75. Gross v. Zorger, 3 Yeates 521, 525, 526 (Pa. 1803).

76. Dixon v. Morehead, Addison 216, 224 (Pa. 1794).

77. *Sampson Against the Philistines, supra* note 51, at 75.

78. H. H. Brackenridge *Law Miscellanies* 434 (1814).

79. 1 *Journal of Law* 24-27 (1833) [reprinting articles from 1830-1831].

80. Benjamin v. Benjamin, 5 W. & S. 562, 563, 564 (Pa. 1843).

81. These comments appear as a note to *Alken v. Bolan,* 3 S.C.L. (1 Brev.) 239, 240 (1803). The case itself enunciates a fairly strict rule against overturning arbitration awards. The note, which first appeared when Brevard's reports were published in 1839, was designed to show that South Carolina practice had changed during the intervening period.

82. 1 D. Hoffman *A Course of Legal Study* 359 (2d ed. 1836).

83. 2 J. Lilly, *The Practical Register* 122 (1719); 3 M. Bacon, *New Abridgment of the Law* 745 (1st Amer. ed., from 6th London ed., 1811).

84. Ch. 720, § 10 [1741], 20th G.A., 3d Sess., *Laws of N.Y.* 216, 220 (H. Gaine, ed. 1774). On the history of special juries in England, see J. B. Thayer, *A Preliminary Treatise on Evidence at the Common Law* 94-97 (1898).

85. *Letter Book of John Watts, supra* note 34, at 285.

86. Ch. 41, § 19 [1786] N. Y. Laws, 9 Sess. 78, 83.

87. Ch. 98 [1801] N. Y. Laws, 24 Sess.

88. Hamilton's *Practical Proceedings, supra,* note 19, at 13, in 1 *Law Practice of Hamilton, supra,* note 3, at 61; W. Wyche, *Treatise on Practice, supra* note 19, at 141-42; G. Caines, *A Summary of the Practice of The Supreme Court of the State of New York* 454-68 (1808). There is one important ambiguity in the above accounts of Hamilton, Wyche, and Caines. Hamilton wrote that the twenty-four unstruck jurors "shall be a Jury for the trial of the Cau[se]." Wyche's account is unclear as to whether the trial jury is twelve or twenty-four. But both the statutes of 1786 and 1801, *supra* notes 86 and 87, on which Caines' account is based, are clear that the trial jury numbers twelve.

89. 2 *Law Practice of Hamilton, supra* note 3, at 73.

90. 1 Cai. R. 217 (N.Y. 1803).

91. *Id.* at 222 (Hamilton's argument for plaintiff).

92. *Id.* at 220, 225 (argument of defendant's counsel).

93. The vote was 2-1. See 2 *Law Practice of Hamilton, supra* note 3, at 561 n. 293.

94. 1 Cai. R. at 236.

95. *Id.* at 243.

96. Livingston v. Columbian Ins. Co., 2 Cai. R. 28 (N.Y. 1804); Manhattan Co. v. Lydig, 2 Cai. R. 380 (N.Y. 1805); Livingston v. Smith, 1 Johns. 141 (N.Y. 1806); Anonymous, 1 Johns. 314 (N.Y. 1806); Wright v. Columbian Ins. Co., 2 Johns. 211 (N.Y. 1807).

97. Genet v. Mitchell, 4 Johns. 186 (N.Y. 1809); Thomas v. Rumsey, 4 Johns. 482 (N.Y. 1809); Thomas v. Croswell, 4 Johns. 491 (N.Y. 1809).

98. No. 1095, § 23 [1769] *Public Laws of S.C.* 272 (Grimke ed. 1790).

99. Law of Dec. 20, 1791, *Acts and Resolutions of the G.A. of S.C., Dec. 1791* 9 (Bowen, ed. 1792).

100. Bay v. Freazer, 1 S.C.L. (1 Bay) 66 (1789); Davis v. Ex'rs of Richardson, 1 S.C.L. (1 Bay) 105 (1790); Scarborough v. Harris, 1 S.C.L. (1 Bay) 177 (1791); Lang v. Brailsford, 1 S.C.L. (1 Bay) 222 (1791); Ash's

Adm'rs v. Brewton's Ex'rs, 1 S.C.L. (1 Bay) 243 (1792); James v. M'Credie, 1 S.C.L. (1 Bay) 294 (1793); Ex'rs of Godfrey v. Forrest, 1 S.C.L. (1 Bay) 300 (1793); Miller v. Russell, 1 S.C.L. (1 Bay) 309 (1793); Winthrop v. Pepoon, Otis & Co., 1 S.C.L. (1 Bay) 468 (1795); Ex'rs of Huger v. Bocquet, 1 S.C.L. (1 Bay) 497 (1792); Payne v. Trezevant, 2 S.C.L. (2 Bay) 23 (1796).

101. Davis v. Ex'rs of Richardson, 1 S.C.L. (1 Bay) 105, 106 (1790).

102. Ex'rs of Huger v. Bocquet, 1 S.C.L. (1 Bay) 497, 498-499 (1792).

103. Winthrop v. Pepoon, Otis & Co., 1 S.C.L. (1 Bay) 468, 469-70 (1795).

VI. THE TRIUMPH OF CONTRACT

1. 1 J. Powell, *Essay upon the Law of Contracts and Agreements* x (1790).

2. 2 *id.* at 229.

3. K. Polanyi, *The Great Transformation: The Political and Economic Origins of Our Time* 115 (Beacon Press ed., 1957).

4. See pp. 173-177 *infra.*

5. See, *e.g.,* L. Fuller and M. Eisenberg, *Basic Contract Law* 121-22 (1972); F. Kessler & G. Gilmore, Contracts 27-28 (1970); T. Plucknett, *A Concise History of the Common Law* 643-44 (5th ed., 1956).

6. T. Plucknett, *supra* note 5, at 643-44.

7. J. Ames, *Lectures on Legal History* 144-45 (1913). See also 3 W. Holdsworth, *A History of English Law* 452 (3d ed., 1923).

8. 2 W. Blackstone, *Commentaries* *440-70.

9. *Id.* at *154-66.

10. *Id.* at *448. The title conception also appears in 1 Z. Swift, *A System of the Laws of the State of Connecticut* 380-81 (1795).

11. 2 J. Powell, *supra* note 1, at 232-33. The last important appearance of the title theory in American contract law occurred in Chancellor Kent's *Commentaries.* His treatment of contracts still focused entirely upon the question of when title passes by delivery, and there was as yet no trace of a discussion of damage remedies for breach of contract. See 2 J. Kent, *Commentaries on American Law* *449-557 (1827). In a world in which markets and speculation were becoming everyday events, Kent's treatment represented the final expression of the eighteenth century view of contract as simply one mode of transferring specific property.

12. 2 W. Black. 1078, 96 Eng. Rep. 635 (C.P. 1776). The decision in *Flureau v. Thornhill* may have been responsible for the widespread adoption in America during the first quarter of the nineteenth century of the rule that for breach of warranty of good title only the purchase price and not expectation damages was recoverable. Although the American decisions adopting this rule may reflect deeper issues concerning the relations between speculative buyers and sellers of land, they also represent the con

tinuing influence of an eighteenth century view of contract that had not yet developed a conception of expectation damages.

13. 2 W. Black. at 1078, 96 Eng. Rep. at 635. I have been able to find only one buyer's action for nondelivery of goods on an executory contract in the English reports of the eighteenth century, and that case did not deal with the measure of damages. In Clayton v. Andrews, 4 Burr. 2101, 98 Eng. Rep. 96 (K.B. 1767), a buyer's action for nondelivery of corn, Lord Mansfield held that the statute of frauds did not apply to executory contracts.

14. 1 Strange 406, 93 Eng. Rep. 598 (K.B. 1760). For a more complete discussion of the case, see Moses v. Macferlan, 2 Burr. 1005, 1010-11, 97 Eng. Rep. 676, 680 (K.B. 1760) (Mansfield, J.).

15. See, e.g., Shepherd v. Johnson, 2 East. 211, 102 Eng. Rep. 350 (K.B. 1802). See also 1 J. Powell, supra note 1, at 137-38.

16. 1 Strange at 406, 93 Eng. Rep. at 598.

17. See Moses v. Macferlan, 2 Burr. 1005, 1011-12, 97 Eng. Rep. 676, 680 (K.B. 1760). See also Clark v. Pinney, 7 Cow. 681, 688-89 (N.Y. 1827; Dutch v. Warren "most manifestly decides nothing which has a bearing upon the question of damages where the action is brought *upon the contract itself,* and not to recover back the money paid").

18. 1 Dall. 15 (Pa. 1767).

19. A. Laussat, *An Essay on Equity in Pennsylvania* 19-27 (1826).

20. 5 Johns. 395 (N.Y. 1810).

21. See, e.g., Carberry v. Tannehill, 1 Har. and J. 224 (Md. 1801); Campbell v. Spencer, 2 Binn. 129, 133 (Pa. 1809); Clitherall v. Ogilvie, 1 Des. 250, 257 (S.C. Eq. 1792); Ward v. Webber, 1 Va. (1 Wash.) 354 (1794). On the other hand, Swift stated that "inadequacy of price, abstracted from all other considerations, seems of itself to furnish no ground on which a court of equity can set aside or relieve a party to a contract." 2 Z. Swift, supra note 10, at 447-48. Swift, however, acknowledged that when inadequacy existed, together with other circumstances, a "court may conclude that the consent of the party was not free, or was conditional, thro [sic] mistake, fear, or misrepresentation, or under the impulse of distress, known to the other party." Id. at 448. In short, even according to Swift, inadequacy of consideration could lead to refusal to enforce a contract without a finding of fraud.

22. Desaussure made this remark as an unnumbered footnote to his report of a case, Clitherall v. Ogilvie, 1 Des. 250, 259 n. (S.C. Eq. 1792).

23. Seymour v. Delanc[e]y, 3 Cow. 445, 447 (N.Y. 1824) (Savage, C.J.).

24. Kent refused to specifically enforce a contract on the grounds of the unfairness of the bargain, but he was to be reversed on appeal. Seymour v. Delanc[e]y, 3 Cow. 445 (N.Y. 1824), rev'g 6 Johns. Ch. 222 (N.Y. Ch. 1822).

25. Seymour v. Delancey, 6 Johns. Ch. 222, 232 (N.Y. Ch. 1822).

26. Holingsworth v. Ogle, 1 Dall. 257, 260 (Pa. 1788).

27. Wharton v. Morris, 1 Dall, 125, 126 (Pa. 1785). See also Conrad v. Conrad, 4 Dall. 130 (Pa. 1793).

28. Gilchreest v. Pollock, 2 Yeates 18, 19 (Pa. 1795) (argument of counsel).

29. Armstrong v. McGhee, Addis. 261 (Pa. C.P. 1795) (argument of counsel).

30. 1 *Legal Papers of John Adams* 9 (L. Wroth and H. Zobel eds. 1965). See also *id.* at 12, 15.

31. Quincy 224 (Mass. 1766).

32. *Id.* at 225 (emphasis deleted).

33. Noble v. Smith, Quincy 254, 255 (Mass. 1767). The case held, by a 3-2 vote, that evidence of inadequate consideration could not be admitted in an action on a promissory note brought by the promisee against the promisor. But it is clear from the case that the court treated notes as an exception to the general rule governing contracts. Indeed, promissory notes soon became the leading example emphasized by those who wished to destroy the doctrine of consideration itself. Although Hutchinson voted to exclude evidence of inadequacy of consideration, his statement does acknowledge the general rule, which he did not contest.

34. 1 *Diary and Autobiography of John Adams* 112 (L. Butterfield, ed., 1961). In recognizing the inconvenience to trade, Adams had presaged later attacks on the substantive doctrine of consideration.

35. 1 Z. Swift, *supra* note 10, at 410, *referring to* Hamlin v. Fitch (Conn. Sup. Ct. Err. 1789).

36. Pledger v. Wade, 1 Bay 35, 37 (S.C. 1786). See also Bourke v. Bulow, 1 Bay 49 (S.C. 1787).

37. Waugh v. Bagg (Va. 1731), *reported in* 1 *Virginia Colonial Decisions,* R77, R78 (R. Barton, ed., 1909).

38. Perit v. Wallis, 2 Dall. 252, 255 (Pa. 1796).

39. Walker v. Smith, 4 Dall. 389, 391 (C.C.D. Pa. 1804).

40. See, *e.g.,* Dean v. Mason, 4 Conn. 428, 434 (1822) (Chapman, J.); Baker v. Frobisher, Quincy 4 (Mass. 1762); Garretsie v. Van Ness, 1 Penning. 20, 27-29 (N.J. 1806) (Rossell, J.) (dictum); Toris v. Long, Tayl. 17 (N.C. Super. Ct. 1799); Whitefield v. M'Leod, 2 Bay 380 (S.C. 1802); Mackie's Ex'r v. Davis, 2 Va. (2 Wash.) 219, 232 (1796); Waddill v. Chamberlayne (Va. 1735), *reported in* 2 *Virginia Colonial Decisions, supra* note 37, at B45; 1 Z. Swift, *supra* note 10, at 384; *cf.* Rench v. Hile, 4 Har. and McH. 495 (Md. 1766). See also Z. Swift, *Digest of the Law of Evidence in Civil and Criminal Cases and a Treatise on Bills of Exchange and Promissory Notes* 341 (1810) ("as in all other cases of the sale of personal property, our law implies a warranty"). W. Wyche's treatise on New York procedure contains an index entry, "Assumpsit for implied warranties." W. Wyche, *Treatise on the Practice of the Supreme Court of Judicature of the State of*

New-York in Civil Actions 339 (1794). The text notes that the action "for deceit in selling unsound horses, or the like" was "especially of late years, usually *declared upon in assumpsit." Id.* at 23.

41. See Parkinson v. Lee, 2 East. 314, 322, 102 Eng. Rep. 389, 392 (K.B. 1802) (Grose, J.); W. Story, *A Treatise on The Law of Contracts Not Under Seal* 333 (1844); G. Verplanck, *An Essay on the Doctrine of Contracts: Being An Inquiry How Contracts Are Affected in Law and Morals* 28-29 (1825). But see 2 W. Blackstone, *Commentaries* *451 (warranties of good title, but not of soundness, are implied by law). An early English manuscript contract treatise had declared that an action lies on an implied warranty of merchantability, "for the party [seller] ought to make them Merchantable goods & see them well delivered without any special provision in the contract." "Of Contracts" (c. 1720) (Hargrave ms. 265, British Museum). I am grateful to Professor John Langbein of the University of Chicago Law School for calling the manuscript to my attention. Professor Langbein believes that the eminent British lawyer, Baron Gilbert, wrote the treatise around 1720.

42. 2 R. Wooddeson, *A Systematical View of the Laws of England* 415 (1792). On the basis of a doubtfully reported seventeenth century case, Chandelor v. Lopus, Cro. Jac. 4, 79 Eng. Rep. 3 (Ex. 1603), noted in 8 *Harv. L. Rev.* 282 (1894), it was supposed by later courts that English law had never allowed an action on an implied warranty. See, *e.g.,* Seixas v. Woods, 2 Cai. R. 48 (N.Y. Sup. Ct. 1804). Yet, like so many other early decisions in English legal history, the court's ruling seems to have been more the product of narrow considerations of pleading than of any direct confrontation with issues of substantive policy. See Hamilton, "The Ancient Maxim Caveat Emptor," 40 *Yale L.J.* 1133, 1166-68 (1931); "Implied Warranty on Sale of Personal Chattels," 12 *Am. Jur.* 311, 315-16 (1834). See also 8 W. Holdsworth, *supra* note 7, at 68-70; McClain, "Implied Warranties in Sales," 7 *Harv. L. Rev.* 213 (1893).

43. See Chapter V.

44. Simpson, "The Penal Bond with Conditional Defeasance," 82 *L.Q. Rev.* 392, 411-12, 415-21 (1966).

45. See Wroth and Zobel, Introduction to 1 *Legal Papers of John Adams, supra* note 30, at xliii n.38.

46. See, *e.g.,* Thompson v. Musser, 1 Dall. 458 (Pa. 1789); Cummings v. Lynn, 1 Dall. 444 (Pa. 1789); Wharton v. Morris, 1 Dall. 125 (Pa. 1785).

47. 6 W. Holdsworth, *supra* note 7, at 663 ("already equity had begun to limit . . . relief to cases in which the sum promised was clearly out of proportion to the loss incurred").

48. Lowe v. Peers, 4 Burr. 2225, 2228, 98 Eng. Rep. 160, 162 (K.B. 1768).

49. Astley v. Weldon, 2 B. & P. 346, 351, 126 Eng. Rep. 1318, 1321 (C.P. 1801).

50. This practice was called to my attention by Professor David Koenig of Washington University who is at work on an edition of eighteenth century Plymouth, Massachusetts court records.

51. 1 *Papers of John Marshall,* 215-18 (H. Johnson ed. 1974).

52. Some of the language of the judges in the case may allow for other interpretations. Judge Tazewell, for example, seems to allow for enforcement of executory contracts without part performance when he states "that in an action upon mutual promise the parties may maintain reciprocal Actions." 1 *Papers of John Marshall, supra* note 51, at 215-18. He also may be recognizing expectation damages when he states that the jury "ought to have assessed Damages according to the differences of price, or any other *Special Damage* which plt. could have proved but here no special Damage appears: the plt. has failed to prove that he paid the whole money. The Damages therefore are excessive & a new Trial must be granted." *Id.*

At the new trial, Tucker reports, John Marshall for the defendant "submitted to the court whether the plt. must not prove paymt. on his part, in order to maintain the present Action." *Id.* The court, with Judge Tazewell dissenting, decided that he must. Thus, it is clear that enforcement of executory contracts in which there was no part performance did not yet exist in Virginia as late as 1787. Whether the requirement of payment was regarded simply as a necessary formality or whether buyers' actions were still conceived of as simply for restitution of money paid is not entirely clear.

53. W. Nelson, *The Americanization of the Common Law* 58 (1975).

54. 1 *Legal Papers of John Adams, supra* note 30, at 4. A modern lawyer would, of course, observe that the condition could be satisfied if the seller had tendered the horse. But in America, the first legal writer explicitly to use the concept of tender for this purpose was Daniel Chipman. See D. Chipman, *An Essay on the Law of Contracts for the Payment of Specifick Articles* 31-40 (1822).

55. 1 Z. Swift *supra* note 10, at 380-81. In Gilchreest v. Pollock, 2 Yeates 18 (Pa. 1795), defendant's counsel reiterated the eighteenth century view that part performance was "a condition precedent to the payment, and the party who is to pay shall not be compelled to part with his money till the thing be performed for which he is to pay." *Id.* at 20. But by enforcing one of the early executory stock contracts the court rejected this view.

56. Given the colonial economy, the only conceivable subject of futures contracts would have been agricultural commodities. However, "the lack of a wide market for farm products was a fundamental characteristic of northern agriculture in the colonial period." P. Bidwell and J. Falconer, *History of Agriculture in the Northern United States, 1620-1860,* at 133 (1941). Lewis Cecil Gray states that there were "occasional instances of future-selling" in colonial Virginia. 1 L. Gray, *History of Agriculture in the Southern United States to 1860* 426 (1941). He offers only one ex-

ample, a contract entered into by George Washington with Alexandria merchants for the sale of his wheat at a uniform price over a period of seven years. And he offers no instances of futures contracts in international trade, which provided the major market for commodities during the colonial period. Nor is any mention made of futures contracts in any other southern colonies. See *id.* at 409-33. In addition, the widespread use of bills of exchange in commercial transactions made executory contracts unnecessary. A 1791 Virginia case noted "that it was the general custom of the English merchants, who solicited tobacco consignments, to appoint agents in this country for that purpose, with power to make advances to the planters, and to draw bills [of exchange] on their principals." Hooe v. Oxley, 1 Va. (1 Wash.) 19, 23 (1791).

57. Blackstone's confused account of assumpsit demonstrates that English lawyers had little occasion to think through the rules governing executory contracts of sale. First, he seemed to deny that executory contracts could be enforced without part performance when he wrote: "If a man agrees with another for goods at a certain price, he may not carry them away before he hath paid for them; for it is no sale without payment, unless the contrary be expressly agreed." 2 W. Blackstone, *Commentaries* *447. Order of performance and contractual obligation, it appears, are confounded. "If neither the money be paid, nor the goods delivered, nor tender made, nor any subsequent agreement be entered into, it is no contract." *Id.* Here Blackstone seems to waver between part performance as a necessary requisite for contractual obligation and a conception of executory contracts made enforceable simply through tender but not delivery.

58. The problem of order of performance, inseparably linked to the idea of executory contracts, had not been worked out until the late eighteenth century. See Kingston v. Preston (K.B. 1773) (Mansfield, C.J.), summarized in Jones v. Barkley, 2 Doug. 685, 689-92, 99 Eng. Rep. 434, 437-38 (K.B. 1781). Even after Mansfield's resolution, the problem continued to confuse American courts for another generation. See, *e.g.,* Havens v. Bush, 2 Johns. 387 (N.Y. 1807); Seers v. Fowler, 2 Johns. 272 (N.Y. 1807).

59. See Graham v. Bickham, 2 Yeates 32 (Pa. 1795) (recovery allowed on a bond in excess of penalty where there had been a sharp market fluctuation).

60. 3 W. Blackstone, *Commentaries* *154-64. Blackstone's discussion of express contracts was brief and essentially uninformative. Its most important break with the past lies in his assertion that, except for the seal, ordinary promises were "absolutely the same" as sealed instruments. 3 W. Blackstone, *Commentaries* *157. Thus, we see the beginnings of a generic conception of contracts united by common principles that transcended the particular form of action under which suits on contracts were brought. But we have yet to see any detailed elaboration of the major categories of nineteenth century contract law: offer and acceptance, consideration, and, most important, rules of contract interpretation.

61. 3 W. Blackstone, *Commentaries* *161. He divided implied contracts into two main headings. The first group consisted of obligations imposed by courts or statutes, which arose, Blackstone thought, from an original social contract. *Id.* at *158-59. A second class, including all of the common counts, arose, he explained, "from natural reason, and the just construction of law." *Id.* at *161. In the latter class, the law assumed "that every man hath engaged to perform what his duty or justice requires." *Id.*

62. *Id.* at *158.

63. 2 Burr. 1005, 97 Eng. Rep. 676 (K.B. 1760).

64. *Id.* at 1012, 97 Eng. Rep. at 681.

65. Reported in W. Cushing, "Notes of Cases Decided in the Superior and Supreme Judicial Courts of Massachusetts, 1772-1789," at 1-2 and App. 1-7 (ms. in Harvard Law School Library).

66. In Griffin v. Lee (Va. 1792), reported in St. George Tucker, "Notes of Cases in the General Court, District Court & Court of Appeals in Virginia, 1786-1811," April 18 and Oct. 15, 1787 (ms. in Tucker-Coleman Collection, Swem Library, College of William & Mary), Judge Tucker protested that the common counts had been "extended far beyond the limits which appear to be reasonable" and "need[ed] no Extending."

67. T. Wood, *An Institute of the Laws of England* 555-56 (9th ed. 1763), quoted in C. Fifoot, *History and Sources of the Common Law: Tort and Contract* 363 (1949).

68. *American Precedents of Declarations* 95 (B. Perham ed. 1802). In Cone v. Wetmore (Mass. 1794). (F. Dana papers, Box 16, "Court Cases A-L," Mass. Historical Society), for example, the plaintiff sued in indebitatus assumpsit for cattle sold and delivered. The Supreme Judicial Court declared that "the Deft may have every advantage of the special [express] agreement in this action which he could have had if it had been special declared on. He might show the appraised value was less than plt. demanded." *Id.* The case thus supports the proposition that a suit on the common counts could be maintained even though an express agreement existed.

69. For an excellent discussion of customary wages in eighteenth century England see E.P. Thompson, *The Making of the English Working Class* 235-237 (Vintage ed. 1966).

70. Quincy 195 (Mass. 1765).

71. Quincy 224 (Mass. 1766).

72. In *Pynchon* Hutchinson also remarked that it was not the practice in England to allow an indebitatus for a customary price. Quincy at 224.

73. See Pynchon v. Brewster, Quincy 224, 225 (Mass. 1766) (Hutchinson, C.J.); 1 *Legal Papers of John Adams, supra* note 30, at 16. This concession to jury discretion may not, however, mean that courts had eroded every practical difference between quantum meruit and indebitatus assumpsit. It was one thing to acknowledge a complete jury power to set "reasonable" prices in quantum meruit; it was another to place a special

burden on the jury to modify a fixed price that the court had established as the standard measuring rod for actions in indebitatus assumpsit. In any case, all of this home-grown lawmaking was swept aside in Glover v. Le-Testue, Quincy 225 n.1 (Mass. 1770), where the Massachusetts court, after hearing extensive citations of English authority, held that only quantum meruit and not indebitatus assumpsit would lie for "Visits, Bleeding [or] Medicines" by a doctor. *Id.* at 226.

74. Nightingal v. Devisme, 5 Burr. 2589, 2592, 98 Eng. Rep. 361, 363 (K.B. 1770).

75. Fitch v. Hamlin (Conn. Sup. Ct. Err. 1789), reported in 1 Z. Swift, *supra* note 10, at 410-12.

76. 2 J. Powell, *supra* note 1, at 232-33.

77. The leading case is Shepherd v. Johnson, 2 East. 211, 102 Eng. Rep. 349 (K.B. 1802). See also M'Arthur v. Seaforth, 2 Taunt. 257, 127 Eng. Rep. 1076 (C.P. 1810); Payne v. Burke (C.P. 1799), discussed at 2 East. 212 n.(a), 102 Eng. Rep. 350 n.(a). While these cases deal explicitly with the question of whether damages should be measured as of the promised date of delivery or as of the date of trial, they are nevertheless also the first cases that recognize any measure of expectation damages.

78. 1 Bay 105 (S.C. 1790).

79. Wiggs v. Garden, 1 Bay 357 (S.C. 1794); Atkinson v. Scott, 1 Bay 307 (S.C. 1793).

80. Atkinson v. Scott, 1 Bay 307 (S.C. 1793) (argument of counsel).

81. 1 Bay 307 (S.C. 1793).

82. 1 Va. (1 Wash.) 1 (1790).

83. 1 Va. (1 Wash.) at 4. This issue remained unsettled ten years later. In Kirtley v. Banks (Va. 1800), reported in Tucker, *supra* note 66 (Dec. 9, 1800), a suit for failure to deliver securities, the court instructed the jury that it "may take the price at either period, but not any higher price at any intermediate period." *Id.* The jury selected the time of delivery as its standard.

84. Marshall v. Campbell, 1 Yeates 36 (Pa. 1791).

85. 2 Yeates 18 (Pa. 1795). Two other cases also granted expectation damages to enforce contracts for the sale of United States securities. See Livingston v. Swanwick, 2 Dall. 300 (C.C.D. Pa. 1793); Graham v. Bickham, 4 Dall. 149 (Pa. 1796).

86. 2 Yeates at 21.

87. Lewis v. Carradan (Pa. 1786), cited in 1 Yeates at 37.

88. The first reported case in Massachusetts involving the measure of damages for nondelivery is also a securities case. Gray v. Portland Bank, 3 Mass. 364, 382, 390-91 (1807).

89. Greening v. Wilkinson, 1 Car. & P. 625, 171 Eng. Rep. 1344 (K.B. 1825); Gainsford v. Carroll, 2 B. & C. 624, 107 Eng. Rep. 516 (K.B. 1824); Leigh v. Paterson, 8 Taunt. 540, 129 Eng. Rep. 493 (C.P. 1818).

90. J. Chitty, *A Practical Treatise on the Law of Contracts, Not under Seal* 132 (1826). Powell's *Essay upon the Law of Contracts* (1790) does not

appear to deal with sales. His only recognition of the effect of changes in the market on contracts of sale is his statement that if, after a contract for delivery of corn, the price *falls* to 5 pounds, the buyer "will be entitled either to . . . [the] corn, or five pounds." 1 J. Powell, *supra* note 1, at 409. He also states the rule that "if one of the parties fail in his part of the agreement, he shall pay the other party such damages as he has sustained by such neglect or refusal." *Id.* at 137. Powell cited the famous case of Dutch v. Warren, 1 Strange 406, 93 Eng. Rep. 598 (K.B. 1720), which, as we have seen, was simply an action for restitution.

In Samuel Comyn's *Treatise on Contracts,* an entire chapter is devoted to contracts for the sale of goods. While Comyn does recognize executory contracts, most of the discussion is devoted either to formation of binding contracts or to sellers' remedies for breach. In his very brief reference to buyers' actions for nondelivery, Comyn concluded only that if the buyer tenders payment, he "may take and recover the things." 2 S. Comyn, *Treatise on Contracts* 212 (1807). For this conclusion he cites only an obscure early seventeenth century treatise. Indeed, this discussion is more in line with Blackstone's title theory analysis of contract as one mode of transfer of property than with a nineteenth century market approach.

Finally, with Joseph Chitty's *Treatise on Contracts,* the rule of expectation damages is announced: "In an action of assumpsit, for not delivering goods upon a given day, the measure of damages is the difference between the contract price, and that which goods of a similar quality and description, bore on or about the day, when the goods ought to have been delivered." J. Chitty, *supra,* at 131-32. Interestingly, he cites only two cases decided in the previous five years.

The chapters on damages in the treatises of Powell, Comyn, and Chitty do not mention the problem of expectation damages. Rather, they address themselves exclusively to the problem of how to distinguish penal clauses from clauses providing for liquidated damages. This emphasis reveals the extent to which commercial transactions were still far more dependent on the use of bonds than on contracts.

91. 16 U.S. (3 Wheat.) 200 (1818). McAllister v. Douglas & Mandeville, 15 F. Cas. 1203 (No. 8657) (C.C.D.D.C. 1805), *aff'd,* 7 U.S. (3 Cranch) 298 (1806), superficially resembles *Shepherd,* but there was no agreed upon contract price.

92. See, *e.g.,* West v. Wentworth, 3 Cow. 82 (N.Y. Sup. Ct. 1824) (salt); Merryman v. Criddle, 18 Va. (4 Munf.) 542 (1815) (corn).

93. Clark v. Pinney, 7 Cow. 681, 687 (N.Y. Sup. Ct. 1827). One earlier case involving a commodity was Sands v. Taylor, 5 Johns. 395 (N.Y. Sup. Ct. 1810).

94. Noble v. Smith, Quincy 254 (Mass. 1767).

95. 3 Burr. 1663, 97 Eng. Rep. 1035 (K.B. 1765). There is no citation of this case in *Noble v. Smith.* The third volume of Burrow's reports was first published in 1771, four years after *Noble v. Smith* was decided.

96. 2 W. Blackstone, *Commentaries* *446.

97. 7 T.R. 350 n.1, 101 Eng. Rep. 1014 n.1 (1778).

98. In the typically chaotic fashion of law reporting of the time, the decision was casually included as a footnote to the report of another case, Mitchinson v. Hewson, 7 T.R. 350, 101 Eng. Rep. 1014 (1797). The earliest recognition of the House of Lords decision in America that I am aware of is St. George Tucker's citation in his 1803 edition of Blackstone. 3 Blackstone's *Commentaries* *446 n.1 (St. G. Tucker ed. 1803). In 1804, William Cranch acknowledged that he just learned of the decision as he was about to publish his elaborate essay on negotiable instruments. 5 U.S. (1 Cranch) 445 n.1.

99. 1 Z. Swift, *supra* note 10, at 373.

100. *Id.* See also Z. Swift, *Digest, supra* note 40, at 339.

101. Livingston v. Hastie, 2 Cai R. 246, 247 (N.Y. Sup. Ct. 1804).

102. 3 Cai. R. 286 (N.Y. Sup. Ct. 1805).

103. *Id.* at 289-91 (Livingston, J., dissenting). This position was also adopted by another judge, William Cranch. See 1 Cranch 445.

104. 3 Cow. 445 (N.Y. 1824), rev'g 6 Johns. Ch. 222 (N.Y. Ch. 1822).

105. *Id.* at 533.

106. Stuart v. Wilkins, 1 Doug. 18, 20, 99 Eng. Rep. 15, 16 (K.B. 1778). Though Mansfield was laying the foundation for the subsequent rejection of the sound price doctrine, his purpose was not clearly understood. Nathan Dane, for one, misread the case as upholding the doctrine and, therefore, attempted in 1823 to show that it was "contrary to most of the settled cases in the books." 2 N. Dane, *A General Abridgement and Digest of American Law* 542 (1823).

107. Parkinson v. Lee, 2 East. 314, 102 Eng. Rep. 389 (K.B. 1802).

108. 2 Cai. R. 48 (N.Y. Sup. Ct. 1804).

109. The case relied upon was Chandelor v. Lopus, Cro. Jac. 4, 79 Eng. Rep. 3 (Ex. 1603).

110. See, *e.g.,* The Monte Allegre, 22 U.S. (9 Wheat.) 616 (1824); Dean v. Mason, 4 Conn. 428 (1822); Bradford v. Manly, 13 Mass. 139 (1816); Curcier v. Pennock, 14 S. & R. 51 (Pa. 1826); Wilson v. Shackleford, 25 Va. (4 Rand.) 5 (1826).

111. Dean v. Mason, 4 Conn. 428, 434-35 (1822) (Chapman, J.).

112. I have not meant to assert that caveat emptor is more conducive to a market economy than the contrary doctrine of caveat venditor, though this might be independently demonstrated. Rather, I have argued that the importance of caveat emptor lies in its overthrow of both the sound price doctrine and the latter's underlying conception of objective value.

113. We can best see the nature of the attack on the "sound price" doctrine in South Carolina, the only state in which it persisted well into the nineteenth century. Urging reversal of the sound price doctrine and adoption in its place of a rule of caveat emptor, the Attorney General of South Carolina argued in 1802 that "such a doctrine . . . if once admitted in the formation of contracts, would leave no room for the exercise of judgment or discretion, but would destroy all free agency; every transaction between

man and man must be weighed in the balance like the precious metals, and if found wanting in . . . adequacy, must be made good to the uttermost farthing." Whitefield v. M'Leod, 2 Bay 380, 382 (S.C. 1802) (argument of counsel). If a court should refuse to enforce a contract made by a man who has had "an equal knowledge of all the circumstances" as well as "an opportunity of informing himself, and the means of procuring information . . . ," he maintained, "good faith and mutual confidence would be at an end. . . . To suffer such a man to get rid of such a contract, under all these circumstances," he concluded, "would establish a principle which would undermine and blow up every contract." *Id.* at 383. According to South Carolina lawyer Hugh Legaré, the rule of caveat emptor was desirable because it rejected the "refined equity" of the civil law in favor of "the policy of society." Though there was "something captivating in the equity of the principle, that a sound price implies a warranty of the soundness of the commodity," he was "certain that this rule is productive of great practical inconveniences." 2 *Writings of Hugh Swinton Legare* 110 (M. Legaré ed. 1845). In South Carolina, he noted, "where we have had ample opportunity to witness its operation, there are very few experienced lawyers but would gladly expunge from our books the case which first introduced it here." *Id.* See also Barnard v. Yates, 1 N. & McC. 142, 146 (S.C. 1818) (noting "the perversion and abuse of [the] rule" which many "thought to have opened a door for endless litigation" in those cases where "the contracting parties had not placed themselves upon a perfect footing of equality in point of value").

114. D. Chipman, *supra* note 54, at 109-11.

115. G. Verplanck, *supra* note 41, at 57.

116. *Id.* at 199.

117. *Id.* at 14 (emphasis deleted).

118. *Id.* at 10.

119. 15 U.S. (2 Wheat.) 178 (1817). The case grew out of a futures contract for sale of tobacco purchased by a merchant who had advance knowledge that the United States and England had signed a peace treaty ending the War of 1812. "The question in this case," Chief Justice Marshall wrote, "is, whether the intelligence of extrinsic circumstances, which might influence the price of the commodity, and which was exclusively within the knowledge of the vendee, ought to have been communicated by him to the vendor?" *Id.* at 195. The Chief Justice held that there was no duty to communicate the information, since "it would be difficult to circumscribe the contrary doctrine within proper limits." *Id.*

120. G. Verplanck, *supra* note 41, at 5.

121. *Id.* at 125-26.

122. Verplanck referred to the issue of fraud as "the [only] purely ethical part of the question." *Id.* at 117.

123. *Id.* at 106.

124. *Id.* at 96.

125. *Id.* at 104.

126. *Id.* at 8.

127. *Id.* at 115. See also *id.* at 133.

128. *Id.* at 225.

129. *Id.* at 135.

130. *Id.* at 120.

131. 1 N. Dane, *supra* note 106, at 100 (emphasis deleted).

132. *Id.* at 107-08.

133. *Id.* at 661.

134. Seymour v. Delanc[e]y, 3 Cow. 445, 533 (N.Y. 1824).

135. 1 J. Story, *Commentaries on Equity Jurisprudence* 249-50 (1836).

136. 1 N. Dane, *supra* note 106, at 223-29.

137. G. Verplanck, *supra* note 41, at 133.

138. Metcalf's lectures were first published between 1839 and 1841, although they were first delivered in 1828 at a law school he had founded in Dedham, Massachusetts. See 1 *U.S. L. Intell. & Rev.* 142 (1829). When, in 1867, Metcalf published his *Principles of the Law of Contracts,* he acknowledged that "the first manuscript of the . . . work was prepared, in the years 1827 and 1828" and was published in *American Jurist* between 1839 and 1841. T. Metcalf, *Principles of the Law of Contracts* iii (1867) [hereinafter cited as *Law of Contracts*]. "That publication," he wrote, "has recently been revised and enlarged by reference to reports and treatises published since 1828; but no change has been made in the original arrangement." *Id.*

139. *Law of Contracts* 4; 20 *Am. Jur.* 5 (1838).

140. Ogden v. Saunders, 25 U.S. (12 Wheat.) 213, 341 (1827); see *Law of Contracts* 4 n.(b); 20 *Am. Jur.* at 5 n.1.

141. *Law of Contracts* 5-6. This passage does not appear in *American Jurist,* though Metcalf did write that "it is manifestly only by a fiction, that a contract or promise is implied. And, indeed, the whole doctrine of implied contracts, in all their varieties, seems to be merely artificial and imaginary." 20 *Am. Jur.* at 9.

142. W. Story, *supra* note 41, at 4. See also 2 S. Greenleaf, *A Treatise on the Law of Evidence* 87 (1850) ("The distinction between general or *implied contracts* and special or *express contracts* lies not in the nature of the undertaking, but in the mode of proof").

143. W. Story, *supra* note 41, at 6.

144. See Annot., 19 *Am. Dec.* 268, 272 (1880).

145. 1 T. Parsons, *The Law of Contracts* 522 n.(1) (1st ed., 1853).

146. 24 Mass. (7 Pick.) 181 (1828), annotated, 19 *Am. Dec.* 268 (1880).

147. 24 Mass. (7 Pick.) at 184, 186, 187.

148. 2 T. Parsons, *The Law of Contracts* 35 & n.(d) (1st ed., 1855). There were two exceptions to this trend. The New York courts applied the express contract theory to building as well as to labor contracts. Smith v. Brady, 17 N.Y. 173, 187 (1858). The second exception is the solitary challenge in New Hampshire to the doctrine against quantum meruit recovery in labor cases. See Britton v. Turner, 6 N.H. 481 (1834).

149. 24 Mass. (7 Pick.) at 185.

150. Even when courts modified in building contracts cases the dominant view of the treatise writers that express contracts barred all recovery on an implied contract, they shared at a deeper level the treatise writers' basic assumption about the relationship between express and implied agreements. The contract price, all agreed, set the limit on recovery in quantum meruit. See, *e.g.,* Hayward v. Leonard, 24 Mass. (7 Pick.) 181, 187 (1828). Similarly, in the great case of Britton v. Turner, 6 N.H. 481 (1834), where New Hampshire Chief Justice Joel Parker stood almost alone in resisting the orthodox view barring quantum meruit recovery on labor contracts, he permitted the employer to deduct from recovery "any damage which has been sustained by reason of the nonfulfillment of the contract." *Id.* at 494.

151. 2 N. Dane, *supra* note 106, at 515-16.

152. *Id.* at 515.

153. *Id.* at 516.

154. C. H. S. Fifoot, *Lord Mansfield* 82-117 (1936).

155. Edie v. East India Co., 96 Eng. Rep. 166, 167 (K.B. 1761).

156. Frith v. Barker, 2 Johns. 327 (N.Y. 1807).

157. Brown v. Jackson, 4 F. Cas. 402, 403 (No. 2,016), 2 Wash. 24, 25 (3d Cir. 1807).

158. Ruan v. Gardner, 20 F. Cas. 1295, 1296 (No. 12,100), 1 Wash. 145, 149 (3d Cir. 1804).

159. [George Caines], *Enquiry into the Law Merchant of the United States, or Lex Mercatoria Americana* 260 (1802).

160. *A Digest of The Law of Evidence . . . and A Treatise on Bills of Exchange and Promissory Notes* 245 (1810).

161. Halsey v. Brown, 3 Day 346 (Conn. 1809).

162. Clark v. Langdon (Mass. 1784) in W. Cushing, "Notes of Cases Received in the Superior and Supreme Judicial Courts of Massachusetts from 1772 to 1789" 47 (ms., Treasure Room, Harvard Law School).

163. Homer v. Dorr, 10 Mass. 26, 28-29 (1813).

164. *Law of Contracts, supra* note 138, at 274-75. 23 *Am. Jur.* at 260-61 (1840).

165. 8 S. & R. 533 (Pa. 1822).

166. *Id.* at 534.

167. *Id.* at 550.

168. *Id.* at 554.

169. *Id.* at 559-60.

170. *Id.* at 561.

171. 1 D. Hoffman, *A Course of Legal Study* 416 (2d ed., 1836). "How far the *lex mercatoria* may be derived from the general usages of merchants of all nations, and thus far claim a place in the law of nations, as being *quasi publici juris;* how far *general usage* among the merchants of England constitutes the *lex mercatoria* of that country; or how far *local usage* may become law, or the entire system be ranked with particular customs, the inquiring student will no doubt duly inform himself."

172. 8 S. & R. at 560.

173. *Law of Contracts, supra* note 138, at 275; 23 *Am. Jur.* at 261 (1840).

174. W. W. Story, *Law of Contracts, supra* note 41, at 161 (emphasis added).

175. Clark v. Baker, 52 Mass. (11 Met.) 186, 188-189 (1846).

176. The Reeside, 20 F. Cas. 458, 459 (No. 11,657), 2 Sumn. 567, 569 (1st Cir. 1837). In Donnell v. Columbian Ins. Co., 7 Fed. Cas. 889, 893 (No. 3,987) (1st Cir. 1836), he had similarly maintained: "I am among those judges who think usages among merchants should be sparingly adopted as rules of court by courts of justice, as they are often founded in mere mistake, and still more often in the want of enlarged and comprehensive views of the full bearing of principles."

177. Bolton v. Colder, 1 Watts 360, 363 (Pa. 1833).

178. 2 T. Parsons, *supra* note 148, at 3-4.

179. *Id.* at 6-9.

180. 2 J. Kent, *Commentaries* 555 (2d ed. 1832) [hereinafter cited as *Commentaries*]. "To reach and carry . . . the mutual intention of the parties . . . into effect," Kent also declared, "the law, when it becomes necessary, will control even the literal terms of the contract, if they manifestly contravene the purpose." *Id.* at 554.

181. G. Verplanck, *supra* note 41, at 114.

182. *Id.* at 115-16.

183. *Id.* at 114.

184. There was a major decline in price differentials among various markets between 1816-20 and 1856-60. See G. Taylor, *The Transportation Revolution, 1815-1860* 333 (1951). This, of course, shows only that markets became more extensive, not that they became more stable. I am assuming, what may prove to be incorrect, that as markets became more extensive, price fluctuations also tended to diminish.

185. See, *e.g.,* Cronk v. Cole, 10 Ind. 485, 489 (1858); Kertz v. Dunlop, 13 Ind. 277, 280-81 (1859).

186. 1 T. Parsons, *supra* note 145, at 460, 466.

187. Taylor, *supra* note 184, at 398.

188. Bradford v. Manly, 13 Mass. 139 (1816).

189. Borrekins v. Bevan, 3 Rawle 23, 43 (Pa. 1831). See also Henshaw v. Robins, 50 Mass. (9 Met.) 83 (1845), which Parsons regarded as establishing the rule as "well settled." 1 T. Parsons, *supra* note 145 at 464, note P.

190. Compare Hastings v. Lovering, 19 Mass. (2 Pick.) 214 (1824) and Osgood v. Lewis, 2 Harr. and Gill 495 (Md. 1829) with Hogins v. Plympton, 28 Mass. (11 Pick.) 97 (1831).

191. Jennings v. Gratz, 3 Rawle 168 (Pa. 1831); see note on exceptions to rule of caveat emptor, 90 *Am. Dec.* 426 (1868).

192. 1 S. Williston, *A Treatise on the Law of Contracts* § 22 (rev. ed. 1936). The argument is identical, though the wording is somewhat different, in Williston's first edition in 1920.

193. E. Patterson, "Equitable Relief for Unilateral Mistake," 28 *Col. L. Rev.* 859, 889-90 (1928).

194. 2 *Commentaries, supra* note 180, at 554.

195. W. Phillips, *Treatise on the Law of Insurance* 158 (1823).

196. 3 *Commentaries* 248 (1st ed., 1828).

197. 13 Johns. 451 (N.Y. 1816) discussed in 3 *Commentaries, supra* note 180, at 252.

198. Since these cases involving insurability for negligence arose in what today is the technically unfamiliar setting of marine insurance doctrines, a word of further explanation is perhaps necessary.

To use contemporary categories, marine insurance policies provided "first party" coverage, *i.e.*, the policy was designed to provide indemnification of personal loss, not to insure against liability to others ("third party" insurance). It was a settled part of marine insurance law that the policy did not insure against the consequences of barratry, which consisted of injury brought about through the design of one's own master or crew. This doctrine was analogous to the general rule that a master was not liable for injuries intentionally brought about by his servant. As the distinction between intentional and negligent action first began to emerge during the early years of the nineteenth century, courts thus confronted the question of whether negligently caused injuries were within the insurer's defense of barratry. The cited cases and opinions before 1825 generally held that the defense of barratry did include negligence, which was therefore uninsurable. In short, the defense of barratry raised virtually the same policy questions as did later efforts to contract out of liability and was analyzed no differently from later questions involving liability to others.

199. 3 *Commentaries, supra* note 180, at 370.

200. Patapsco Ins. Co. v. Coulter, 28 U.S. (3 Pet.) 222, 232-38 (1830); Columbian Ins. Co. v. Lawrence, 35 U.S. (10 Pet.) 507, 517-18 (1836); Waters v. Merchants' Louisville Ins. Co., 36 U.S. (11 Pet.) 213, 220-25 (1837). See also Catlin v. Springfield Fire Ins. Co., 5 F. Cas. 310 (No. 2,522), 1 Sumn. 434 (1st Cir. 1833).

201. J. Angell, *A Treatise on The Law of Fire and Life Insurance* 158-59 (1854).

202. 4 *Am. Jur.* 28, 35 (1830).

203. *Id.* at 54, 41.

204. *Id.* at 38.

205. *Id.* at 56.

206. *Id.* at 62-63.

207. Barney v. Prentiss, 4 H & J 317 (Md. 1818); Gordon v. Little, 8 S. & R. 533, 550, 558 (Pa. 1822); Schieffelin v. Harvey, 6 Johns. 170, 177 (N.Y. 1810).

208. Gordon v. Little, 8 S. & R. 533, 550 (Pa. 1822); Aymar v. Astor, 6 Cow. 266 (N.Y. 1826).

209. Nicholson v. Willan, 102 Eng. Rep. 1164 (K.B. 1804). See also Forward v. Pittard, 99 Eng. Rep. 953 (K.B. 1785).

210. 2 J. Kent, *Commentaries* 471 (1st ed. 1827).

211. J. Story, *Commentaries on the Law of Bailments* § 549 (1st ed., 1832).

212. Beckman v. Shouse, 5 Rawle 179, 189 (Pa. 1835).

213. Atwood v. Reliance Transportation Co., 9 Watts 87, 88 (Pa. 1839).

214. Laing v. Colder, 8 Pa. st. 479, 484 (1848).

215. Camden & Amboy R.R. v. Baldauf, 16 Pa. 67, 76-77 (1851).

216. Patton v. Magrath, 23 S.C.L. (Dudley) 159, 163 (1838).

217. Swindler v. Hilliard, 31 S.C.L. (2 Rich. L.) 286, 303 (1846).

218. Orange County Bank v. Brown, 9 Wend. 85 (1832).

219. Cole v. Goodwin, 19 Wend. 251 (N.Y. 1838); Hollister v. Nowlen, 19 Wend. 234 (N.Y. 1838); Clark v. Faxton, 21 Wend. 153 (N.Y. 1839); Camden & Amboy R.R. v. Belknap, 21 Wend. 354 (N.Y. 1839).

220. Cole v. Goodwin, 19 Wend. 251, 280-81 (N.Y. 1838).

221. 2 Hill 623 (N.Y. 1842).

222. New Jersey Steam Navig. Co. v. Merchants Bank, 47 U.S. (6 How.) 344, 383-85 (1848).

223. Mercantile Mutual Ins. Co. v. Chase, 1 Smith 115 (Ct. Com. Pl. 1850); Dorr v. N.J. Steam Navig. Co., 6 N.Y. Super. Ct. Rep. (4 Sanf.) 136 (Super. Ct. N.Y. 1850); Stoddard v. Long Island R.R., 7 N.Y. Super. Ct. Rep. (5 Sanf.) 180 (Super. Ct. N.Y. 1851); Parsons v. Monteath, 13 Barb. 353 (Sup. Ct. N.Y. 1851); Moore v. Evans, 14 Barb. 524 (Sup. Ct. N.Y. 1852); Dorr v. N.J. Steam Navig. Co., 11 N.Y. 485 (N.Y. 1854).

224. Dorr v. N.J. Navig. Co., 6 N.Y. Super. Ct. Rep. (4 Sanf.) 136, 141-42 (Super. Ct. N.Y. 1850).

225. Dorr v. N.J. Navig. Co., 11 N.Y. 485, 493 (1854).

226. New Jersey Steam Navig. Co. v. Merchants Bank, 47 U.S. (6 How.) 344, 383-85 (1848). The court, however, held that "the intent of the parties" to the contract was not to immunize the carrier from liability for negligence. *Id.* at 383.

227. Wells v. Steam Navig. Co., 8 N.Y. 375 (1853).

228. Welles v. N.Y. Central R.R., 26 Barb. 641, 644 (1858).

229. Smith v. N.Y. Central R.R., 29 Barb. 132, 138 (1859).

230. N.Y. Central R.R. v. Lockwood, 84 U.S. (17 Wall.) 357 (1873).

231. 2 S. Greenleaf, *Law of Evidence,* 205-06 (1st ed., 1846).

232. J. Angell, *A Treatise on the Law of Carriers* 241 (1849); 1 T. Parsons, *supra* note 145 at 711 n.(h).

233. I. Redfield, *A Practical Treatise upon The Law of Railways* 280-82 (2d ed., 1858).

234. *Id.* at 278-79.

235. J. Angell, *Law of Carriers, supra* note 232 at 241.

236. B. Bailyn, *Education in the Forming of American Society* 17 (1960).

237. *Id.* at 30-31.

238. T. Reeve, *The Law of Baron and Femme* 374 (1816).

239. Commonwealth ex rel. Gear v. Conrow, 2 Pa. 402, 403 (1845).

240. Preistley v. Fowler, 150 Eng. Rep. 1030 (1837).

241. Murray v. South Carolina R.R., 26 S.C.L. (1 McMul.) 385 (1841).

242. 45 Mass. (4 Met.) 49 (1842).

243. 26 S.C.L. (1 McMul.) at 399.

244. *Id.* at 400.

245. *Ibid.*

246. *Id.* at 402.

247. 45 Mass. (4 Met.) at 51.

248. *Id.* at 56-57.

VII. THE DEVELOPMENT OF COMMERCIAL LAW

1. 2 R. Wooddeson, *A Systematical View of the Laws of England* 387, 388 (1792).

2. Clerke v. Martin, 2 Ld. Raym. 757, 758 (1702).

3. J. Holden, *The History of Negotiable Instruments in English Law* 79 (1955). J. Reeder, "Corporate Loan Financing in the Seventeenth and Eighteenth Centuries," 2 *Anglo-American L. Rev.* 487, 515-16 (1973).

4. New Hampshire, North and South Carolina. The Crown refused to give its assent to a 1767 New York law adopting the English statute, Ch. 1327 [1767], *Laws of N.Y.* 497 (Gaine, ed. 1774). Another New York law, "An Act for giving Relief on Promissory Notes" passed in 1773, probably did not make notes negotiable. Ch. 1612 [1773], *Laws of N.Y.* 772 (Gaine, ed. 1774). See Lewis v. Burr, 2 Cai. Cas. 195, 197 (N.Y. 1796).

5. See, *e.g.,* A. Jensen, *The Maritime Commerce of Colonial Philadelphia* 14-16 (1963).

6. Most of the existing writing on the history of negotiable instruments in colonial America seems to have stretched the evidence much too far in attempting to find early precedents that recognized negotiability. In his "Colonial Sources of the Negotiable Instruments Law of the United States," 34 *Ill. L. Rev.* 137 (1939), Fredrick K. Beutel recognized "the distinction between assignability and negotiability" and correctly saw that assignability "is still erroneously used by many as one of the infallible evidences of negotiability." *Id.* at 140. As was the fashion in the legal history of the last generation, he sought to show that as "the reactionary . . . English common law" gained a dominant foothold in America during the eighteenth century, it exercised a "retarding influence" on "the liberal development" of negotiability in the seventeenth and early eighteenth centuries. *Id.* at 148, 150. Yet, the evidence that Beutel has adduced for an early colonial recognition of negotiability seems to go no further than to demonstrate the existence of a more limited principle of assignability. In any case, whether or not there was an anticommercial retrogression during the eighteenth century — a dubious proposition — we do not substantially disagree about the state of the law immediately before the American Revolution.

Herbert Alan Johnson also maintains that by 1730 "promissory notes were well established in the law of New York." *The Law Merchant and Negotiable Instruments in Colonial New York, 1664 to 1730,* at 36 (1963). It is true, as he observes, that after the passage of the English statute of

Anne, "many [New York] suits thereafter specifically allege the existence of the statute." *Id.* But several questions remain. How many of these notes were actually negotiated? Of the five eighteenth century declarations on promissory notes published in Richard B. Morris, ed., *Select Cases of the Mayor's Court of New York City, 1674-1784* (1935), four represent suits between the original parties to the note — the note was never negotiated — and only one represents an action by an indorsee. *Id.* at 132-137, 513-526. Johnson cites, I believe, only two additional unpublished cases of actions brought by an "assignee." *Id.* at 71, notes 43 and 44.

Elwin L. Page, in his *Judicial Beginnings in New Hampshire, 1640-1700* (1959), carries the argument one step further. "It seems to have been a New England invention," he writes "to treat promissory notes as if they were bills of exchange. In our earliest practice, the assignment of a note had the same consequences as the assignment of a bill. . . . It is perfectly clear that our seventeenth century law recognized the same negotiable quality in a note as in a bill of exchange." *Id.* at 88-89.

Unlike Johnson, Page does cite a number of cases in which promissory notes were assigned. But, as I argue in the text, even evidence of assignment does not illuminate the crucial characteristic of negotiability: limitations on the personal defenses of the promisor. We know, for example, that eighteenth century Massachusetts still refused to allow an endorsee to recover against a promisor who had paid the value of the note to his original promisee. Indeed, one of the majority opinions expressly recognized the difference between bills and notes. Russell v. Oakes, Quincy 48 (Mass. 1763). We also know that this question of personal defenses was still very much unsettled in America at the beginning of the nineteenth century, and none of the eighteenth century pleadings really sheds any further light on this issue. Neither Johnson nor Page is able to indicate from the pleadings whether the New York or New Hampshire courts recognized that the endorsee received a better title than the endorsor had to grant.

7. Quincy 48 (Mass. 1763).

8. Quincy 335 (Mass. 1772).

9. 1 Dane *Abridgement* 378.

10. "The establishment of banking institutions . . . in different sections of the country meant a considerably broader potential market for commercial paper." A. Greef, *The Commercial Paper House in the United States* 6 (1938). Greef goes too far in stating that "it seems more than a coincidence that the earliest dealings in promissory notes of which any record is readily available cannot be traced back much farther than 1793, a year later than the establishment of the last of the [first] eleven banks [in America]." *Id.* at 6-7. In fact, there are many earlier instances — some of them during the colonial period — of the endorsement of promissory notes, though we do not know whether they were negotiable. See *supra* note 6.

11. New Hampshire, North and South Carolina had adopted the English negotiability statute during the colonial period. New York did not enact a clear negotiability statute until 1794, Ch. 48 [1794], *N.Y. Laws,*

17th Sess. 140 (Greenleaf, ed. 1794), although colonial lawyers regularly pleaded that suits on notes were "by force of the Statute . . . Lately made," referring, apparently, to the English statute. Mills v. Richardson in Joseph Murray, "Form Book" (1740-41) (ms. Columbia Law Library) at 137; Peter Sander, "Forms of Sundry Presedents" (1764) at 87, 89 (ms. N.Y. Historical Society). Georgia passed a negotiability statute in 1799. Though Delaware and Massachusetts eventually adopted the principle of negotiability entirely through judicial decision, they still recognized major barriers to negotiability as of 1800. McKnight v. Welsh, 1 Del. Cas. 451 (1797); Pierce v. McIntire, 1 Dane Abr. 111 (Mass. 1785), Webster v. Lee, 5 Mass. 334 (1809).

12. Z. Swift, *A Digest of the Law of Evidence . . . And A Treatise on Bills of Exchange and Promissory Notes* 345 (1810).

13. Norton v. Rose, 2 Wash. 233 (Va. 1796); Mackie's Exec. v. Davis, 2 Wash. 219 (Va. 1796); cf. Buckner v. Smith, 1 Wash. 296 (Va. 1794). M'Cullough v. Houston, 1 Dall. 441 (Pa. 1789); Wheeler v. Hughes, 1 Dall. 23 (Pa. 1776).

14. Bibb v. Prather, 2 Ky. (Sneed) 136 (1802); Drake v. Johnson, 3 Ky. (Hardin) 218 (1808); Spratt v. M'Kinney, 4 Ky. (1 Bibb) 595 (1809); Duncan v. Littell, 5 Ky. (2 Bibb) 424 (1811).

15. In Pennsylvania, however, the legislature made important concessions to negotiability. In 1793 it made notes discounted at the Bank of Pensylvania negotiable. Ch. 147, §13 (1793), 3 *Laws of Pa., 1790-1795* 329-30 (A. Dallas, ed. 1795). In 1797 notes dated in the city or county of Philadelphia were made negotiable.

16. Pierce v. McIntire, 1 Dane Abr. 111 (Mass. 1785).

17. Webster v. Lee, 5 Mass. 334 (1809). In his manuscript notes of Jones v. Alexander, 2 Mass. 36 (1806) (ms. Francis Dana papers, Mass. Historical Society), Justice Dana revealed how unsettled the question of negotiability still was. In a case involving the assignment of an insurance contract, he noted that the "assignee must take the thing subject to all the Equity to which *the original party was* subject" but added: "If this rule applied to bills & promissory Notes, it would stop their currency."

18. 1 S.C.L. (1 Bay) 398.

19. *Id.* at 399-440.

20. *Id.* at 415 (Waties, J., dissenting).

21. *Id.* at 428-29.

22. *Id.* at 402.

23. Only five states in the nineteenth century recognized the negotiability of bonds: Georgia, Illinois, Iowa, North Carolina, and Tennessee. For the English cases involving negotiability of bonds, see Reeder, *supra* note 3 at 517-19.

24. 9 Hening's *Laws of Virginia* 431. See 3 Tucker's *Blackstone* 469 n. 26 (1803); 1 *The Papers of George Mason, 1725-1792,* at 423 (R. Rutland, ed., 1970).

25. Rhodes v. Risley, 1 Chip. Rep. 52, 53 (Ver. 1791).

26. "A Dissertation on The Negotiability of Notes" (1792) in N. Chipman, *Reports and Dissertations* 89, 95 (2d ed. 1871).

27. *Id.* at 106.

28. Rhodes v. Risley, 1 Chip. Rep. 52, 53 (Ver. 1791).

29. Norton v. Rose, 2 Wash. 233, 252 (1796) (Carrington, J.).

30. 3 William Griffith, *Annual Law Register* [7]. [This page appears in the Harvard Law School Library's copy, though it does not appear in the Arno Press reprint of the copy at the University of Illinois Law Library.]

31. 99 Eng. Rep. 1104 (K.B. 1786).

32. Jordaine v. Lashbrooke, 7 T.R. 603 (K.B. 1798).

33. Parker v. Lovejoy, 3 Mass. 565 (1795)

34. 3 Johns. Cas. 185, 197 (N.Y. 1802) overruled in Stafford v. Rice, 5 Cow. 23, 25 (N.Y. 1825) and Bank of Utica v. Hillard, 5 Cow. 153, 160 (N.Y. 1825).

35. Churchill v. Suter, 4 Mass. 156, 161 (1808).

36. 1 S. Greenleaf, *Treatise on The Law of Evidence* 430-31 n. 2 (1st ed. 1842).

37. Bank of the United States v. Dunn, 31 U.S. (6 Pet.) 51, 57 (1832).

38. Smyth v. Strader, 4 How. 404, 417, 418-20 (1846).

39. Norton v. Rose, 2 Wash. 233 (Va. 1796).

40. Mackie's Exec. v. Davis, 2 Wash. 219 (Va. 1796).

41. Dunlop v. Harris, 9 Va. (5 Call.) 16 (Va. 1804).

42. Dunlop v. Silver, 1 D.C. (1 Cranch) 27 (D.C. Cir. 1801); Riddle v. Mandeville, 1 D.C. (1 Cranch) 95 (D.C. Cir. 1802).

43. Mandeville v. Riddle, 5 U.S. (3 Cranch) 290 (1803).

44. Appendix to Mandeville v. Riddle, 1 Cranch 367, reprinted in 3 *Select Essays in Anglo-American Legal History* 72 (A.A.L.S. ed., 1909).

45. 1 Cranch, *supra* note 44, at 374, 3 *Select Essays, supra,* note 44, at 74-75 [part of the quote has been omitted in *Select Essays.*]

46. 1 *Works of James Wilson* 279 (McCloskey ed., 1967).

47. 1 Cranch, *supra* note 44 at 461.

48. Harris v. Johnson, 7 U.S. (3 Cranch) 322, 331 (1809).

49. Riddle v. Mandeville, 9 U.S. (5 Cranch) 322, 331 (1809).

50. 27 U.S. (2 Pet.) 331 (1829).

51. In his opinion Justice Johnson acknowledged the Kentucky cases opposing the endorsee's action but pretended that they were simply holdings that there was no action at law. "The conclusion, then, results from our own decisions," he wrote, "that he must be let into equity. . . . And this we understand to be consistent with the received opinions and practice of Kentucky" 27 U.S. (2 Pet.) 331, 348 (1829). In fact, there was no such "received opinion" in Kentucky. Even after the decision in *Riddle v. Mandeville* (1809), the Kentucky high court held that an endorsee could not sue a remote endorsee because of absence of privity and hence failure of consideration. Duncan v. Littell, 2 Bibb. 424 (Ky. 1811). Such a defense undoubtedly would have prevailed in equity as well.

52. J. Story, *Commentaries on the Law of Promissory Notes* 10 (1845).

53. The states were Alabama (Law of Dec. 18, 1812, as cited in Robinson v. Crenshaw, 2 S. & P. 276, 312 [1832]); Arkansas (Law of Dec. 8, 1818, *Ark. Terr. Laws* 24); Indiana (Law of Jan. 29, 1818, ch. 12, [1838] *Rev'd Stat. of Ind.*); Maryland (see below); Mississippi (Law of June 25, 1822, § 9, *Laws of Miss.* 385); Missouri (Law of Feb. 11, 1825, § 9 [1825] 1 *Laws of Mo.* 143); New Jersey (Law of Jan. 30, 1799, § 4 [1820] *Rev'd Laws of N.J.* 395); Pennsylvania (Law of Feb. 27, 1797, ch. 1920, 1 *Stat. at Large of Pa.* 484); and Virginia (Ch. 68, [1786] *Laws of Virginia* 358). *Cf.* Story, *Promissory Notes* 10 n. 1. In 1829 the Maryland high court seemed to assume that the English statute of 3 & 4 Anne ch. 9 was received in Maryland. Bowie v. Duvall, 1 Gill & Johns. 175, 179 (1829). But a statute the next year preserved all defenses to a promisor in an action on a note. Ch. 51 [1830], *Laws of Md., 1829-1830 Gen. Ass.* (J. Hughes, ed., 1830). As late as 1839 the Maryland court declared that "the same defenses . . . are open to the debtor, as if the action was brought in the name of the assignor." Harwood v. Jones, 10 Gill & Johns. 404, 405 (1839). In Missouri and New Jersey, negotiability could be accomplished in a note by use of words specially prescribed by their statutes.

54. 50 U.S. (9 How.) 213 (1850).

55. 1 Cranch, *supra* note 44, at 422.

56. 50 U.S. (9 How.) 213, 222 (1850).

57. Watson v. Tarpley, 59 U.S. (18 How.) 517, 521 (1855).

58. Oates v. National Bank, 100 U.S. 239, 246, 249 (1879). In 1873 Alabama changed its statute, appearing no longer to preserve all of the promisor's defenses. Law of Apr. 8, 1873 [1872-73] *Acts of Ala.* 111. The United States Supreme Court, while holding that the new provision had apparently received the above construction by the Alabama Supreme Court, declared nevertheless that it did not matter whether it was "mistaken in our interpretation of the decision of the Supreme Court of Alabama," since the general commercial law would govern in any event. *Id.* at 246.

59. *Id.* at 249. The Supreme Court expressly refused to take account of "a direct violation of the statutes against usury" in Alabama. In determining the validity of such defenses, the court held, "we are not bound by the decision of the State court." *Id.* Compare Gaither v. Bank, 1 Pet. 37 (1828).

60. C. Binney, *The Life of Horace Binney* 60-61 (1903).

61. These figures should be taken as only gross approximations. There are no New York reports for 1802. Moreover, I have calculated each yearly percentage independently and then simply averaged them. I have therefore not taken account of the fact that a larger number of cases may have been decided (or reported) in any one year. For the changes in New York marine insuring after 1790, see 2 *Law Practice of Alexander Hamilton* 404-06 (J. Goebel, ed., 1964).

62. J. Park, *A System of the Law of Marine Insurances* xl (1st ed., 1787).

63. W. Vance, "The Early History of Insurance Law," 3 *Select Essays in Anglo-American Legal History* 115 (1909).

64. Wright and Fayle, *A History of Lloyd's* 39, 56 (1928).

65. D. Gibb, *Lloyd's of London* 45 (1957).

66. Quoted in B. Supple, *The Royal Exchange Assurance* 189 (1970).

67. Gibb, *supra* note 65, at 55.

68. 2 *Law Practice of Alexander Hamilton, supra* note 61, at 397. See also Heubner, "The Development and Present Status of Marine Insurance in the United States" in 26 *Annals of the American Academy of Political and Social Science* 432-33 (1905).

69. M. McLaughlin, "Marine Insurance in Boston, 1724-1781" 31 (Harvard College senior thesis).

70. Silva v. Low, 1 Johns. Cas. 184, 199 (N.Y. 1799); Barnewall v. Church, 1 Caines 217 (N.Y. 1803); Mumford v. Smith, 1 Cai. R. 520 (N.Y. 1804); Byrnes v. Alexander, 1 Brev. 213 (S.C. 1803); Wallace v. DePau, 1 Brev. 252, 2 Bay 503 (S.C. 1803).

71. See B. Supple *supra* note 66 at 54-55 for a discussion of the development of actuarial theory before 1800. See also J. Cassedy, *Demography in Early America: Beginnings of the Statistical Mind, 1600-1800* (1969).

72. "The trading part of the community have generally ranged themselves" against insolvency laws. [J. Gallison], *Considerations on an Insolvency Law* 4 (1814).

73. P. Coleman, *Debtors and Creditors in America* 283-85 (1974).

74. C. Warren, *Bankruptcy in United States History* 25-45 (1935). Gallison, *supra* note 72, at 36, 6, may be the first to have written of the need to provide relief to the "fair and blameless debtor" whose failure was "caused by manufacture, which could not be guarded against by ordinary prudence and foresight." See also "On a National Bankrupt Law," 1 *Am. Jurist* 35 (1829); "A Bill to Establish a Uniform System of Bankruptcy Throughout the U.S.," 7 *North Am. Rev.* 25 (1818).

75. J. Dorsey, *Treatise on the American Law of Insolvency* 16 (1832).

76. By the late 1850s Philadelphia insurance companies insured greater values under fire than under marine insurance policies. J. Fowler, *History of Insurance in Philadelphia* 217 (1888).

77. Quoted in 2 C. Warren, *History of the Harvard Law School* 244 (1908).

78. Scripture v. Lowell Mutual Fire Ins. 10 Cush. 356, 363 (1852).

79. G. Taylor, *The Transportation Revolution, 1815-1860* 322 (1951); F. Oviatt, "Historical Study of Fire Insurance in the United States," 26 *Annals supra* note 68, at 342-43.

80. 3 J. Kent, *Commentaries* 320 (2d ed. 1832); J. Angell, *A Treatise on the Law of Fire and Life Insurance* 41 (1854).

81. Stetson v. Mass. Mutual Fire Ins. Co., 4 Mass 330 (1808). "We find in the books but few cases in which the subject of insurance against loss by fire, has come under consideration." Harris v. Eagle Fire Co., 5 Johns. 368, 373 (1810).

82. Burritt v. Saratoga County Mutual Fire Ins. Co., 5 Hill 188, 192 (N.Y. 1843); Holmes v. Charlestown Mutual Fire Ins. Co., 10 Met. 211, 214 (Mass. 1845).

83. Jolly's Admin. v. Baltimore Equitable Society, 1 Har. and Gill 295, 300, 302 (1827).

84. 2 Hall 632 (N.Y. 1829).

85. 1 Cai. R. 217 (N.Y. 1803).

86. *Id.* at 227.

87. *Id.* at 225.

88. *Id.* at 245 (opinion of Lewis, C.J.).

89. J. Duer, *The Law and Practice of Marine Insurance* 654 (1846).

90. P. DuPonceau, *A Dissertation on the Nature and Extent of the Jurisdiction of the Courts of the United States,* 22 (1824).

91. Fleming v. Marine Ins. Co. 3 Watts & Serg. 144, 153 (Pa. 1842).

92. Peters v. Warren Ins. Co., 14 Pet. 99, 109 (1840).

93. Farmers' Ins. Co. v. Snyder, 16 Wend. 481, 493 (1836).

94. Barnewall v. Church, 1 Cai. R. 217 (1803); Talcot v. Commercial Ins. Co., 2 Johns. 124 (1807). Justice Livingston tried to reverse the rule in Patrick v. Hallett, 1 Johns. 241 (1806) but was overruled *sub silentio* in Talcot v. Commercial Ins. Co., *supra.* Merchant opposition to the rule governing seaworthiness appears in statement of counsel for insurer Barnewall v. Church, *supra* at 225 and in the unwillingness of juries to find unseaworthiness despite four court-ordered new trials in Talcot v. Commercial Ins. Co., 2 Johns. 467 (1807).

95. M. Howe, "The Creative Period in the Law of Massachusetts," 69 *Proceedings of the Massachusetts Historical Society* 232, 237 (1947-50).

96. T. Parsons, *Treatise on the Law of Marine Insurance* 380 (1868); Walsh v. Washington Marine Ins. Co., 32 N.Y. 427, 436-37 (1865); Deshon v. Merchants Ins. Co., 11 Met. 199, 207 (Mass. 1846); Snethen v. Memphis Ins. Co., 3 La. Ann. 474 (1848).

97. T. Parsons, *supra* note 96, at 6-10.

98. Holmes, "The Path of the Law," 10 *Harv. L. Rev.* 455, 469 (1897).

99. W. Sheldon, *Cursory Remarks on the Laws Concerning Usury,* iii-iv, 28-29 (1798).

100. F. Gilmer, *Vindication of the Laws Limiting the Rate of Interest on Loans,* 43, 62 (1820). For an American Roman Catholic priest's religiously based attack on usury, see J. O'Callaghan, *Usury, Funds and Banks* (1824).

101. Fitch v. Hamlin (1789) reported in 1 Z. Swift, *A System of the Laws of the State of Connecticut,* 410-12 (1795), reversing Hamlin v. Fitch, Kirby 260 (1787). A second trial of the same cause appears in 2 Kirby 42 (1789). For similar views on usury, see the opinion of Virginia Chancellor George Wythe reported and reversed in Groves v. Graves, 1 Wash. 1 (Va. 1790).

102. *Memoir of Jeremiah Mason,* 46-49 (1873).

103. J. McCulloch, *Interest Made Equity* 30 (1826).

104. T. Cooper, *Lectures on the Elements of Political Economy* 111-12 (1826).

105. W. Phillips, *A Manual of Political Economy* 66 (1828).

106. 1 *The Journal of Law* 49, 64, 81, 97 (1833).

107. *Id.* at 52.

108. *Id.* at 88.

109. T. Dew, *Essay on the Interest of Money, and the Policy of Laws Against Usury* 8 (1834).

110. "Usury Law," 6 *Am. Jur.* 282, 295 (1831).

111. *A Familiar View of the Operation and Tendency of Usury Laws,* 58 (1837).

112. The Petition and Committee Report appear in A. Everett [and J. Bowles], "Usury and the Usury Laws," 39 *No. Am. Rev.* 68 (1834). [While the index ascribes the article to Everett, there is a separate index at Harvard University's Widener Library that gives both Everett and Bowles as authors.]

113. J. Bowles, *Treatise on Usury and Usury Laws,* 28-29, 33, 47 (1837).

114. 2 *Statutes at Large of South Carolina* 713 (T. Cooper., ed., 1837); 3 *id.* at 784-785 (1838).

115. "On the Usury Law," 17 *Am. Jur.* 331, 343, 348, 359, 362 (1837).

116. *Opinion of Chancellor Kent on the Usury Laws,* 9-11 (1837).

117. This information is derived from William Griffith, *Annual Law Register of the United States,* vols. 3 and 4 (1822). [Vols 1 and 2 were never published.] Griffith provided a summary of the law of each state based on questionnaires sent to prominent lawyers. Questions 122 and 123 dealt with usury. In selecting 1819 as the dividing line I have modified Griffith's breakdown only to the extent of counting Kentucky as a state that voided usurious contracts. Griffith's summary takes account, as mine does not, of the 1819 Kentucky statute eliminating a provision voiding usurious contracts.

118. A Pennsylvania statute of 1722-23 provided that persons who "receive or take" unlawful interest on any bond or contract should "upon conviction thereof . . . forfeit the money and other things lent," one half to the government, "and the other half to the person who shall sue for the same." Statutes of Pennsylvania ch. 262. But in Wycoff v. Longhead, 2 Dall. 92 (1785), the Pennsylvania Supreme Court, in an action on a usurious promissory note, held that the creditor could recover the debt plus legal interest; that only where the creditor "actually receives" unlawful interest — instead of merely contracting for future receipt — the statute requires that he incur a forfeiture of the entire amount lent; "but if an action is brought to recover the amount of the loan, a verdict ought not to be given for the [debtor] as that would, in effect, be putting the money into his pocket, instead of working a forfeiture to the Commonwealth." Thus, not only did the court ignore the terms of the statute; it also made forfeiture depend on a proceeding brought by governmental authority, which entirely divorced enforcement from private self-interest. In *A Summary Review of the Laws of the United States of North America* 14 (1788), an anonymous "Barrister of the State of Virginia" wrote: "The laws against usury are in force in Virginia but not in Pennsylvania, and some other States. They are certainly not in practice."

119. Connecticut, Delaware, Maryland, New Jersey, New York, North Carolina, Virginia.

120. This summary is derived from a chart in J. Murray, *History of Usury* ix-x (1866). He provides no information on penalties, if any, in Minnesota and I have, therefore, excluded it from the summary.

121. Iowa, New Hampshire, Maryland, and New York so provided by statute. Hackley v. Sprague, 10 Wend. 113 (N.Y. 1833), Sauerwein v. Brunner, 1 Har. and Gill. 477, 481-82 (Md. 1827) (note by reporter). Courts in Kentucky, Chiles v. Coleman, 9 Ky. 296 (1820), and Illinois, Conkling v. Underhill, 4 Ill. 388 (1842) reached the same result. On the question of whether the maker of a note could plead that an endorsee had received the instrument through a usurious discount, see "Law of Usury," 2 *U.S. Law Jo.* 51 (1826).

122. Five states—Georgia, Illinois, Iowa, North Carolina, and Tennessee—permitted bonds to be negotiated, although the first three also barred inquiry into usurious transactions involving even unsealed negotiable instruments.

123. Barnewall v. Church, 1 Cai. R. 217, 225 (1803) (argument of counsel). See also Wilkie v. Roosevelt, 3 Johns. Cas. 206 (1802) where the court orders a third trial after two consecutive juries ignored the trial judge's instructions and found in favor of an innocent purchaser of a note despite the usurious character of the original transaction.

124. Hackley v. Sprague, 10 Wend. 113 (1833).

125. 16 Pet. 1 (1842).

126. Southern Pacific v. Jensen 244 U.S. 205, 222 (1917). See also G. Dunne, *Joseph Story and the Rise of the Supreme Court* 406-08 (1971).

127. 1 W. Crosskey, *Politics and the Constitution* 573 (1953).

128. See Chapter I, at pp. 15-16.

129. J. Goodenow, *Historical Sketches of the Principles and Maxims of American Jurisprudence in Contrast with the Doctrines of the English Common Law on the Subject of Crimes and Punishments* 36.

130. J. Story, *Commentaries on Conflict of Laws,* sec. 7 (1st. ed., 1834).

131. *Id.* at secs. 7 and 8.

132. *Id.* at sec. 28, quoting Saul v. Creditors, 17 Martin 569, 595-96 (La. 1827).

133. *Id.* at sec. 36, 22, and 28.

134. *Id.* at sec. 242.

135. See Meigs, "Decisions of the Federal Courts on Questions of State Law," 45 *Am. L. Rev.* 47, 68-73 (1911).

136. See *supra* at pp. 220-223.

137. Watson v. Tarpley, 59 U.S. (18 How.) 517, 521 (1855); Pease v. Peck, 59 U.S. (18 How.) 595, 598-99 (1855); Oates v. National Bank, 100 U.S. 239, 246, 249 (1879).

138. 7 Fed. Cas. 418 (No. 3,776) (CCD. Mass. 1815).

139. 1 *Life and Letters of Joseph Story* 270 (W. Story, ed., 1851).

140. Story himself remarked in Peele v. Merchants Ins. Co., 19 Fed.

Cas. 98 (1822) that he had "hitherto supposed the point [decided in *DeLovio*] rather of theoretical than practical importance, presuming that from private convenience, the benefit of trial by jury, and the confidence that is so justly placed in our state tribunals, the insured would almost universally elect a domestic forum." However, when he declared privately immediately after *DeLovio* that the decision "is rather popular among merchants" because "they are not fond of juries," *supra* note 139, he was not quite so enthusiastic about the "benefits" of the jury system. Actually, since the insurance companies were usually not plaintiffs, they were not generally able to choose the federal forum and take advantage of nonjury trials in admiralty.

141. Ramsay v. Allegre, 12 Wheat. 611, 638 (1827) (Johnson, J.); Bains v. The James and Catherine, 2 Fed. Cas. 410, 416 (C.C.D., Pa. 1832) (Baldwin, J.); Taylor v. Carryl, 20 How. 583, 615 (1857) (Taney, C., dissenting).

142. New England Mutual Marine Ins. Co. v. Dunham, 78 U.S. (11 Wall.) 1 (1870). See also [F. Loring], "Of the Jurisdiction of Admiralty over Contracts of Marine Insurance," 3 *Am. L. Rev.* 666 (1869); "History of Admiralty Jurisdiction in the Supreme Court of the United States," 5 *Am. L. Rev.* 581, 582-83, 617-19 (1871).

143. Genesee Chief v. Fitzhugh, 12 How. 443 (1851). See also The Daniel Ball, 10 Wall. 557 (1870).

VIII. THE RISE OF LEGAL FORMALISM

1. O. Holmes, "Privilege, Malice and Intent," 8 *Harv. L. Rev.* 1, 7 (1894); K. Llewellyn, *The Common Law Tradition: Deciding Appeals* 38 (1960).

2. See C. Warren, *Bankruptcy in United States History* 25, 45 (1935).

3. See M. Horwitz, "The Conservative Tradition in the Writing of American Legal History," 17 *Am. J. L. Hist.* 275, 281-82 (1973).

4. P. Miller, *The Life of the Mind of America: From the Revolution to the Civil War* 156-64 (1965).

5. Wythe's lectures appear to be lost. Wilson lectures, delivered 1790-91, appear in 1, 2 *The Works of James Wilson* (R. McCloskey, ed. 1967). For Kent, see *An Introductory Lecture to a Course of Law Lectures, Delivered November 17, 1794* (1794).

6. E. Corwin, "The Extension of Judicial Review in New York: 1793-1905," 15 *Mich. L. Rev.* 281 (1917). I have counted only those cases listed by Corwin *id.* at 306-13, in which a court actually set aside a statute and not those in which it simply construed a statute to avoid constitutionality.

7. M'Clenachan v. Curwin, 3 Yeates 362 (Pa. 1802); Beekman v. Saratoga & Schenectady R.R., 3 Paige 45 (N.Y. 1831); Harvey v. Thomas, 10 Watts 63 (Pa. 1840).

8. Taylor v. Porter, 4 Hill 140 (N.Y. 1843).

9. See Chapter II.

10. Beekman v. Saratoga & Schenectedy R.R., 3 Paige 45, 57 (N.Y. 1831) (argument of counsel); Harvey v. Thomas, 10 Watts 63 (Pa. 1840) (Gibson, C. J.).

11. Taylor v. Porter, 4 Hill 140, 143 (N.Y. 1843).

12. West River Bridge Co. v. Dix, 6 How. 507, 545-48 (1848) (Woodbury, J.); People v. Salem, 20 Mich. 452, 470 (1870) (Cooley, J.); Memphis Freight Co. v. Mayor, 44 Tenn. 419, 425-26, 429-30 (1867); Reeves v. Treasurer, 8, Ohio St. 333, 345-47 (1858). Compare Sharpless v. Mayor of Philadelphia, 21 Pa. St. 147, 167, 169-71 (Black, C. J.) (1853) with 2 *Am. L. Reg.* 34-35, 92-93 (Lowrie and Lewis, JJ., dissenting). *Cf.* Boston & Roxbury Mill Dam Corp. v. Newman, 12 Pick. 467, 475-77 (Mass. 1832).

13. Stowell v. Flagg, 11 Mass. 364 (1814).

14. Fiske v. Framingham Mfg. Co., 12 Pick. 68, 70 (1832); Murdock v. Stickney, 8 Cush. 113, 116 (1851); Chase v. Sutton Mfg. Co. 4 Cush. 152, 169-70 (1849).

15. Head v. Amoskeag Mfg. Co., 113 U.S. 9 (1884).

16. R. Hale, "Bargaining, Duress, and Economic Liberty," 43 *Col. L. Rev.* 603, 616 (1943).

17. J. Dawson, "Economic Duress—An Essay in Perspective," 45 *Mich. L. Rev.* 253, 266-67 (1947). See also J. Dalzell, "Duress by Economic Pressure," 20 *N.C.L. Rev.* 237, 240 (1942): "The consent to a contract is probably far more real than the typical contractual consent."

18. J. Dawson, *supra* note 17, at 266.

19. Bostwick v. Lewis, 1 Day 250, 254 (Conn. 1804); Foley v. Cowgill, 5 Blackf. 18, 20 (Ind. 1838); Page v. Bent, 2 Metc. 371, 372-73, 374 (Mass. 1841); Gatling v. Newell, 9 Ind. 572, 576 (1857); Gordon v. Parmelee, 2 Allen 212, 213-14 (Mass. 1861).

20. The first English case to create a sharp distinction between fact and opinion was Pasley v. Freeman, 3 Durf. and East. 51 (1784). This distinction "came down from a period," Professor Keeton observed " . . . in which the law was characterized by extreme individualism, and by its unmoral attitude. Each person was assumed to be capable of looking out after his own interests. The stock argument of the courts was that the situation was produced by the party's own folly, and that he must abide the consequences." W. P. Keeton, "Fraud: Misrepresentation of Opinion," 21 *Minn. L. Rev.* 643, 651 (1937). See also Harper and McNeely, "A Synthesis of the Law of Misrepresentation," 22 *Minn. L. Rev.* 939, 956-57 (1938).

21. See Chapter VI at pp. 197-201, and G. Gilmore, *The Death of Contract* 41-43 (1974).

22. On the effort to make equity "scientific," see Story's attack on the "just price" doctrine and his approval of the doctrine of Seymour v. Delanc[e]y, 3 Cow. 445 (N.Y. 1824).

23. 5 W. Holdsworth, *A History of English Law,* 330-32 (2d. ed., 1937).

24. [J. Lansing, Jr.], *An Essay on the Law of Mortgages in the State of*

New York (1824); J. Holcombe, *An Introduction to Equity Jurisprudence* 186 n.1. (1846).

25. L. Friedman, *A History of American Law* 217-18 (1973).

26. Compare Otis v. Wood, 3 Wend. 498 (N.Y. 1830) with Strong v. Taylor, 2 Hill 326 (N.Y. 1842). See also Barrett v. Pritchard, 2 Pick. 512, 13 *Am. Dec.* 449 (Mass. 1824); Dresser Mfg. Co. v. Waterston, 3 Met. 9 (Mass. 1841); Herring v. Hoppock, 15 N.Y. 409 (1857); "Bailments and Conditional Sales," 44 *Am. L. Reg.* 335 (1896); "Question Whether Transaction is Mortgage or Conditional Sale," note to Palmer v. Howard, 1 *Am. St. R.* 63 (1887); 2 J. Kent. *Commentaries* 496-98, note 1 (12th ed., O. Holmes, ed. 1873).

Index